SYMMETRIES OF ALGEBRAS

VOLUME 1

by

Chelsea Walton

619 Wreath

Copyright © 2024 Chelsea Walton

All rights reserved. No part of this publication may be reproduced, stored or transmitted in any form or by any means, electronic, mechanical, photocopying, recording, scanning, or otherwise without written permission from the publisher. It is illegal to copy this book, post it to a website, or distribute it by any other means without permission.

First edition, 2024

ISBN 978-1-958469-20-0

Published by 619 Wreath

To Matt & The Zoo

PREFACE

Welcome! I am happy that you are here, and I am excited to present to you a topic that I have enjoyed over the last several years. My journey towards landing on this topic began in graduate school, where I studied material related to the first chapter of this book (namely, algebras over a field). Later as a faculty member, I encountered the material comprising the remaining chapters here, discovering the world of Algebraic Quantum Symmetry. I've found all of the structures in this field quite beautiful and important in their own way.

This book is geared for newcomers who would love to learn about intriguing algebraic structures in nature beyond their first Abstract Algebra course(s). By now, you might have skimmed the table of contents and thought to yourself, "I know a few of these words, but certainly not all." If you are concerned about this, fret not. To be perfectly honest with you, I did not know most of those words as a student, but this is the way that it is supposed to work. Mathematics is certainly not a 'young man's game'; it is for everyone who simply wants an adventure in discovering new knowledge.

So, I wish to serve as your guide in finding and understanding some fascinating algebraic structures, motivated by the concept of symmetry. Let's proceed!

– C. Walton, 2024

Acknowledgments

This is my first book project. So, I would like to first acknowledge key mentors who were pivotal in my development as a mathematician: my undergraduate advisor, Jeanne Wald; my Ph.D. advisors, Toby Stafford and Karen Smith; and my postdoctoral mentors, James Zhang, Ellen Kirkman (unofficial), Sarah Witherspoon, and Pavel Etingof. Wald and Stafford introduced me to noncommutative ring theory, and Stafford and Smith taught me its connections to various guises of geometry. They were also instrumental in my learning how to write mathematics formally, and I sincerely thank them for their time during those stages of my training. Zhang, Kirkman, and Witherspoon introduced me to Hopf algebras and their symmetries on algebras, which launched my research program. They were also my first collaborators and I am quite grateful for our interactions during those times. Etingof was very influential in my career path by exposing me to the world of tensor categories and by helping me further develop my taste in mathematics.

"You can do everything with tensor categories!"   – P. Etingof.

This sentiment inspired the last decade of my research program, and it is my hope that this book contributes as proof of the statement.

This book project was primarily written during my long-term stays at the University of Hamburg, Germany, in 2022 and 2023. I thank my host, Christoph Schweigert, the Hamburg Algebra and Mathematical Physics research group, and staff member Karin Zimmer for hosting my stays. I am also grateful to the Alexander von Humboldt Foundation for sponsoring these visits.

I am also grateful for the partial support provided by the US National Science Foundation (grants #DMS-2100756, 2348833) and by the American Mathematical Society's Claytor–Gilmer Fellowship.

I would like to thank Candice Price and Miloš Savić, the leaders of the 619 Wreath Publishing Company, for agreeing to take on this book project. They have been supportive throughout the entire writing and production process, which has boosted my creativity and motivation to finish this book.

I thank Rice students, Bingbing Hu, Evan Huang, and Jacob Kesten for participating in reading courses at Rice University: Hu and Huang covered a draft of Chapters 1 and 2 in Fall 2022, and Kesten reviewed a draft of Chapters 2–4 during the 2023 calendar year. I am greatly appreciative of the careful comments they provided during these reading courses.

I would like to thank the students of my Spring 2024 courses at Rice University: Justin Cruz-Ortiz, Stian du Preez, Luis Gomez Gorgonio, Zhengyi Han, Thomas Kaldahl, Anna Lowery, Abiel Quinonez, and Owen Silberg. We covered the topics in this book and more, and their feedback prompted key improvements.

I am grateful to the following individuals who provided comments that also improved the quality of this book: Fernando Liu Lopez, Jo Nelson, David Reutter, Christoph Schweigert, Vincent Thompson, Abigail Watkins, and Harshit Yadav.

The members of my 2022–2024 research team (Fabio Calderón, Amanda Hernandez, Hongdi Huang, Jacob Kesten, Fernando Liu Lopez, Anh Tuong Nguyen, Vincent Thompson, and Harshit Yadav) provided a great amount of inspiration for this endeavor, and I thank them for our interactions.

I sincerely thank Vincent Thompson for his hired assistance and careful work in providing the Tikz code for several initial figures in this book.

I am grateful to Matilde Marcolli for her genuine encouragement in completing this book and for providing insightful advice on the publishing process. I also thank Pavel Etingof for his helpful comments and for providing several important tips on the textbook writing process.

I send a special acknowledgment to Mohamed Omar for providing encouraging professional support throughout this book project.

Finally, I am grateful to my husband, Matthew Abbatessa, who has given me a steady and unconditional source of support in my personal life. And for the record, my pups, Mr. Mischief Maker, Dr. Boom-Boom, and Belly Belly Cavatelly, acknowledge themselves as sources of support as well.

Contents

List of Figures

$$\cdot\ \text{C{\small HAPTER}}\ 0\ \cdot$$

I{\small NTRODUCTION}

In this part, we provide some motivation for the material in this book series, as well as give a summary of its contents and features. Some reasons for studying symmetries of algebras are presented in Section 0.1, but it is not required to understand this in its entirety. It is enough to gather the flavor of the content at this time. Section 0.2 discusses the contents and organization of the book series, and Section 0.3 lists the book series' features. Note that *italicized words* throughout are terms that will either be defined vaguely, be explained later in the text, or are outside of the scope of the text. **Bold words** throughout are either headers, result labels, terms highlighted for emphasis, or terms that are defined formally.

§0.1. Motivation

What is a symmetry? Symmetry is a ubiquitous concept that we have all encountered since childhood– it appears intuitively in art, music, science, etc. Mathematically, let us consider the following definition. A *symmetry* of an object X is a transformation from X to itself. If X has certain features or a structure, then we may require symmetries to preserve such aspects of X. The set of symmetries of X, denoted by $\mathrm{Sym}(X)$ here, can form an interesting structure when we equip it with the operation of composition. For instance, if the composition of two symmetries of X is again a symmetry of X, then $\mathrm{Sym}(X)$ is a *monoid*, that is, a set that has an associative binary operation and an identity element with respect to the operation. Here, the identity element of $\mathrm{Sym}(X)$, with respect to composition, is "do nothing to X". If, further, all symmetries of X are reversible (or invertible), then $\mathrm{Sym}(X)$ is a *group*, that is, $\mathrm{Sym}(X)$ is a monoid where each element is accompanied by an inverse element.

Let us consider a couple of examples, which are depicted in Figure 0.1. Take X to be two states (or entities) a and b. Leaving the states alone, and swapping the

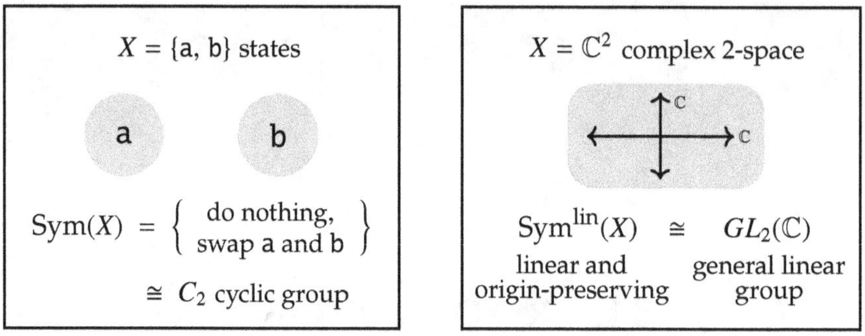

$X = \{a, b\}$ states

a   b

$\text{Sym}(X) = \left\{ \begin{array}{c} \text{do nothing,} \\ \text{swap a and b} \end{array} \right\}$

$\cong C_2$ cyclic group

$X = \mathbb{C}^2$ complex 2-space

$\text{Sym}^{\text{lin}}(X) \cong GL_2(\mathbb{C})$

linear and   general linear
origin-preserving   group

Figure 0.1: Examples of classical symmetry for: (1) two states; and (2) $\mathbb{C}^2$.

states, are both reversible symmetries of X. In this case, these two symmetries can be identified with the abstract cyclic group of order 2, yielding $\text{Sym}(\{a, b\}) \cong C_2$.

Next, take the complex numbers $\mathbb{C}$, and let X be the complex 2-plane,

$$\mathbb{C}^2 = \left\{ \begin{pmatrix} x \\ y \end{pmatrix} \;\middle|\; x, y \in \mathbb{C} \right\}.$$

Here, let us consider the symmetries of X that are linear (i.e., that send lines to lines), that preserve the origin $\begin{pmatrix} 0 \\ 0 \end{pmatrix}$, and that are reversible. These symmetries, denoted by $\text{Sym}^{\text{lin}}(\mathbb{C}^2)$, are captured by the general linear group $GL_2(\mathbb{C})$ of invertible 2×2 matrices in $\mathbb{C}$, and the transformations are given by matrix multiplication,

$$GL_2(\mathbb{C}) \times \mathbb{C}^2 \to \mathbb{C}^2, \quad \left(\begin{pmatrix} a & b \\ c & d \end{pmatrix}, \begin{pmatrix} x \\ y \end{pmatrix} \right) \mapsto \begin{pmatrix} ax + by \\ cx + dy \end{pmatrix}.$$

In short, $\text{Sym}^{\text{lin}}(\mathbb{C}^2) \cong GL_2(\mathbb{C})$ as groups.

So far, this discussion all lies within the framework of *Classical Symmetry*, in which symmetries of objects are those that we can see, or that we can 'observe' in the sense of classical physics. But there are many interesting objects that we cannot see nor observe, namely those within the quantum world.

Generalizing symmetry. One way of constructing an object in the quantum world is by altering an object in the classical world; we will illustrate this by using the processes of *linearization* and *deformation* below.

Keep in mind that a good framework for symmetry should remain unchanged under such alterations. Namely, taking alterations and symmetries in either order, saying forming $\text{Sym}(X_{\text{alt}})$ and $\text{Sym}(X)_{\text{alt}}$, should yield the same result. In this case, we say that these processes *commute*. See Figure 0.2. Even if we are accustomed to deformed shapes having more or less symmetry than their original counterparts (e.g., scalene triangles being less symmetric than regular triangles), it does not mean that the symmetry framework that we are used to is ideal. Let us expand our perception of symmetry here...

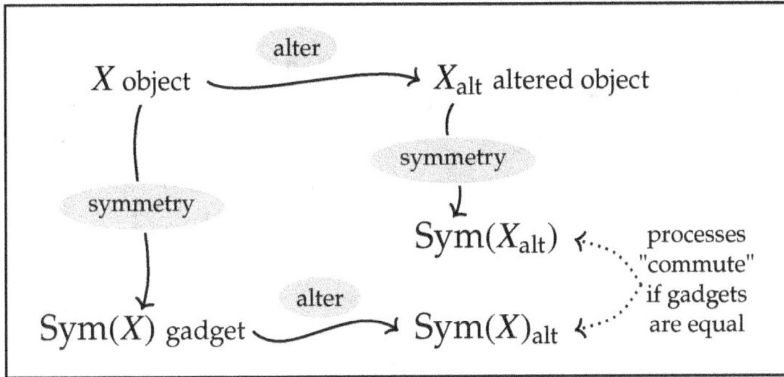

Figure 0.2: Ideal framework: Symmetries preserved after alteration.

In Classical Symmetry, the outputs $\mathrm{Sym}(X_{\mathrm{alt}})$ and $\mathrm{Sym}(X)_{\mathrm{alt}}$ are groups. But often for $\mathrm{Sym}(X_{\mathrm{alt}})$ to equal $\mathrm{Sym}(X)_{\mathrm{alt}}$, we have to work in a framework larger than group symmetries, using more sophisticated algebraic structures. This is our entrance into the realm of *(Algebraic) Quantum Symmetry*.

Structures that pop up frequently in generalized symmetry frameworks are algebras, which are the key algebraic structures of this book series. An *algebra* is, in short, a vector space that has a compatible structure of a unital ring. Here, a *vector space* is an algebraic structure in which we can add and subtract tuples of elements, and scalar-multiply tuples of elements by a number. We will take the numbers to be complex numbers $\mathbb{C}$ in the examples below, forming $\mathbb{C}$-*vector spaces*. Moreover, a *unital ring* is an algebraic structure in which we can perform addition, subtraction, and multiplication, with having additive and multiplicative identity elements. Algebras that have the underlying structure of a $\mathbb{C}$-vector space are called $\mathbb{C}$-*algebras*.

Linearizing symmetries. Let us now extend the first example in Figure 0.1 to see how Classical Symmetry is generalized when incorporating *linearization*, and to see how symmetries of algebras arise naturally.

Motivated by the notion of *quantum superposition*, we linearize the pair of states in Figure 0.1 and create a *quantum state space*, $\mathbb{C}\mathrm{a} \oplus \mathbb{C}\mathrm{b}$, which is a $\mathbb{C}$-vector space with basis given by the states a and b. The group of invertible, linear, origin-preserving symmetries of this space is the general linear group $GL_2(\mathbb{C})$.

However, if we were to capture symmetries of the quantum state space by simply linearizing the group of symmetries in the first example of Figure 0.1, we would obtain a small subset of the symmetries obtained in the above procedure. Namely, we would get $\mathbb{C}$-linear combinations of the identity matrix and the anti-identity matrix in $GL_2(\mathbb{C})$. So, linearizing the group symmetry framework is not sufficient for capturing all linear symmetries of the quantum state space. See Figure 0.3.

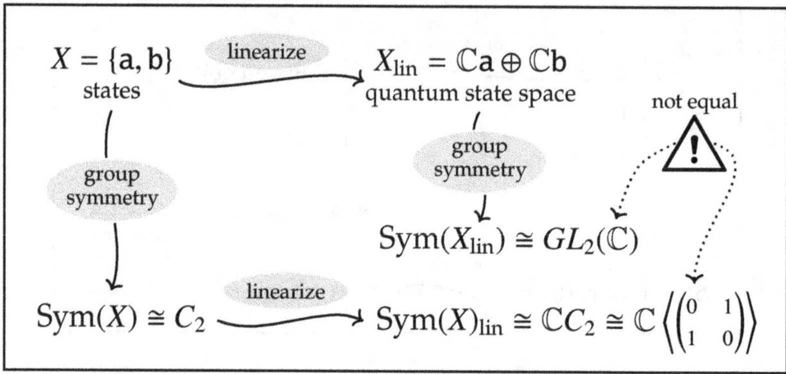

Figure 0.3: Group symmetries breaking after linearizing, for Fig. 0.1(1).

To capture all of the symmetries of quantum state spaces, one can use symmetries of algebras. In fact, linear, origin-preserving symmetries of $\mathbb{C}a \oplus \mathbb{C}b$ correspond to degree-preserving symmetries of the *tensor algebra* $T(\mathbb{C}a \oplus \mathbb{C}b)$ on this space. This algebra is also known as the *free algebra* $\mathbb{C}\langle a, b\rangle$ on variables a and b, which has the following structure:

- a $\mathbb{C}$-vector space basis of words in a and b (e.g., a, ba, aabab...), with
- multiplication given by concatenation (e.g., ba $*$ aabab $=$ baaabab), and
- unit given by the empty word.

One can also recover the quantum state space from $T(\mathbb{C}a \oplus \mathbb{C}b)$ by taking $\mathbb{C}$-linear combinations of words of degree (or length) one, which is its *generating space*. See Figure 0.4. This puts symmetries of $\mathbb{C}$-linearized states in the framework of symmetries of $\mathbb{C}$-algebras.

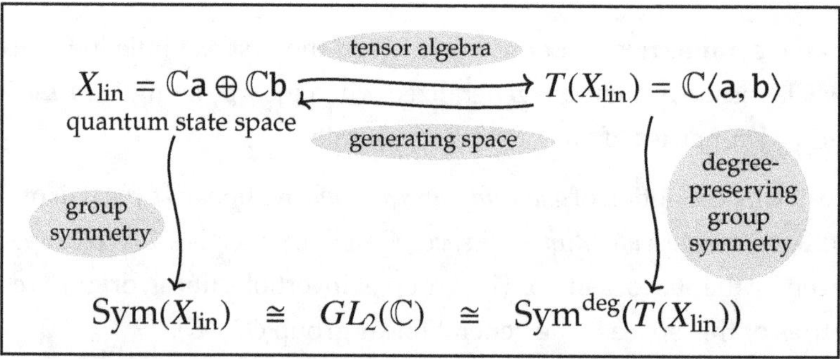

Figure 0.4: Connection to symmetries of algebras, for Figure 0.1(1).

Deforming symmetries. Next, let us build on the second example in Figure 0.1 to see how the framework for symmetries needs to be generalized when incorporating *deformation*. This procedure alters some features of an object while other features remain unchanged.

Since we cannot visualize deformations of the space $\mathbb{C}^2$, it is best to replace this geometric object with an algebraic one. This involves taking the *coordinate algebra* $\mathcal{O}(\mathbb{C}^2)$ of $\mathbb{C}^2$, which is a polynomial algebra with variables being the basis of the space, e.g., $\mathbb{C}[x, y]$ in this case. This procedure matches *ideals I* of $\mathbb{C}[x, y]$ with shapes in $\mathbb{C}^2$ cut out by setting elements of I equal to zero. For instance, the ideal (x) of $\mathbb{C}[x, y]$ corresponds to the y-axis, the ideal (x, y) of $\mathbb{C}[x, y]$ corresponds to the origin, and the ideal (0) of $\mathbb{C}[x, y]$ corresponds to all of $\mathbb{C}^2$. The bigger the ideal of $\mathbb{C}[x, y]$, the smaller the shape in $\mathbb{C}^2$. In fancier terms, this algebro-geometric correspondence is said to be *contravariant*– the directions of maps between objects get reversed when going between the geometric and algebraic settings. Retrieving the geometric shape from its coordinate algebra involves taking the *spectrum* of the algebra, but we will not go into the details here.

Returning to symmetries, let us consider symmetries of $\mathbb{C}^2$ in the geometric setting by taking symmetries of its coordinate algebra $\mathcal{O}(\mathbb{C}^2) = \mathbb{C}[x, y]$ in the algebraic setting. Linear, origin-preserving symmetries of $\mathbb{C}^2$ correspond to degree-preserving algebra maps of $\mathbb{C}[x, y]$, and both groups of symmetries are the general linear group $GL_2(\mathbb{C})$. See Figure 0.5.

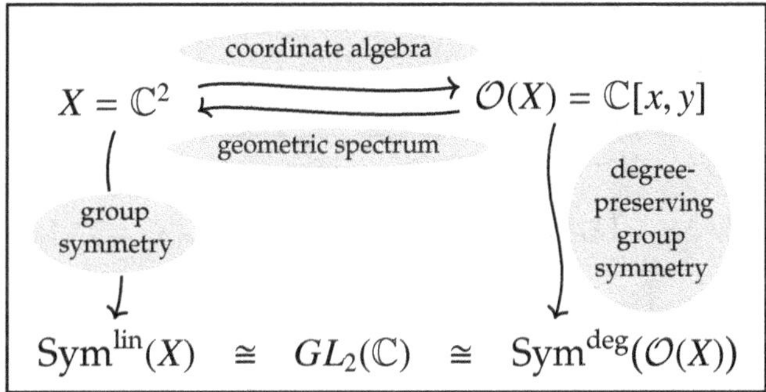

Figure 0.5: Connection to symmetries of algebras, for Figure 0.1(2).

On the other hand, to investigate deformations of the space $\mathbb{C}^2$ (which reside in the quantum world, and thus, cannot visualize), we deform its coordinate algebra $\mathbb{C}[x, y]$ and proceed algebraically. *Algebraic deformation* is a process that creates from one algebra A another algebra A_{def} that has the same vector space basis as A, but has a different multiplication. One way of deforming $\mathbb{C}[x, y]$ is by taking a nonzero complex number q, and creating the *q-polynomial algebra*, $\mathbb{C}_q[x, y]$, generated by non-commuting variables x and y, with $yx = qxy$ as a relation. This is a well-behaved algebra that serves as the coordinate algebra for a deformation of $\mathbb{C}^2$, commonly known as the *quantum 2-space*, $\mathbb{C}_q^2$.

Do the processes taking of a q-deformation and taking group symmetries commute? Unfortunately not. We show this next. To compute the group of degree-preserving symmetries of $\mathbb{C}_q[x, y]$, it suffices to determine the matrices $\begin{pmatrix} a & b \\ c & d \end{pmatrix}$ in $GL_2(\mathbb{C})$ such that the transformations $x \mapsto ax + by$ and $y \mapsto cx + dy$ preserve the relation space $\mathbb{C}(yx - qxy)$ of $\mathbb{C}_q[x, y]$. Under the transformations,

$$yx - qxy \mapsto (1 - q)acx^2 + (bc - qad)xy + (ad - qbc)yx + (1 - q)bdy^2.$$

When $q = 1$, then

$$yx - xy \mapsto (ad - bc)(yx - xy),$$

which lies in the relation space $\mathbb{C}(yx - xy)$; thus, there are no restrictions on a, b, c, d in this case. But when $q = -1$, then

$$yx + xy \mapsto 2acx^2 + (ad + bc)(yx + xy) + 2bdy^2,$$

and we need $a = d = 0$ or $b = c = 0$ for the output to lie in the relation space $\mathbb{C}(yx+xy)$. Thus, the group symmetries in the $q = -1$ case consist of transformations by diagonal or anti-diagonal matrices. Likewise, we can only have transformations by diagonal matrices when $q \neq \pm 1$. So, the group of symmetries of $\mathbb{C}_q[x, y]$ shrinks greatly when we move away from the $q = 1$ case, yet there is no way we can q-deform the group $GL_2(\mathbb{C})$ to mirror this shrinkage. We need to expand beyond the framework of group symmetries for deformations to be properly taken into account. See Figure 0.6.

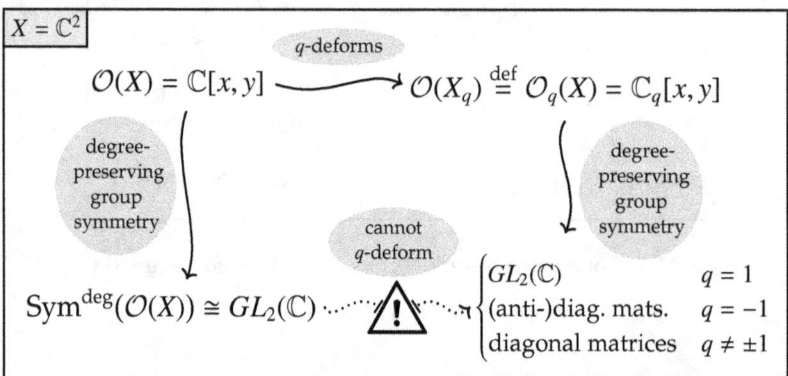

Figure 0.6: Group symmetries breaking after deforming, for Fig. 0.1(2).

To land in a setting in which symmetry commutes with deformation, we exploit the contravariant relationship between $\mathbb{C}^2$ and its coordinate algebra $\mathcal{O}(\mathbb{C}^2) = \mathbb{C}[x, y]$. Recall that the geometric symmetries of $\mathbb{C}^2$ are given by matrix multiplication:

$$GL_2(\mathbb{C}) \times \mathbb{C}^2 \to \mathbb{C}^2, \quad \left(\begin{pmatrix} a & b \\ c & d \end{pmatrix}, \begin{pmatrix} x \\ y \end{pmatrix} \right) \mapsto \begin{pmatrix} ax + by \\ cx + dy \end{pmatrix}.$$

The algebraic symmetries of $\mathcal{O}(\mathbb{C}^2)$ are then given by *matrix comultiplication*:

$$\mathcal{O}(\mathbb{C}^2) \to \mathcal{O}(\mathbb{C}^2) \otimes_{\mathbb{C}} \mathcal{O}(GL_2(\mathbb{C})), \quad \begin{pmatrix} x \\ y \end{pmatrix} \mapsto \begin{pmatrix} x \otimes_{\mathbb{C}} a + y \otimes_{\mathbb{C}} c \\ x \otimes_{\mathbb{C}} b + y \otimes_{\mathbb{C}} d \end{pmatrix}.$$

Here, $GL_2(\mathbb{C})$ is not only a group, it is also realized as a geometric shape in complex 5-space with coordinates a, b, c, d, t, subject to the relation $(ad - bc)t = 1$. Namely, the variable t is used to force the determinant $ad - bc$ to be nonzero. Then, the coordinate algebra $\mathcal{O}(GL_2(\mathbb{C}))$ of $GL_2(\mathbb{C})$ is the quotient algebra of $\mathbb{C}[a, b, c, d, t]$ by the ideal generated by the relation $(ad - bc)t = 1$. We also have that the commutative algebra $\mathcal{O}(GL_2(\mathbb{C}))$ inherits the rich structure of being a *Hopf algebra* (or a *quantum group*) due to $GL_2(\mathbb{C})$ being a group, but we will not elaborate on this here.

Fortunately, the Hopf algebra $\mathcal{O}(GL_2(\mathbb{C}))$ can q-deform. Its q-deformation $\mathcal{O}_q(GL_2(\mathbb{C}))$ is also a Hopf algebra, which as an algebra, is the quotient algebra of a q-polynomial algebra $\mathbb{C}_q[a, b, c, d, t]$ with certain q-commutation relations by the ideal generated by the relation $(ad - qbc)t = 1$. This relation encodes that a matrix in coordinates a, b, c, d has nonzero q-*determinant*. Most importantly, the symmetries of the quantum space $\mathbb{C}_q^2$, and its coordinate algebra $\mathbb{C}_q[x, y]$, are also given by the same matrix comultiplication rule as above:

$$\mathcal{O}_q(\mathbb{C}^2) \to \mathcal{O}_q(\mathbb{C}^2) \otimes_{\mathbb{C}} \mathcal{O}_q(GL_2(\mathbb{C})), \quad \begin{pmatrix} x \\ y \end{pmatrix} \mapsto \begin{pmatrix} x \otimes_{\mathbb{C}} a + y \otimes_{\mathbb{C}} c \\ x \otimes_{\mathbb{C}} b + y \otimes_{\mathbb{C}} d \end{pmatrix}.$$

In short, the processes of taking q-deformation does not commute with group symmetries, but it does commute with *Hopf symmetries*, also called *quantum group symmetries*. This is a clean way of q-deforming symmetries from the classical setting to the quantum setting, and it requires a framework beyond group symmetry. See Figure 0.7.

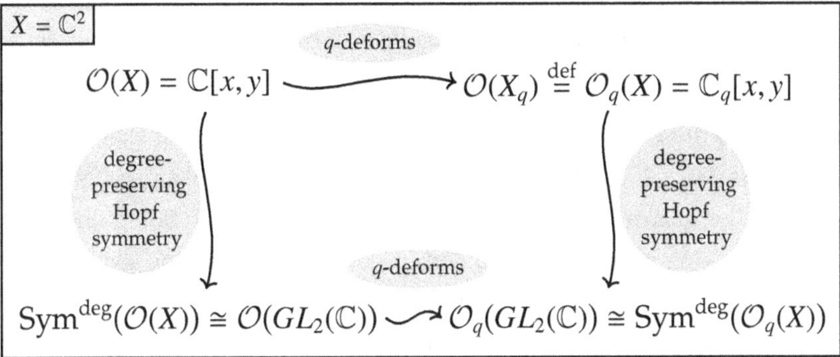

Figure 0.7: Hopf symmetries preserved after deforming, for Fig. 0.1(2).

Framework of Category Theory. What is beautiful about the algebraic framework for capturing symmetries is that it can all be expressed in the language of Category Theory. This provides us with powerful tools for translation between mathematical frameworks (e.g., algebraic, geometric, topological, combinatorial, etc.). A *category* is simply a collection of objects and structure-preserving maps between the objects, subject to certain predictable axioms (e.g., there is always an identity map between an object and itself). For instance, the collection of $\mathbb{C}$-vector spaces and $\mathbb{C}$-linear maps forms a category, denoted by $\mathsf{Vec}_{\mathbb{C}}$. Related to the examples above, the category of *modules* over a group G, denoted by $G\text{-Mod}$, and the category of *comodules* over the coordinate algebra $\mathcal{O}(G)$, denoted by $\mathsf{Comod}\text{-}\mathcal{O}(G)$, will play a vital role in this book series.

Moreover, what makes the categories $\mathsf{Vec}_{\mathbb{C}}$, $G\text{-Mod}$, $\mathsf{Comod}\text{-}\mathcal{O}(G)$ especially useful here are that they admit a monoidal structure. Namely, a category $\mathcal{C}$ is *monoidal* if it comes equipped with an associative binary operation $\otimes$ on $\mathcal{C}$ and an object $\mathbb{1}$ of $\mathcal{C}$ for which $(\mathcal{C}, \otimes, \mathbb{1})$ mimics the structure of a monoid. This allows us to combine two objects (resp., maps) of $\mathcal{C}$ to build an object (resp., a map) in $\mathcal{C}$. For instance, $\mathsf{Vec}_{\mathbb{C}}$ is a monoidal category, with $\otimes$ being the tensor product $\otimes_{\mathbb{C}}$ over $\mathbb{C}$, and with $\mathbb{1} = \mathbb{C}$. Similarly, both $G\text{-Mod}$ and $\mathsf{Comod}\text{-}\mathcal{O}(G)$ are monoidal categories.

Note that if we consider (co)modules over an arbitrary algebra H, then we need extra structure on H to make $H\text{-(Co)Mod}$ a monoidal category. This entails putting a *bialgebra*, or a Hopf algebra, structure on H, as we will see later in the book series. Again, indeed, the coordinate algebras $\mathcal{O}(GL_2(\mathbb{C}))$ and $\mathcal{O}_q(GL_2(\mathbb{C}))$ both admit the structure of a Hopf algebra.

One can also build algebraic structures within monoidal categories. For instance, an *algebra* in $(\mathsf{Vec}_{\mathbb{C}}, \otimes_{\mathbb{C}}, \mathbb{C})$ is a $\mathbb{C}$-vector space A that comes equipped with a $\mathbb{C}$-linear *multiplication map* $m : A \otimes_{\mathbb{C}} A \to A$ and a $\mathbb{C}$-linear *unit map* $u : \mathbb{C} \to A$ such that the triple (A, m, u) mimics the structure of a unital ring. In fact, algebras in the monoidal category $\mathsf{Vec}_{\mathbb{C}}$ are the same as $\mathbb{C}$-algebras. Likewise, one can build algebras, and other interesting algebraic structures (such as *coalgebras* and *Frobenius algebras*) in general monoidal categories.

Returning to the running examples above, recall for the first example that the tensor algebra $T(\mathbb{C}a \oplus \mathbb{C}b)$ has invertible, degree-preserving, algebra symmetries captured by the group $GL_2(\mathbb{C})$. This amounts to $T(\mathbb{C}a \oplus \mathbb{C}b)$ being an algebra in the monoidal category $GL_2(\mathbb{C})\text{-Mod}$. On the other hand, for the second example pertaining to symmetries of q-polynomial algebra $\mathbb{C}_q[x, y]$, we are able to remedy the problem with using group symmetries in Figure 0.5 by using Hopf symmetries, as seen Figure 0.6. This amounts to changing the monoidal categories in which we work, using $\mathsf{Comod}\text{-}\mathcal{O}(GL_2(\mathbb{C}))$ instead of $GL_2(\mathbb{C})\text{-Mod}$, as illustrated in Figure 0.8.

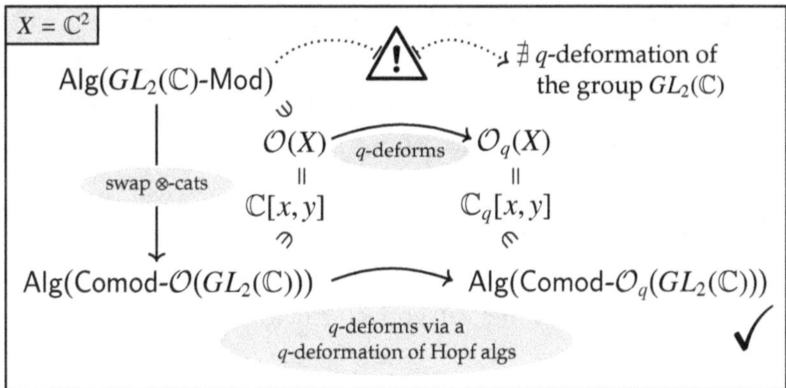

Figure 0.8: Monoidal categories and deformed symmetries, for Fig. 0.1(2).

Even though the majority of monoidal categories in this book series are algebraic in nature, there are many important monoidal categories that are inherently non-algebraic, e.g., topological, geometric, or even information-theoretic. Thus, once we understand monoidal categories and algebraic structures within them, the tools in this book series can be adapted to study symmetries of algebras in other mathematical settings!

§0.2. Contents of the book series

The contents of this three-volume book series are guided by notions that one would like to understand fully when learning any mathematical concept.

Terminology .... (Definitions)

Examples .... (Models)

Constructions .... (How to build more)

Key theorems .... (Tools)

History .... ("Why did they care?")

Applications .... ("Why care now?")

We center the notions above on the algebraic structures featured in Figure 0.9, and within the framework of category theory as illustrated in Figure 0.10. The three volumes of the book series are divided as follows.

Volume 1 pertains to algebras and categories: Chapter 1 is on algebras over a field; Chapter 2 is on category theory; Chapter 3 is on monoidal categories; and Chapter 4 is on algebras in monoidal categories.

See Figure 4.13 for a summary of the results we will start with here in Chapter 1, and how we plan to end in Chapter 4, via the contents in Chapters 2 and 3.

Volume 2 will be on coalgebras and Frobenius algebras: Chapter 5 will be on coalgebras over a field; Chapter 6 will be on coalgebras in monoidal categories; Chapter 7 will be on Frobenius algebras over a field; and Chapter 8 will be on Frobenius algebras in monoidal categories.

Volume 3 will be on Hopf algebras and beyond: Chapter 9 will be on bialgebras and Hopf algebras over a field; Chapter 10 will be on braided monoidal categories; Chapter 11 will be on Hopf algebras and related structures in monoidal categories; and Chapter 12 will be on higher categorical structures.

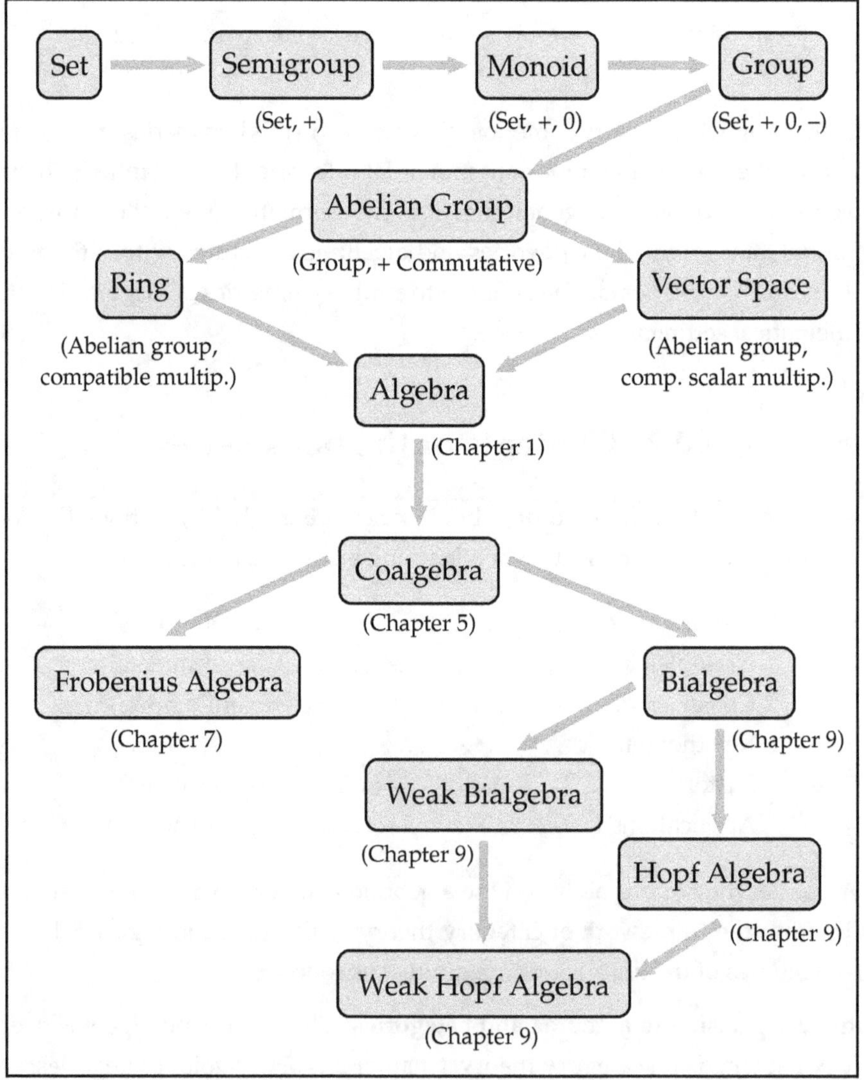

Figure 0.9: Algebraic structures with underlying set structure.

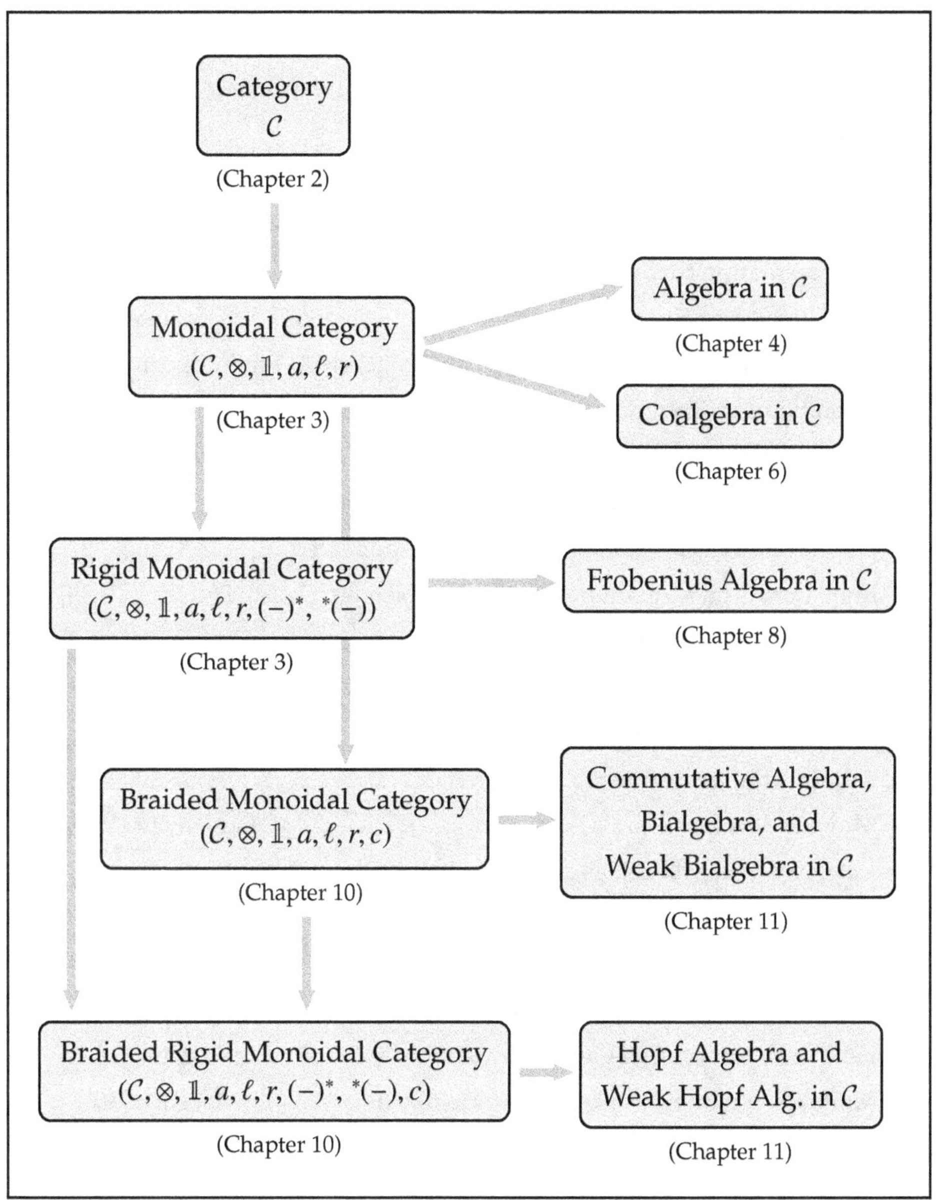

Figure 0.10: Categorical (algebraic) structures.

Timeline.　　The topics in this book series will be presented out of the order in which they were introduced historically. But here is a snapshot of when the structures in Volume 1 first arose in the literature.

Historical timeline of the algebraic and categorical structures in Volume 1

- Axiomatic **vector spaces** appeared . . . . . Grassmann [1844], Peano [1888]
- **Matrix algebras** investigated . . . . Cayley [1858]
- **Quaternion algebra** introduced . . . . Hamilton [1866]
- Finite dim., associative **algebras** over a field first studied . . Peirce [1881]
- **Representation theory of groups** investigated . . . . Maschke [1899]
- Definition of an **algebra** over a field first developed . . . . . Dickson [1903]
- First structure results for assoc. **fin. dim. algebras** . . Wedderburn [1908]
- Definition of a (commutative, assoc.) **ring** appeared . . . . Fraenkel [1915]
- **Abstract ring theory** established . . . . Noether [1921]
- Foundations for **algebras** over a field published . . . . Dickson [1923]
- Structure results for **semisimple algebras** appeared . . . . Artin [1927]
- More results on **semisimple algebras** . . Noether [1929], Hopkins [1939]
- **Categories** and **functors** introduced . . . . Eilenberg and MacLane [1945]
- **Abelian categories** arose . . . . . . Buchsbaum [1955], Grothendieck [1957]
- Foundations in **Homological Algebra** set . . Cartan and Eilenberg [1956]
- **Adjoint functors** studied . . . . Kan [1958]
- **Monads** developed . . . . Godement [1958]
- **Category theory** began as an independent subject . . . . . . Lawvere [1963]
- **Monoidal categories** introduced . . . . Bénabou [1963]
- **Algebras** in monoidal categories introduced . . . . Bénabou [1964]
- **Monoidal categories** axiomatized . . . . Mac Lane [1963], Kelly [1964]
- Landmark **text on Category Theory** appeared . . . . MacLane [1971]
- **Modules** in monoidal categories formalized . . . . MacLane [1971]
- **Rigid monoidal categories** developed . . . . Saavedra Rivano [1972]
- Monoidal **algebras and modules** examined . . . . Pareigis [1977a]
- **Applications** of monoidal algebras arose . . . . Fuchs et al. [2002]
- **Algebraic structures in fusion categories** studied . . . . Ostrik [2003c]
- Key results on **fusion categories** established . . . . Etingof et al. [2005]
- Vital text on **tensor categories** published . . . . Etingof et al. [2015]

§0.3. Features of the book series

Prerequisites. It is best if the reader is familiar with abstract vector spaces, groups, rings and ideals, as covered in many undergraduate mathematics courses. These concepts and other important concepts from linear algebra (that are sometimes not offered in a course) are covered at the beginning of Chapter 1. It is important that the reader has a solid understanding of this material before proceeding with the rest of the text. Depth is more important than speed.

Reading paths. Volume 1 (Chapters 1–4) serves as a representative sample of what the book series has to offer, and it alone could be the subject of a course or a source for independent study.

Considering all of the volumes together after Volumes 2 and 3 are released, one can proceed in order, starting at Chapter 1 and ending with Chapter 12. But, as you can see in Figure 0.9, one could skip the categorical material and proceed by only studying Chapters 1, 5, 7, and 9. On the other hand, as shown in Figure 0.10, the focus could be purely on the categorical material, with the aim of studying Chapters 2, 3, 4, 6, 8, 10, 11, and 12 in depth; some material from previous non-categorical chapters may be needed for context.

In any case, one may also want to just read the first few sections of each chapter to gather the basic terms, tools, and examples of the material.

Highlighted references. This book series is not intended to be reference books containing an extensive bibliography for all items needed to do research. Instead, this is a learning book series, and a carefully curated list of additional textbooks and articles will be featured at the end of each chapter for further exploration.

Modern applications. To add to the "Why care now?" component mentioned above, there are sections dedicated to modern applications of and research themes for the algebraic and categorical structures introduced here. This will also include references for material beyond the scope of the book series, such as web addresses to lectures publicly available at the time of this publication.

Diagrammatic arguments. Many of the arguments in the book series involve showing that two sets of composed functions are equal. This can be done with the familiar line-by-line method of using equalities. On the other hand, these arguments can be executed using more visually appealing methods, namely, using *commutative diagrams* or *graphical calculus*. The latter methods will be explained in detail in this volume, and all three techniques will be used throughout the series.

Exercises.   There will be numerous exercises in this book series; in fact, there are close to 200 exercises in this volume alone. This will aid the reader in exploring the material further. The exercises will have varying degrees of difficulty, but the tougher problems are not marked with stars or labeled as "challenges", nor are the straightforward problems labeled as "easy". Full solutions could vary greatly in length depending on the difficulty of the problem and the interest of the reader. Moreover, some problems are intentionally open-ended (and labeled as such) to prompt the reader to discover their taste in this subject. This may inspire potential research directions!

The exercises are collected at the end of each chapter, and they are cited within the body of the chapters roughly in order. So, read carefully to not only hunt for the problems but to also hunt for some of their solutions. Have fun!

$\cdot$ CHAPTER 1 $\cdot$

ALGEBRAS OVER A FIELD

History

An *algebra* is a mathematical structure that is a combination of a unital ring and a vector space. A special case of these structures first appeared in the literature as "hypercomplex numbers", which are algebras over the real numbers that have multiplicative inverses (i.e., $\mathbb{R}$-division algebras). Examples of such $\mathbb{R}$-division algebras include the real numbers $\mathbb{R}$, the complex numbers $\mathbb{C}$, and the quaternions $\mathbb{H}$ introduced by Hamilton [1866]. Other examples of algebras, namely matrix algebras, were examined by Cayley [1858], predating the work on hypercomplex numbers. The first comprehensive study of hypercomplex numbers was carried out by Peirce [1881], and the first general structure results for algebras were achieved by Wedderburn [1908], building on the work of Cartan [1898] and Molien [1892]. The introduction of abstract ring theory by Noether [1921], along with the ring axioms presented by Fraenkel [1915], set the foundation for the study of abstract algebras including the classes above. The foundations for the theory of algebras over a field were then established by Dickson [1923], twenty years after the first postulates for these structures were posed by Dickson [1903].

Overview

This chapter on algebras over a field sets the foundation for most of the material in this book, as illustrated in Figures 0.9 and 0.10. Basic terminology and notions about algebras over a field are discussed in §1.1, building on results from the theory of groups, rings, and vector spaces. Many examples of algebras are provided in §1.2. How algebras *act* on vector spaces, i.e., to form representations and modules, is discussed in §1.3. Constructions of algebras and their representations and modules are provided are §1.4. Structure results are then given in §§1.5–1.7 for algebras that are *simple*, *semisimple*, and *separable*, respectively. A diagram summarizing the structure results and examples is given in §1.8. The chapter ends with a discussion of modern applications in §1.9, and references in §1.10, along with several exercises.

Standing hypotheses.   Throughout the book, $\mathbb{N}$ is the collection of natural numbers including 0. Moreover, $\mathbb{Z}, \mathbb{Q}, \mathbb{R}$, and $\mathbb{C}$ are the collections of integers, rational numbers, real numbers, and complex numbers, respectively.

All linear algebraic structures are over a **ground field** $\Bbbk$, i.e., are $\Bbbk$-vector spaces. Here, $\Bbbk$ has characteristic 0 and is algebraically closed. That is, $1_\Bbbk + \cdots + 1_\Bbbk$ is never $0_\Bbbk$ for nonzero linear structures, and polynomials over $\Bbbk$ have all roots in $\Bbbk$. For instance, take $\Bbbk = \mathbb{C}$. Several results do not require these assumptions on $\Bbbk$, but we will emphasize when they are needed.

---

§1.1. Algebras over a field

We review various notions about groups, rings, and vector spaces over a field, and then build on these notions to study algebras over a field.

§1.1.1. Groups

A **monoid** is a set G equipped with an associative binary operation,

$$\star := \star_G : G \times G \to G, \quad (g,h) \mapsto gh,$$

(that is, $(gh)\ell = g(h\ell)$ for all $g, h, \ell \in G$), and with an identity element $e := e_G$ with respect to the operation $\star$ (that is, $ge = g = eg$, for all $g \in G$).

A **group** is a monoid $(G, \star, e)$ such that, for each element $g \in G$, there exists an inverse element g^{-1} (that is, $gg^{-1} = g^{-1}g = e$). Sometimes, we denote the underlying set of G by G_{set}.

The **order** of a group G is the cardinality of G_{set}, denoted by $|G|$. We say that G is **(in)finite** if $|G|$ is (in)finite.

A group G is **abelian** if $gh = hg$, for all $g, h \in G$. In this case, we denote the operation $\star_G$ by $+$, the identity element e_G by 0, and inverse elements by $-g$.

Given groups $(G, \star, e)$ and $(G', \star', e')$, a function $\phi : G \to G'$ is a **group (homo)morphism** or a **group map** if preserves the operations of G and G':

$$(\phi(g \star h) =) \quad \phi(gh) = \phi(g)\phi(h) \quad (= \phi(g) \star' \phi(h)),$$

for all $g, h \in G$. Homomorphisms provide a way of comparing two groups.

Commutative diagrams. Notice that the equation, $\phi(gh) = \phi(g)\phi(h)$, above is captured by the diagram of composed morphisms below. Namely, the compositions of morphisms (going $\rightarrow$ then $\downarrow$, or going $\downarrow$ then $\rightarrow$) from the starting object $G \times G$ to the finishing object G' yield the same outcome. In this case, we say that the diagram **commutes**.

$$
\begin{array}{ccc}
G \times G & \xrightarrow{\ \star\ } & G \\
{\scriptstyle \phi \times \phi}\big\downarrow & & \big\downarrow{\scriptstyle \phi} \\
G' \times G' & \xrightarrow{\ \star'\ } & G'
\end{array}
\qquad
\begin{array}{ccc}
(g,h) & \longmapsto & gh \\
\big\downarrow & & \big\downarrow \\
(\phi(g),\phi(h)) & \longmapsto & \phi(g)\phi(h) = \phi(gh)
\end{array}
$$

We will give many more examples of commutative diagrams throughout this book, and they will serve as a tool to illustrate when compositions of morphisms are equal.

Observe that we can combine elements of groups by using commutative diagrams. To see this, take the one-point set, $\{\cdot\}$, and note that the cartesian product $\{\cdot\} \times \{\cdot\}$ is isomorphic to $\{\cdot\}$, as sets (i.e., there exists a bijection between the two sets). Next, identify each element $g \in G$ with a function,

$$
\vec{g} : \{\cdot\} \rightarrow G.
$$

Then, one can identify a product of elements, gh, with the composition of morphisms, $\star \circ (\vec{g} \times \vec{h})$. For instance, the equations $ge = g = eg$ are encoded by the commutative diagrams below.

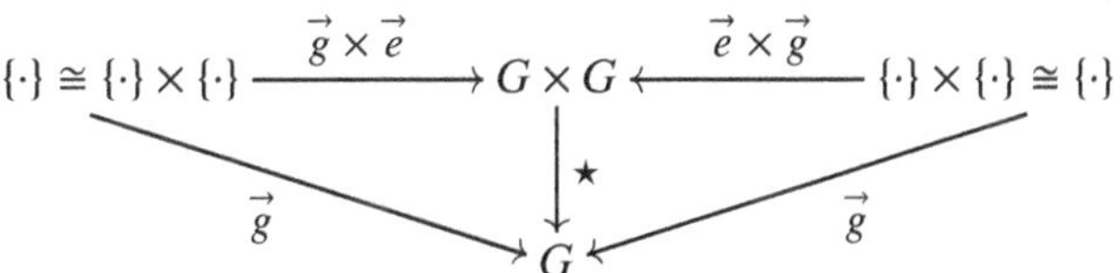

A group (homo)morphism $\phi : G \rightarrow G'$ is called a **group monomorphism** if ϕ is injective; a **group epimorphism** if ϕ is surjective; a **group isomorphism** if ϕ is bijective; a **group endomorphism** if $G' = G$; and a **group automorphism** if ϕ is bijective and $G' = G$. For example, the **identity map** id_G, where $\mathrm{id}_G(g) := g$ for all $g \in G$, is a group automorphism.

Sometimes, monomorphisms and epimorphisms are referred to as **embeddings** and **projections**, respectively.

We also have that ϕ is a **group isomorphism** if, equivalently, there exists a group homomorphism $\psi : G' \rightarrow G$ such that $\psi \circ \phi = \mathrm{id}_G$ and $\phi \circ \psi = \mathrm{id}_{G'}$. In this case, we say that G is **isomorphic** to G' and write $G \cong G'$. Two groups G and G' are considered to be the same abstractly when $G \cong G'$.

Structure versus property. Observe that a set G becoming a group involves equipping G with extra **structure**– features that are attached. However, a group G being abelian requires that G has a certain **property**– a condition that either holds or does not hold. In other words, structures are to nouns, as properties are to adjectives. Moreover, morphisms preserve structure, and properties are preserved under isomorphisms for free. For instance, we have morphisms between groups that do not need to preserve the property of being abelian. But isomorphic groups must simultaneously be abelian, or simultaneously not be abelian. See Exercise 1.1.

Given a group G, a nonempty subset H of G is a **subgroup** if H is a group under the same operation and identity element of G. In this case, there is a canonical group monomorphism $H \to G$ given by inclusion.

A subgroup N of G is called **normal** if $gng^{-1} \in N$ for all $g \in G$ and $n \in N$. Subgroups of abelian groups are always normal.

Given a subgroup H of G and $g \in G$, the set $gH = \{gh \mid h \in H\}$ is a **left coset of H in G**, and the collection of left cosets of H in G is denoted by G/H.

The set G/H can be upgraded to a group if and only if $H =: N$ is a normal subgroup of G. Here,

$$(gN)(g'N) := (gg')N, \qquad e_{G/N} := eN = N \qquad (gN)^{-1} := g^{-1}N,$$

for all $g, g' \in N$. In this case, G/N is called a **quotient group**, and there is a canonical group epimorphism $G \to G/N$ defined by $g \mapsto gN$.

Let $\phi : G \to G'$ be a group homomorphism. Then, the **kernel** of ϕ is the set

$$\ker(\phi) := \{g \in G \mid \phi(g) = e_{G'}\},$$

and the **image** of ϕ is

$$\mathrm{im}(\phi) := \{\phi(g) \in G' \mid g \in G\}.$$

In fact, $\ker(\phi)$ is a normal subgroup of G, but $\mathrm{im}(\phi) \subseteq G'$ is a subgroup that is not necessarily normal. We also have the following isomorphisms of groups:

$$G/\ker(\phi) \cong \mathrm{im}(\phi).$$

Now let $\mathrm{im}(\phi)_{\mathrm{norm}}$ be the smallest normal subgroup of G' containing $\mathrm{im}(\phi)$ (which exists because the intersection of normal subgroups is normal). Then, the **cokernel** of ϕ is the quotient group,

$$\mathrm{coker}(\phi) := G'/\mathrm{im}(\phi)_{\mathrm{norm}}.$$

Note that if G' is abelian, then $\mathrm{im}(\phi)_{\mathrm{norm}} = \mathrm{im}(\phi)$.

§1.1.2. Rings

A **ring** is an abelian group $(R, +, 0)$ that comes equipped with an associative binary operation

$$\cdot : R \times R \to R, \quad (r, s) \mapsto rs,$$

such that $r(s + t) = rs + rt$ and $(r + s)t = rt + st$ for all $r, s, t \in R$. The compatibility conditions between the two operations, + (**addition**) and · (**multiplication**), can be expressed by the following commutative diagrams.

Here, ▷. is the following left multiplication of R on $R \times R$: $r \triangleright. (s, t) = (r \cdot s, r \cdot t)$. Likewise, ◁. is defined by $(r, s) \triangleleft. t = (r \cdot t, s \cdot t)$.

A ring $(R, +, 0, \cdot)$ is **unital** if there is an identity element with respect to multiplication, which is denoted by $1 := 1_R$ in this case.

Given a ring $(R, +, 0, \cdot)$, its **opposite ring** is $R^{\mathrm{op}} := (R, +, 0, \cdot^{\mathrm{op}})$, where for $r, s \in R$, we have that $r \cdot^{\mathrm{op}} s := sr$. A ring R is **commutative** if $R = R^{\mathrm{op}}$, or if $rs = sr$ for all $r, s \in R$.

A ring R is a **domain** if it does not have zero divisors, that is, if $r, s \in R$ with $rs = 0$, then $r = 0$ or $s = 0$.

Given rings R and R', a function $\phi : R \to R'$ is a **ring (homo)morphism** or a **ring map** if addition and multiplication are preserved, that is, if

$$\phi(r + s) = \phi(r) + \phi(s) \qquad \text{and} \qquad \phi(rs) = \phi(r)\phi(s),$$

for all $r, s, \in R$. If R and R' are unital, then a ring homomorphism $\phi : R \to R'$ is **unital** when $\phi(1_R) = 1_{R'}$.

A unital ring $(R, +, 0, \cdot, 1)$ is called a **division ring** or **skew field** if there exists a two-sided inverse element r^{-1} with respect to ·, for each nonzero $r \in R$ (that is, $rr^{-1} = r^{-1}r = 1_R$). A commutative division ring is called a **field**.

The terminology for the various types of morphisms (e.g., **mono-**, **epi-**, **iso-**, **endo-**, **auto-**) from the group setting also applies in the ring setting.

Given a ring R, a nonempty subset S of R is a **subring** if S is a ring under the same operations of R. If R is unital, then a subring S is a **unital subring** if it has a multiplicative identity and $1_S = 1_R$. In this case, there is a canonical (unital) ring monomorphism $S \to R$ given by inclusion.

One example of a subring of a ring R is given by the **center** of R:

$$Z(R) := \{r \in R \mid rs = sr, \ \forall s \in R\}.$$

A subring $S =: I$ of R is called a **left** (resp., **right**) **ideal** if $rl \in I$ (resp., $lr \in I$) for all $r \in R$ and $l \in I$. A **(two-sided) ideal** of R is a left ideal that is also a right ideal. When R is unital, an ideal is a unital subring of R if and only if $I = R$.

Given a subring S of R, the set R/S of additive cosets can be upgraded to a group as the underlying group structure of R is abelian and S is a normal subgroup. The quotient group R/S can be upgraded further to a ring if and only if $S =: I$ is an ideal of R. Here,

$$(r + I)(r' + I) := (rr') + I,$$

for all $r, r' \in R$, and R/I is called a **quotient ring**. If R is unital, then so is R/I with

$$1_{R/I} := 1_R + I.$$

There is a canonical (unital) ring epimorphism, $R \to R/I$, defined by $r \mapsto r + I$.

Let $\phi : R \to R'$ be a ring homomorphism. Then, the **kernel** of ϕ is the set $\ker(\phi) = \{r \in R \mid \phi(r) = 0_{R'}\}$, and the **image** of ϕ is $\mathrm{im}(\phi) = \{\phi(r) \mid r \in R\}$. In fact, $\ker(\phi)$ is an ideal of R, but $\mathrm{im}(\phi) \subseteq R'$ is a subring that is not always an ideal. We also have that, as rings:

$$R/\ker(\phi) \cong \mathrm{im}(\phi).$$

Similar to the group setting, let $\mathrm{im}(\phi)_{\mathrm{ideal}}$ be the smallest ideal of R' containing $\mathrm{im}(\phi)$ (which exists because the intersection of ideals is an ideal). Then, the **cokernel** of ϕ is the quotient ring,

$$\mathrm{coker}(\phi) := R'/\mathrm{im}(\phi)_{\mathrm{ideal}}.$$

However, it can be difficult to understand $\mathrm{im}(\phi)_{\mathrm{ideal}}$. For this reason and others, the collection of rings behaves pathologically (namely, the *category* of rings is quite weird). Therefore, we will primarily work with the collection of vector spaces over a field (discussed next) to construct algebras over a field later.

§1.1.3. Vector spaces

A **vector space over a field** $\Bbbk$ is an abelian group $(V, +, 0)$ that comes equipped with a binary operation

$$* : \Bbbk \times V \to V, \quad (\lambda, v) \mapsto \lambda v,$$

such that, for all $\lambda, \mu \in \Bbbk$ and $v, w \in V$, we get:

$$\lambda(v + w) = \lambda v + \lambda w, \qquad (\lambda + \mu)v = \lambda v + \mu v, \qquad (\lambda\mu)v = \lambda(\mu v), \qquad 1_{\Bbbk}v = v.$$

The first three compatibility conditions between the operations, + (**addition**) and ∗ (**scalar multiplication**), can be expressed by the commutative diagrams below.

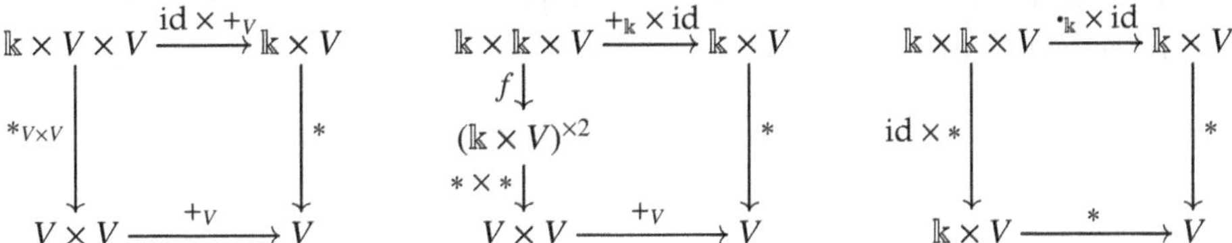

Here, $*_{V \times V}$ is scalar multiplication on $V \times V$ given by $\lambda(v, w) := (\lambda v, \lambda w)$. Moreover, f is the composition of $\mathrm{id}_\Bbbk \times \mathrm{id}_\Bbbk \times \mathrm{diag}$, for $\mathrm{diag}(v) := (v, v)$, with $\mathrm{id} \times \mathrm{flip} \times \mathrm{id}$.

Elements of vector spaces are called **vectors**. Each vector comes equipped with a (non-unique) collection of vectors $\mathcal{B} := \{b_i\}_i$ such that for each $v \in V$, we have that $v = \sum_i^{\text{finite}} \lambda_i b_i$ for some scalars $\lambda_i \in \Bbbk$ (i.e., $\mathcal{B}$ **spans** V), and such that the only way to express 0_V in terms of $\mathcal{B}$ is as $0_V = \sum_i 0_\Bbbk b_i$ (i.e., $\mathcal{B}$ is **linearly independent**). Here, $\mathcal{B}$ is called a **basis** of V. The cardinalities of any two bases of V are equal, and this value is the **dimension** of V, denoted by $\dim_\Bbbk V$. By convention, the zero vector space is only vector space of dimension 0.

Given vector spaces V and V', a function $\phi : V \to V'$ is a (**$\Bbbk$-)linear map** or a **vector space (homo)morphism** if

$$\phi(v + w) = \phi(v) + \phi(w) \quad \text{and} \quad \phi(\lambda v) = \lambda \phi(v),$$

for all $\lambda \in \Bbbk$ and $v, w \in V$. The terminology for the various types of morphisms (namely, **mono-**, **epi-**, **iso-**, **endo-**, **auto-**) from the group setting also applies in the vector space setting.

To any linear map $\phi : V \to V'$, with fixed bases $\mathcal{B} = \{b_j\}_j$ of V and $\mathcal{B}' = \{b_i'\}_i$ of V', we have that $\phi(b_j) = \sum_i \lambda_{i,j} b_i'$, for some scalars $\lambda_{i,j} \in \Bbbk$. The **matrix of ϕ with respect to $\mathcal{B}$ and $\mathcal{B}'$**, is the $\dim_\Bbbk V' \times \dim_\Bbbk V$ matrix given by $\mathrm{Mat}_{\mathcal{B}, \mathcal{B}'}(\phi) = (\lambda_{i,j})_{i,j}$. This matrix is invertible if and only if ϕ is an isomorphism.

Given a vector space V, a nonempty subset U of V is a **subspace** if U is a vector space under the same operations of V. There is a canonical linear embedding $U \to V$ in this case given by inclusion.

Given a subspace U of V, the set V/U of additive cosets is a group because V_{group} is abelian and U is a normal subgroup of V. The quotient group V/U is also a vector space, where $\lambda(v + U) := (\lambda v) + U$, for $\lambda \in \Bbbk$, $v \in V$. In this case, V/U is called a **quotient space**, and there is a linear projection, $V \to V/U$, given by $v \mapsto v + U$.

Let $\phi : V \to V'$ be a linear map. The **kernel** of ϕ is $\ker(\phi) = \{v \in V \mid \phi(v) = 0_{V'}\}$, which is also known as the **nullspace** of ϕ. The dimension of the nullspace of ϕ is called the **nullity** of ϕ.

The **image** of a linear map $\phi : V \to V'$ is $\mathrm{im}(\phi) = \{\phi(v) \mid v \in V\}$, also known as the **range** of ϕ. The dimension of the range of ϕ is called the **rank** of ϕ.

For a linear map $\phi : V \to V'$, we have that $\ker(\phi)$ is a subspace of V, $\mathrm{im}(\phi)$ is a subspace of V', and as vector spaces: $V/\ker(\phi) \cong \mathrm{im}(\phi)$. So, if $\dim_{\Bbbk} V < \infty$, then

$$\dim_{\Bbbk} V = \mathrm{nullity}(\phi) + \mathrm{rank}(\phi),$$

which is known as the **Rank-Nullity Theorem**.

Moreover, the **cokernel** of ϕ is the quotient space, $\mathrm{coker}(\phi) := V'/\mathrm{im}(\phi)$, and its dimension is called the **corank** of ϕ.

§1.1.4. Operations on vector spaces

To combine algebras later in §1.4, we need the vector space operations below.

i. Direct products, sums, and direct sums

The **direct product** of vector spaces $V_1, \ldots, V_r$ is given by

$$V_1 \times \cdots \times V_r = \{(v_1, \ldots, v_r) \mid v_i \in V_i, \ i = 1, \ldots, r\},$$

a vector space with component-wise addition and scalar multiplication.

For a vector space V, the **sum** of subspaces of $V_1, \ldots, V_r$ of V is given by

$$V_1 + \cdots + V_r = \{\textstyle\sum_{i=1}^{r} v_i \mid v_i \in V_i, \ i = 1, \ldots, r\},$$

which is a subspace of V with summand-wise addition and scalar multiplication.

If every element of $V_1 + \cdots + V_r$ is a sum of vectors $\sum_i v_i$ for a *unique* choice of $v_i \in V_i$, then $V_1 + \cdots + V_r$ is a **direct sum**, denoted by $V_1 \oplus \cdots \oplus V_r$.

We also have that

$$\dim_{\Bbbk}(V_1 \times \cdots \times V_r) = \dim_{\Bbbk}(V_1 \oplus \cdots \oplus V_r) = \textstyle\sum_{i=1}^{r} \dim_{\Bbbk} V_i,$$

but $\dim_{\Bbbk}(V_1 + \cdots + V_r)$ could be strictly less than $\sum_{i=1}^{r} \dim_{\Bbbk} V_i$.

Remark 1.1. There is a subtle difference between direct products and direct sums of vector spaces. Both constructions have component-wise operations. However, the components of a direct sum are substructures (summands), and this is not required for direct products. Also, note that for a direct product of vector spaces $V_1 \times V_2$, we have that $V_1 \times \{0_{V_2}\}$ and $\{0_{V_1}\} \times V_2$ are subspaces of $V_1 \times V_2$, and that $V_1 \times \{0_{V_2}\} \cong V_1$ and $\{0_{V_1}\} \times V_2 \cong V_2$ as vector spaces. Therefore,

$$V_1 \times V_2 \cong (V_1 \times \{0_{V_2}\}) \oplus (\{0_{V_1}\} \times V_2) \cong V_1 \oplus V_2,$$

where the first isomorphism is given by $(v_1, v_2) \mapsto (v_1, 0_{V_2}) + (0_{V_1}, v_2)$.

Moreover, we leave it to the reader to verify the result below.

Lemma 1.2. *We have a linear embedding and projection, for each $j = 1, \ldots, r$, given by:*

$$V_j \to V_1 \times \cdots \times V_r, \quad v_i \mapsto (0_{V_1}, \ldots, 0_{V_{j-1}}, v_j, \, 0_{V_{j+1}}, \ldots, 0_{V_r}),$$

$$V_1 \times \cdots \times V_r \to V_j, \quad (v_1, \ldots, v_j, \ldots, v_r) \mapsto v_j.$$

Similar statements hold for direct sums. $\qquad\qquad\square$

ii. Bilinear maps and multilinear maps

For vector spaces V, W, Z, a function $T : V \times W \to Z$ is called a **bilinear map** if it is linear in each slot, that is, for each $v \in V$ and $w \in W$,

$$T(-, w) : V \to Z, \;\; v \mapsto T(v, w) \quad \text{and} \quad T(v, -) : W \to Z, \;\; w \mapsto T(v, w)$$

are linear maps. Bilinear maps of the type $T : V \times V \to \Bbbk$ are called **bilinear forms on V**. A bilinear form T on V is said to be **nondegenerate** if $T(v, v') = 0_{\Bbbk}$ for all nonzero $v' \in V$ implies that $v = 0_V$. Likewise, one can define **multilinear maps**, **multilinear forms**, and define when such forms are **nondegenerate**.

iii. Tensor products

Now we will define a product of two vector spaces first by using bases. If $\{b_i\}_i$ is a basis of a vector space V, and $\{c_j\}_j$ is a basis of a vector space W, then one can form symbols $b_i \otimes c_j$, which we refer to as **simple tensors**. Sometimes, we write $\otimes_{\Bbbk}$ to emphasize the ground field $\Bbbk$. The **tensor product of V and W** is the collection of finite linear combinations of simple tensors:

$$V \otimes W \; := \; V \otimes_{\Bbbk} W \; := \; \{ \textstyle\sum_{i,j}^{\text{finite}} \lambda_{i,j} \, (b_i \otimes c_j) \mid \lambda_{i,j} \in \Bbbk \}.$$

It is a $\Bbbk$-vector space with basis $\{b_i \otimes c_j\}_{i,j}$, via the definitions below:

$$\lambda(b_i \otimes c_j) := \lambda b_i \otimes c_j := b_i \otimes \lambda c_j,$$

$$(b_i \otimes c_j) + (b_i' \otimes c_j) := (b_i + b_i') \otimes c_j, \quad (b_i \otimes c_j) + (b_i \otimes c_j') := b_i \otimes (c_j + c_j'),$$

for $\lambda \in \Bbbk$, and $b_i, b_i' \in V$, and $c_i, c_i' \in W$. Here, $\dim_{\Bbbk}(V \otimes W) = (\dim_{\Bbbk} V)(\dim_{\Bbbk} W)$.

In other words, let $F(V \times W) := \operatorname{span}_{\Bbbk}\langle(v, w) \mid v \in V, w \in W\rangle$, for $\Bbbk$-vector spaces V and W. Then, $V \otimes W$ is the quotient space $F(V \times W)/R$, where R is the subspace of $F(V \times W)$ spanned by the following vectors (for $v, v' \in V;\; w, w' \in W;\; \lambda \in \Bbbk$):

$$\begin{aligned} \lambda(v, w) - (\lambda v, w), \qquad & \lambda(v, w) - (v, \lambda w), \\ (v + v', w) - (v, w) - (v', w), \qquad & (v, w + w') - (v, w) - (v, w'). \end{aligned} \tag{1.3}$$

There is another way of defining the $\Bbbk$-vector space $V \otimes W$, which is particularly helpful when comparing $V \otimes W$ with other vector spaces. This uses the next notion.

Universal property.   *"It's as easy as α, β, γ..."*

Take a gadget X. A **universal structure attached to** X is a structure, $\mathrm{Univ}(X)$, that is connected to X, such that for any arbitrary structure, $\mathrm{Arb}(X)$, connected to X in the same way, there is a unique structure-preserving map between $\mathrm{Univ}(X)$ and $\mathrm{Arb}(X)$ that respects the connections. Universal structures have one of the two forms below.

Form I.   The connection between X and $\mathrm{Univ}(X)$ is of the form,

$$\alpha : X \to \mathrm{Univ}(X).$$

The **universal property** of $\mathrm{Univ}(X)$ is that for any structure $\mathrm{Arb}(X)$ connected to X via $\beta : X \to \mathrm{Arb}(X)$, there is a unique structure map $\gamma : \mathrm{Univ}(X) \to \mathrm{Arb}(X)$ with $\beta = \gamma \circ \alpha$. This is depicted via the commutative diagram below.

$$
\begin{array}{ccc}
X & \xrightarrow{\;\alpha\;} & \mathrm{Univ}(X) \\
& \forall\beta \searrow & \Big\downarrow{\scriptstyle \exists!\,\gamma} \\
& & \mathrm{Arb}(X)
\end{array}
$$

Loosely speaking, the connection α *feeds into* $\mathrm{Univ}(X)$ making it *rule over* all $\mathrm{Arb}(X)$. There is a bijection between the connections β and the structure maps γ above.

$$\{\beta : X \to \mathrm{Arb}(X)\} \quad \overset{1\text{-}1}{\longleftrightarrow} \quad \{\gamma : \mathrm{Univ}(X) \to \mathrm{Arb}(X)\}$$

Form II.   The connection between X and $\mathrm{Univ}(X)$ is of the form,

$$\alpha' : \mathrm{Univ}(X) \to X.$$

The **universal property** of $\mathrm{Univ}(X)$ is that for any structure $\mathrm{Arb}(X)$ is connected to X via $\beta' : \mathrm{Arb}(X) \to X$, then there is a unique structure map $\gamma' : \mathrm{Arb}(X) \to \mathrm{Univ}(X)$ with $\beta' = \alpha' \circ \gamma'$. This is depicted via the commutative diagram below.

$$
\begin{array}{ccc}
\mathrm{Arb}(X) & & \\
\Big\downarrow{\scriptstyle \exists!\,\gamma'} & \searrow{\scriptstyle \forall\beta'} & \\
\mathrm{Univ}(X) & \xrightarrow{\;\alpha'\;} & X
\end{array}
$$

Here, the connection α' *takes away from* $\mathrm{Univ}(X)$ making it *ruled under* all $\mathrm{Arb}(X)$. Likewise, we have a bijection:

$$\{\beta' : \mathrm{Arb}(X) \to X\} \quad \overset{1\text{-}1}{\longleftrightarrow} \quad \{\gamma' : \mathrm{Arb}(X) \to \mathrm{Univ}(X)\}.$$

Universal structures need not exist. But if they do exist, they are unique up to structure-preserving isomorphism; see Exercise 1.2.

Universal property of $V \otimes W$. Now take vector spaces V and W, and observe that we can also define the tensor product of V and W via the result below.

Definition-Proposition 1.4. The vector space $V \otimes W$ is equipped with a bilinear map

$$\alpha : V \times W \to V \otimes W, \quad (v, w) \mapsto v \otimes w,$$

such that for any bilinear map $\beta : V \times W \to Z$, there exists a unique linear map $\gamma : V \otimes W \to Z$ with $\beta = \gamma \circ \alpha$. Namely, the following diagram commutes.

$$
\begin{array}{ccc}
V \times W & \xrightarrow{\ \alpha\ } & V \otimes W \\
& \searrow & \downarrow \exists! \text{ linear map } \gamma \\
\forall \text{ bilinear map } \beta & & Z \text{ vs}
\end{array}
$$

Here, we have a bijection:

$$\{\beta : V \times W \to Z, \text{ bilinear map}\} \xleftrightarrow{1\text{-}1} \{\gamma : V \otimes W \to Z, \text{ linear map}\}.$$

Proof. To start, one can check that α is a bilinear map. Also, recall from above that $V \otimes W \cong F(V \times W)/R$, for the subspace R of $V \times W$ defined in (1.3).

Now take an arbitrary bilinear map $\beta : V \times W \to Z$. Then, we get a linear map

$$\hat{\beta} : F(V \times W) \to Z, \qquad \hat{\beta}(\textstyle\sum_i \lambda_i(v_i, w_i)) := \sum_i \lambda_i \beta(v_i, w_i),$$

for $\lambda_i \in \Bbbk$ and $(v_i, w_i) \in V \times W$. Since β is bilinear, R is contained in the kernel of $\hat{\beta}$. Thus, $\hat{\beta}$ factors through a linear map $\gamma : V \otimes W \to Z$. We also obtain that $\beta = \gamma \circ \alpha$.

To show that γ is unique, suppose that $\tilde{\gamma} : V \otimes W \to Z$ is a linear map such that $\beta = \tilde{\gamma} \circ \alpha$. Then, we have that $\gamma(v \otimes w) = \beta(v, w) = \tilde{\gamma}(v \otimes w)$, for all $(v, w) \in V \times W$. Since the elements $\{v \otimes w\}_{v \in V, w \in W}$ span $V \otimes W$ as a vector space, the linear maps γ and $\tilde{\gamma}$ must be equal. $\qquad\square$

Observe that the definition of $V \otimes W$ above is independent of a choice of bases for V and W. See Exercise 1.3 for practice, and consider the result below.

Proposition 1.5. *For vector spaces V and W, we have an isomorphism of vector spaces:*

$$V \otimes W \cong W \otimes V.$$

Proof. Consider the map $\beta : V \times W \to W \otimes V$ defined by $\beta(v, w) = w \otimes v$ for $v \in V$ and $w \in W$. One can check that β is bilinear. Thus, by the universal property of $V \otimes W$, we get a unique linear map $\gamma : V \otimes W \to W \otimes V$ given by

$$\gamma(v \otimes w) = \gamma\alpha(v, w) = \beta(v, w) = w \otimes v.$$

Likewise, we can define a bilinear map $\bar{\beta} : W \times V \to V \otimes W$ by $\bar{\beta}(w, v) = v \otimes w$, and this yields a linear map $\bar{\gamma} : W \otimes V \to V \otimes W$ given by $\bar{\gamma}(w \otimes v) = v \otimes w$. So, γ and $\bar{\gamma}$ are mutually inverse linear maps, and thus, $V \otimes W \cong W \otimes V$ as vector spaces. $\quad\square$

Universal property of $V_1 \otimes \cdots \otimes V_n$. Likewise, for vector spaces $V_1, \ldots, V_n$, the vector space $V_1 \otimes \cdots \otimes V_n$ can be defined via the universal property attached to a multilinear map,

$$\alpha : V_1 \times \cdots \times V_n \to V_1 \otimes \cdots \otimes V_n, \quad (v_1, \ldots, v_n) \mapsto v_1 \otimes \cdots \otimes v_n.$$

Namely, we require that the following diagram commutes.

$$
\begin{array}{ccc}
V_1 \times \cdots \times V_n & \xrightarrow{\;\;\alpha\;\;} & V_1 \otimes \cdots \otimes V_n \\
 & \searrow{\scriptstyle \forall \text{ multilinear map } \beta} & \big\downarrow{\scriptstyle \exists! \text{ linear map } \gamma} \\
 & & Z \text{ vs}
\end{array}
$$

Here, we have a bijection:

$$\{\beta : V_1 \times \cdots \times V_n \to Z, \text{ multilinear map}\} \overset{1\text{-}1}{\longleftrightarrow} \{\gamma : V_1 \otimes \cdots \otimes V_n \to Z, \text{ linear map}\}.$$

iv. Homs and duals

For fixed vector spaces V and W, the collection of linear maps $V \to W$ forms a vector space called a **Hom space**, denoted by $\mathrm{Hom}_{\Bbbk}(V, W)$.

Namely, if ϕ and ϕ' are elements of $\mathrm{Hom}_{\Bbbk}(V, W)$, then we define the operations:

$$(\phi + \phi')(v) := \phi(v) + \phi'(v) \quad \text{and} \quad (\lambda * \phi)(v) := \lambda *_W \phi(v) \; (= \phi(\lambda *_V v)),$$

for all $\lambda \in \Bbbk$ and $v \in V$. We have that $\dim_{\Bbbk} \mathrm{Hom}_{\Bbbk}(V, W) = (\dim_{\Bbbk} V)(\dim_{\Bbbk} W)$.

In particular, the Hom space,

$$V^* := \mathrm{Hom}_{\Bbbk}(V, \Bbbk),$$

is the **dual space** to V. Here, $\dim_{\Bbbk} \mathrm{Hom}(V, \Bbbk) = (\dim_{\Bbbk} V)(\dim_{\Bbbk} \Bbbk) = \dim_{\Bbbk} V$. Vectors of V^* are referred to as **linear functionals**, or as **linear forms**, on V. If $\{b_i\}_i$ is a basis of V, then V^* has the **dual basis** $\{b_i^*\}_i$ given by $b_i^*(b_j) = \delta_{i,j} 1_{\Bbbk}$, with $\delta_{i,j}$ being the Kronecker delta (which is $= 1$ if $i = j$, and $= 0$ if $i \neq j$). See Exercise 1.4.

There is a key relationship between the Hom space and tensor product constructions above, given by vector space isomorphisms below.

$$\mathrm{Hom}_{\Bbbk}(U \otimes V, W) \cong \mathrm{Hom}_{\Bbbk}(U, \mathrm{Hom}_{\Bbbk}(V, W)), \quad \begin{cases} \phi \mapsto [\, u \mapsto (v \mapsto \phi(u \otimes v)) \,] \\ [\, u \otimes v \to \psi(u)(v) \,] \mapsfrom \psi \end{cases}$$

$$\mathrm{Hom}_{\Bbbk}(U \otimes V, W) \cong \mathrm{Hom}_{\Bbbk}(V, \mathrm{Hom}_{\Bbbk}(U, W)), \quad \begin{cases} \phi \mapsto [\, v \mapsto (u \mapsto \phi(u \otimes v)) \,] \\ [\, u \otimes v \to \psi(v)(u) \,] \mapsfrom \psi \end{cases}$$

This relationship is called **Tensor-Hom adjunction**, and it admits generalizations for modules and in the categorical setting discussed later in the book.

v. Operations on linear maps

The direct product, the sum, and the direct sum of linear maps are defined component-wise in a natural way. Also, given linear maps $f : V \to W$ and $f' : V' \to W'$, along with a vector space U, we have the following linear maps built from the tensor product and Hom constructions.

$$f \otimes f' : V \otimes V' \to W \otimes W', \qquad\qquad v \otimes v' \mapsto f(v) \otimes f'(v')$$

$$\mathrm{Hom}_{\Bbbk}(f, U) : \mathrm{Hom}_{\Bbbk}(W, U) \to \mathrm{Hom}_{\Bbbk}(V, U), \qquad g \mapsto g \circ f$$

$$\mathrm{Hom}_{\Bbbk}(U, f') : \mathrm{Hom}_{\Bbbk}(U, V') \to \mathrm{Hom}_{\Bbbk}(U, W'), \qquad g \mapsto f' \circ g$$

§1.1.5. Algebras

As mentioned at the beginning of the chapter, an algebra is a mathematical structure that is a combination of a unital ring and a vector space over $\Bbbk$. One version of an algebra is given as follows.

Definition 1.6. A unital ring $(A, +, 0, \cdot, 1)$ is called a **($\Bbbk$-)algebra** if it comes equipped with a unital ring homomorphism $\phi : \Bbbk \to A$ such that $\mathrm{im}(\phi) \subset Z(A)$.

But as mentioned at the end of §1.1.2, the collection of (unital) rings is not well-behaved. So, we define algebras in an alternative way, first using vector spaces, instead of starting with unital rings. Linear Algebra is a nice setting to work in overall.

Definition 1.7. A $\Bbbk$-vector space $(A, +, 0, *)$ is called a **($\Bbbk$-)algebra** if it comes equipped with two linear maps

$$m := m_A : A \otimes A \to A, \quad m(a \otimes b) =: ab \qquad \text{and} \qquad u := u_A : \Bbbk \to A, \quad u(1_{\Bbbk}) =: 1_A$$

such that the following axioms hold:

- **(associativity)** $\quad m(m \otimes \mathrm{id}_A) = m(\mathrm{id}_A \otimes m)$, and
- **(unitality)** $\quad m(u \otimes \mathrm{id}_A) = \mathrm{id}_A = m(\mathrm{id}_A \otimes u)$.

These structure axioms can be visualized via the commutative diagrams below.

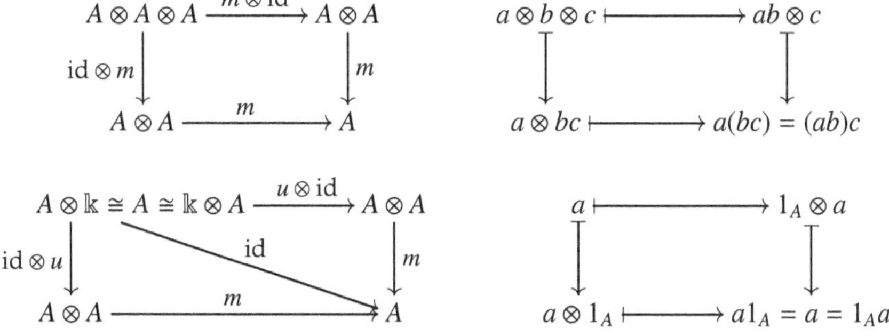

Exercise 1.5 entails showing that Definitions 1.6 and 1.7 are equivalent. Indeed, the map ϕ in Definition 1.6 gives A the structure of a $\Bbbk$-vector space for Definition 1.7, where $\lambda * a := \phi(\lambda) \cdot a$, for $\lambda \in \Bbbk$ and $a \in A$.

Standing hypothesis. All algebras in this book are assumed to be associative and unital, unless stated otherwise.

Examples of $\mathbb{R}$-algebras include $\mathbb{R}$ and $\mathbb{C}$, and $\Bbbk$ is a $\Bbbk$-algebra. Moreover, the **zero $\Bbbk$-algebra** is the zero $\Bbbk$-vector space with m and u being zero maps.

Features from rings and vector spaces also get adapted to algebras. For instance, an algebra is **commutative** if it is commutative as a ring. The **dimension** of an algebra is the dimension of its underlying vector space.

Moreover, a **division algebra** is a unital algebra $(A, +, 0, \cdot, 1, *)$ such that each nonzero element a of A has a two-sided inverse element a^{-1} with respect to $\cdot$.

Given algebras A and A', a function $\phi : A \to A'$ is an **algebra (homo)morphism** or an **algebra map** if ϕ is a ring morphism between the underlying ring structures and a linear map between the underlying vector space structures.

We denote the set of algebra morphisms from A to A' by $\mathrm{Hom}_{\mathrm{Alg}_{\Bbbk}}(A, A')$, which is a subspace of the Hom space between underlying vector spaces, $\mathrm{Hom}_{\Bbbk}(A_{\mathrm{vs}}, A'_{\mathrm{vs}})$.

The terminology for the various types of morphisms (namely, **mono-**, **epi-**, **iso-**, **endo-**, **auto-**) from the group setting also applies in the algebra setting.

Given an algebra A, a nonempty subset B of A is a **subalgebra** if B is an algebra under the same operations of A, and $1_B = 1_A$. There is a canonical algebra monomorphism $B \to A$ in this case.

If we drop the condition that $1_B = 1_A$ (or if 1_B does not exist), then we call B a **nonunital subalgebra** of A; the linear embedding $B \to A$ here is multiplicative, but not necessarily unital. For instance, if A is the *direct product* of two nonzero algebras A_1 and A_2 (defined later in §1.4.1), then A_1 and A_2 are nonunital subalgebras of A.

For a nonunital subalgebra B of A, the set A/B of additive cosets is a quotient space, and is a **quotient algebra** if and only if $B =: I$ is an ideal of A. Here, there is a canonical algebra epimorphism, $A \to A/I$, defined by $a \mapsto a + I$.

Let $\phi : A \to A'$ be an algebra morphism. Then, the following statements hold.

- The **kernel** of ϕ, namely, the set $\ker(\phi) := \{a \in A \mid \phi(a) = 0_{A'}\}$, is an ideal of A.
- The **image** of ϕ, namely, the set $\mathrm{im}(\phi) := \{\phi(a) \mid a \in A\}$, is just a subalgebra of A' and not necessarily an ideal of A'.

- We have an isomorphism of algebras: $A/\ker(\phi) \cong \operatorname{im}(\phi)$.

- Let $\operatorname{im}(\phi)_{\text{ideal}}$ denote the smallest ideal of A' containing $\operatorname{im}(\phi)$. Then, the **cokernel** of ϕ is defined as the quotient algebra: $\operatorname{coker}(\phi) := A'/\operatorname{im}(\phi)_{\text{ideal}}$.

§1.2. Examples of algebras over a field

We now provide several examples of algebras over a field that are ubiquitous throughout the literature.

§1.2.1. Matrix algebras and endomorphism algebras

Let n be a positive integer. The collection of $n \times n$ matrices with entries in $\Bbbk$,

$$\operatorname{Mat}_n(\Bbbk) = \{(c_{i,j}) \mid c_{i,j} \in \Bbbk,\ i, j = 1, \ldots, n\},$$

is an algebra under matrix addition and matrix multiplication, and scalar multiplication given by $\lambda * (c_{i,j}) := (\lambda c_{i,j})$ for $\lambda \in \Bbbk$. This is a **matrix algebra over** $\Bbbk$.

For indices k, ℓ, an **elementary matrix** is of the form $E_{k,\ell} := (\delta_{k,i}\, \delta_{\ell,j}\, 1_{\Bbbk})_{i,j=1}^n$, and elementary matrices $\{E_{k,\ell}\}_{k,\ell=1}^n$ form a vector space basis of $\operatorname{Mat}_n(\Bbbk)$.

Let V be a vector space. The Hom space of linear maps $V \to V$,

$$\operatorname{End}_{\Bbbk}(V) := \operatorname{Hom}_{\Bbbk}(V, V),$$

can be upgraded to an algebra with composition as multiplication, and with id_V as the unit. This is called the **endomorphism algebra of the vector space** V.

If $\dim_{\Bbbk} V = n$, then we have an isomorphism between the algebras above:

$$\operatorname{End}_{\Bbbk}(V) \cong \operatorname{Mat}_n(\Bbbk).$$

Here, the linear endomorphism, $\phi : V \to V$, corresponds to the matrix, $\operatorname{Mat}_{\mathcal{B}}(\phi)$, for a choice of a basis $\mathcal{B}$ of V.

Likewise, one can define a **matrix algebra over an algebra** A, denoted by $\operatorname{Mat}_n(A)$, and one gets that it is isomorphic to a certain endomorphism algebra; cf. Exercise 1.26 after reading §1.4.3ii. See also Exercise 1.23 after reading §1.4.2v.

§1.2.2. Free algebras, tensor algebras, and quotient algebras

Our next set of examples are algebras which can be defined with a universal property. First, we define the algebras in terms of elements. Let $\{v_i\}_{i \in I}$ be a collection of variables for an index set I.

The **free algebra on** $\{v_i\}_{i \in I}$, denoted by $\Bbbk\langle v_i \rangle_{i \in I}$, is the algebra with vector space basis being finite (possibly empty) products of variables

$$v_{i_1} \cdots v_{i_r}, \quad (\textbf{words})$$

with multiplication given by concatenation of the basis elements,

$$m(v_{i_1} \cdots v_{i_r} \otimes v_{j_1} \cdots v_{j_s}) := v_{i_1} \cdots v_{i_r} v_{j_1} \cdots v_{j_s},$$

and with unit $1_{\Bbbk\langle v_i \rangle_{i \in I}}$ being the empty product of the variables. If $|I| = n < \infty$, then we write $\Bbbk\langle v_i \rangle_{i \in I}$ as $\Bbbk\langle v_1, \ldots, v_n \rangle$. Elements of $\Bbbk\langle v_i \rangle_{i \in I}$ are linear combinations of finitely many words in $\{v_i\}_{i \in I}$.

Next, let $V = \bigoplus_{i \in I} \Bbbk v_i$ be the vector space with basis $\{v_i\}_{i \in I}$. The **tensor algebra of** V, denoted by $T(V)$, is the algebra with vector space basis being simple tensors,

$$v_{i_1} \otimes \cdots \otimes v_{i_r} \in V^{\otimes r},$$

for all $r \in \mathbb{N}$. Here, $V^{\otimes 0}$ is $\Bbbk$ by convention. In other words, as a vector space,

$$T(V) = \Bbbk \oplus V \oplus (V \otimes V) \oplus (V \otimes V \otimes V) \oplus \cdots .$$

Since $V^{\otimes r} \otimes V^{\otimes s} \cong V^{\otimes (r+s)}$, multiplication is given by concatenation:

$$m((v_{i_1} \otimes \cdots \otimes v_{i_r}) \otimes (v_{j_1} \otimes \cdots \otimes v_{j_s})) := v_{i_1} \otimes \cdots \otimes v_{i_r} \otimes v_{j_1} \otimes \cdots \otimes v_{j_s}.$$

The unit morphism, $u : \Bbbk \to T(V)$, of $T(V)$ is given by embedding.

We have an isomorphism between free algebras and tensor algebras above:

$$\Bbbk\langle v_i \rangle_{i \in I} \cong T(V),$$

for $V = \bigoplus_{i \in I} \Bbbk v_i$, via the identification of $v_{i_1} \cdots v_{i_r}$ with $v_{i_1} \otimes \cdots \otimes v_{i_r}$.

Universal property of $T(V)$. We will see here that tensor algebras (and, thus, free algebras) can be defined universally. For a vector space V, we obtain that the tensor algebra $T(V)$ satisfies the following result.

Definition-Proposition 1.8. The tensor algebra $T(V)$ is equipped with a linear embedding:

$$\alpha : V \to T(V)_{\mathrm{vs}} \sqsubset T(V),$$

where $\sqsubset$ denotes taking the underlying vector space, such that for any algebra Z equipped with a linear map β from V to the underlying vector space of Z, there exists a unique algebra map $\gamma : T(V) \to Z$ such that $\beta = \gamma \circ \alpha$. Namely, the following diagram commutes.

$$
\begin{array}{ccc}
V & \xrightarrow{\;\;\alpha\;\;} & T(V)_{\mathrm{vs}} \sqsubset T(V) \\
& {\scriptstyle \forall\, \text{linear map } \beta} \searrow & \big\downarrow {\scriptstyle \exists!\, \text{algebra map } \gamma} \\
& & Z_{\mathrm{vs}} \sqsubset Z \text{ alg}
\end{array}
$$

Here, we get a bijection between sets: $\mathrm{Hom}_{\Bbbk}(V, Z_{\mathrm{vs}}) \cong \mathrm{Hom}_{\mathrm{Alg}_{\Bbbk}}(T(V), Z)$.

Proof. To start, α is the linear embedding defined in Lemma 1.2. Next, given an arbitrary linear map $\beta : V \to Z_{vs}$, define a map

$$\hat{\beta}_i : V^{\times i} \to Z, \quad (v_1, \ldots, v_i) \mapsto \beta(v_1) \cdots \beta(v_i),$$

where we use the multiplication of the algebra Z for the right-hand side. One can check that $\hat{\beta}_i$ is multilinear, so this yields a unique linear map

$$\gamma_i : V^{\otimes i} \to Z, \quad \gamma(v_1 \otimes \cdots \otimes v_i) \mapsto \beta(v_1) \cdots \beta(v_i),$$

due to the universal property of the i-fold tensor product. Now we obtain the desired algebra map γ as the sum of linear maps $\sum_{i \in \mathbb{N}} \gamma_i$. We leave it to the reader to verify that γ is multiplicative and unital (i.e., is an algebra map and not just a linear map), and to verify the uniqueness of γ. $\square$

Quotients of free (or tensor) algebras. Arbitrary structures in algebra and elsewhere often arise as a 'quotient' of a free structure. Let us discuss the quotient algebras of free (and tensor) algebras here. These are algebras that are defined via a collection of independent variables (*generators*), subject to certain expressions in those variables set equal to zero (*relations*).

In a free algebra $\mathbb{k}\langle v_i \rangle_{i \in I}$, take a collection of elements $\{f_j\}_{j \in J}$, for index sets I and J. Then, the set $\{f_j\}_{j \in J}$ generates an ideal of $\mathbb{k}\langle v_i \rangle_{i \in I}$, denoted by $(f_j)_{j \in J}$. So, we obtain that the quotient space,

$$A := \mathbb{k}\langle v_i \rangle_{i \in I} \ / \ (f_j)_{j \in J},$$

is an algebra. Here, we say that A is **generated** by $\{v_i\}_{i \in I}$, subject to the **relations** $\{f_j\}_{j \in J}$, which forms a **presentation** of A. If $|I| < \infty$ (resp., $|I| < \infty$ and $|J| < \infty$), then A is **finitely generated** (resp., **finitely presented**).

Presentations of algebras are not unique. One way to understand algebras built with generators and relations in detail is to use the *Diamond Lemma* by Bergman [1978], which we will not discuss further here.

---

§1.2.3. Polynomial algebras and symmetric algebras

Now we impose commutativity relations on free/tensor algebras to get the algebras below. Let $\{v_i\}_{i \in I}$ be a collection of variables for an index set I.

The **polynomial algebra**, $\mathbb{k}[v_1, \ldots, v_n]$, is the algebra with vector space basis

$$\{v_1^{i_1} \cdots v_n^{i_n} \mid i_r \in \mathbb{N}, \ r = 1, \ldots, n\},$$

with unit $1_{\mathbb{k}[v_1,\ldots,v_n]} := v_1^0 \cdots v_n^0$, and with multiplication given by

$$m(v_1^{i_1} \cdots v_n^{i_n} \otimes v_1^{j_1} \cdots v_n^{j_n}) := v_1^{i_1+j_1} \cdots v_n^{i_n+j_n}.$$

In fact, we have an algebra isomorphism,

$$\Bbbk[v_1,\ldots,v_n] \cong \Bbbk\langle v_1,\ldots,v_n\rangle/(v_iv_j - v_jv_i)_{1\leq i<j\leq n}.$$

For instance, the polynomial algebra $\Bbbk[v_1, v_2]$ is finitely presented by generators v_1 and v_2, and subject to the relation $v_1v_2 - v_2v_1$.

On the other hand, we can form a universal commutative algebra as follows. The **symmetric algebra of** V, denoted by $S(V)$, is defined to be the unique algebra attached to a vector space V that comes equipped with an embedding,

$$\alpha : V \to S(V)_{\mathrm{vs}} \sqsubset S(V),$$

such that for any commutative algebra Z, the commutative diagram below holds.

$$
\begin{array}{ccc}
V & \xrightarrow{\ \alpha\ } & S(V)_{\mathrm{vs}} \sqsubset S(V) \\
& \searrow{\scriptstyle \forall \text{ linear map } \beta} & \big\downarrow{\scriptstyle \exists! \text{ algebra map } \gamma} \\
& & Z_{\mathrm{vs}} \sqsubset Z \text{ comm alg}
\end{array}
$$

By $Z_{\mathrm{vs}} \sqsubset Z$, we mean take Z_{vs} the underlying vector space of Z.

In fact, one can realize $S(V)$ as a quotient algebra of $T(V)$ by the ideal generated by the set, $\{v \otimes w - w \otimes v\}_{v,w\in V}$. Thus, $S(V)$ is a commutative algebra.

If $V = \bigoplus_{i=1}^n \Bbbk v_i$, then $S(V) \cong \Bbbk[v_1,\ldots,v_n]$ as algebras; see Exercise 1.6.

Note that $\mathrm{Hom}_{\Bbbk}(V, Z_{\mathrm{vs}}) \cong \mathrm{Hom}_{\mathrm{CommAlg}_{\Bbbk}}(S(V), Z)$ as sets. Namely, maps of commutative algebras are just algebra maps.

§1.2.4. Exterior algebras

A counterpart to the commutative algebras discussed in the previous section are the universal anti-commutative algebras below.

Take a vector space V. The **exterior algebra of** V is defined to be the unique algebra $\Lambda(V)$ attached to V that comes equipped with an embedding,

$$\alpha : V \to \Lambda(V)_{\mathrm{vs}} \sqsubset \Lambda(V),$$

such that $\alpha(v)\alpha(v) = 0$ for each $v \in V$, and such that for any algebra Z, the diagram below commutes.

$$
\begin{array}{ccc}
V & \xrightarrow{\ \alpha\ } & \Lambda(V)_{\mathrm{vs}} \sqsubset \Lambda(V) \\
& \searrow{\scriptstyle \forall \text{ linear } \beta \text{ with } \atop \beta(v)\beta(v)=0 \text{ for each } v\in V} & \big\downarrow{\scriptstyle \exists! \text{ algebra map } \gamma} \\
& & Z_{\mathrm{vs}} \sqsubset Z \text{ alg}
\end{array}
$$

In fact, one can realize $\Lambda(V)$ as a quotient algebra of $T(V)$ by the ideal generated by the set, $\{v \otimes v\}_{v\in V}$. See Exercise 1.7. In particular, if $V = \bigoplus_{i=1}^n \Bbbk v_i$, then as algebras:

$$\Lambda(V) := \Lambda(v_1,\ldots,v_n) \cong \Bbbk\langle v_1,\ldots,v_n\rangle/(v_iv_j + v_jv_i)_{1\leq i\leq j\leq n}.$$

§1.2.5. Path algebras

Next, we turn our attention to algebras built from graphs. A **quiver** is a synonym for a directed graph used in algebraic settings; it is given by the data below:

$$Q := (Q_0, Q_1, s, t).$$

Here, Q_0 is the vertex set of Q, and Q_1 is the arrow set of Q, and $s, t : Q_1 \to Q_0$ are the source and target maps for Q, respectively.

For example, for the arrow below,

$$\bullet_1 \xrightarrow{\ a\ } \bullet_2$$

we have that $s(a) = 1$ and $t(a) = 2$.

We say that Q is **finite** if both $|Q_0|$ and $|Q_1|$ are finite sets.

We say that Q is **connected** if its underlying undirected graph is connected.

A **path** in Q is a composition of arrows in Q, where composition is read left-to-right here. That is, if $a, b \in Q_1$, then ab is a path in Q where $t(a) = s(b)$. Likewise, we can compose paths: If $p = a_1 \cdots a_n$ and $q = b_1 \cdots b_m$ are paths for $a_i, b_j \in Q_1$, then we can define the path, pq as $a_1 \cdots a_n b_1 \cdots b_m$ precisely when $t(a_n) = s(b_1)$.

A **cycle** is a path of the form $a_1 \cdots a_n$ with $t(a_n) = s(a_1)$. A quiver is **acyclic** if it does not contain a cycle.

A path of the form, $a_1 \cdots a_n$, for $a_i \in Q_1$, has **length** n. The **trivial path** at a vertex $i \in Q_0$ is the path, denoted e_i, with $s(e_i) = i = t(e_i)$ of length 0. The set of paths of Q of length n is denoted by Q_n.

Now the **path algebra**, $\Bbbk Q$, attached to a quiver Q is a not-necessarily-unital algebra with vector space basis being paths in Q; that is,

$$\Bbbk Q := \Bbbk Q_0 \oplus \Bbbk Q_1 \oplus \Bbbk Q_2 \oplus \Bbbk Q_3 \oplus \cdots$$

as vector spaces. The multiplication of $\Bbbk Q$ is given by the composition of paths when this makes sense, and 0 otherwise. Also, $\Bbbk Q$ is unital if and only if $|Q_0| < \infty$; in this case, $1_{\Bbbk Q} = \sum_{i \in Q_0} e_i$. Note that $\Bbbk Q$ is rarely commutative, and that $\dim_{\Bbbk} \Bbbk Q < \infty$ if and only if Q is finite and acyclic.

Examples of path algebras are provided in Figure 1.1 below.

- If Q is a quiver with one vertex $\{1\}$ and no arrows, then the basis of $\Bbbk Q$ is $\{e_1\}$. Here, the product on the basis element is $e_1^2 = e_1$ and $1_{\Bbbk Q} = e_1$. In terms of generators and relations, $\Bbbk Q = \Bbbk\langle e_1 \rangle / (e_1^2 - e_1) \cong \Bbbk e_1 \cong \Bbbk$, as algebras.

- If Q is a loop a on a vertex labeled by 1, then paths of Q are $\{a^i\}_{i \in \mathbb{N}}$, including $a^0 = e_1$. In this case, $a^i a^j = a^j a^i$ and $1_{\Bbbk Q} = e_1$. So, $\Bbbk Q$ is isomorphic to the free algebra $\Bbbk\langle a \rangle$ (or to the polynomial algebra $\Bbbk[a]$) in the variable a.

- Likewise, if Q is an n-loop with arrows $a_1, \ldots, a_n$ on vertex 1, then $\Bbbk Q$ is isomorphic to the free algebra $\Bbbk\langle a_1, \ldots, a_n \rangle$ in the variables $a_1, \ldots, a_n$ with $1_{\Bbbk Q} = e_1$.

- If Q is an arrow a with $s(a) = 1$ and $t(a) = 2$, then $\Bbbk Q$ has vector space basis e_1, e_2, a, and $\Bbbk Q$ is isomorphic to the algebra of 2×2 upper triangular matrices with entries in $\Bbbk$. The correspondence below extends to an algebra isomorphism:

$$e_1 \leftrightarrow \begin{pmatrix} 1 & 0 \\ 0 & 0 \end{pmatrix} =: E_{1,1}, \qquad e_2 \leftrightarrow \begin{pmatrix} 0 & 0 \\ 0 & 1 \end{pmatrix} =: E_{2,2}, \qquad a \leftrightarrow \begin{pmatrix} 0 & 1 \\ 0 & 0 \end{pmatrix} =: E_{1,2}.$$

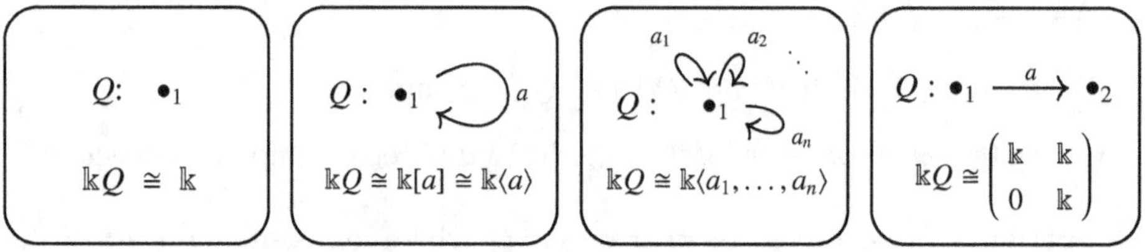

Figure 1.1: Examples of path algebras of quivers.

Universal property of $\Bbbk Q$. Path algebras of finite quivers satisfy a universal property. Take Q to be a finite quiver. Then, the **path algebra** $\Bbbk Q$ is the unique algebra attached to $Q_0 \times Q_1$ that comes equipped with a function,

$$\alpha = (\alpha_0, \alpha_1) : Q_0 \times Q_1 \to \Bbbk Q, \quad i \in Q_0 \mapsto \alpha_0(i) = e_i, \quad a \in Q_1 \mapsto \alpha_1(a) = a,$$

that satisfies the following conditions,

- $\alpha_0(i)\, \alpha_0(j) = \delta_{i,j}\, \alpha_0(i)$ for each $i, j \in Q_0$,

- $1_{\Bbbk Q} = \sum_{i \in Q_0} \alpha_0(i)$,

- $\alpha_0(s(a))\, \alpha_1(a)\, \alpha_0(t(a)) = \alpha_1(a)$ for each $a \in Q_1$,

such that for any algebra Z, it satisfies the commutative diagram below.

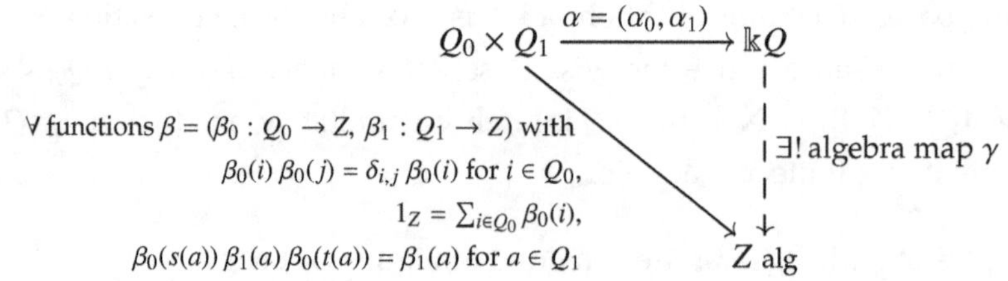

$$\begin{array}{ccc} & \alpha = (\alpha_0, \alpha_1) & \\ Q_0 \times Q_1 & \longrightarrow & \Bbbk Q \\ & & \downarrow \exists!\ \text{algebra map } \gamma \\ & & Z \text{ alg} \end{array}$$

$\forall$ functions $\beta = (\beta_0 : Q_0 \to Z,\ \beta_1 : Q_1 \to Z)$ with
$\beta_0(i)\,\beta_0(j) = \delta_{i,j}\,\beta_0(i)$ for $i \in Q_0$,
$1_Z = \sum_{i \in Q_0} \beta_0(i)$,
$\beta_0(s(a))\,\beta_1(a)\,\beta_0(t(a)) = \beta_1(a)$ for $a \in Q_1$

That is, $\gamma(e_i) = \beta_0(i)$ for all $i \in Q_0$, and $\gamma(a) = \beta_1(a)$ for all $a \in Q_1$. (Yes, yes, there are a lot of conditions on β in the diagram above. But the point is that path algebras do satisfy a universal property.)

§1.2.6. Group algebras

Next, we focus on algebras built from groups. Let G be a group.

The **group algebra**, $\Bbbk G$, attached to G is the algebra with vector space basis consisting of elements of G, with multiplication being the group operation applied to the basis elements, and with unit $1_{\Bbbk G}$ equal to the identity element e of G.

Note that $\Bbbk G$ is commutative if and only if G is abelian. Moreover, $\Bbbk G$ is finite-dimensional if and only if G is a finite group.

Universal property of $\Bbbk G$. Group algebras satisfy a universal property. To see this, take an algebra A, and consider its **group of units** :

$$A^\times := \{ a \in A \mid ab = ba = 1_A, \text{ for some } b \in A \}.$$

Note that G is a subgroup of $(\Bbbk G)^\times$, and consider the inclusion $\subset$ of the group $(\Bbbk G)^\times$ in $\Bbbk G$. Then, $\Bbbk G$ is the unique algebra attached to G that comes equipped with a group homomorphism,

$$\alpha : G \to (\Bbbk G)^\times \subset \Bbbk G, \quad g \mapsto g,$$

such that for any algebra Z, it satisfies the commutative diagram below.

$$
\begin{array}{ccc}
G & \xrightarrow{\ \ \alpha\ \ } & (\Bbbk G)^\times \subset \Bbbk G \\
 & {}_{\forall \text{ group map } \beta}\searrow & \big\downarrow {\scriptstyle \exists!\, \text{alg. map } \gamma} \\
 & & Z^\times \subset Z \text{ alg}
\end{array}
$$

Here, we get a bijection between sets: $\mathrm{Hom}_{\mathrm{Group}}(G, Z^\times) \cong \mathrm{Hom}_{\mathrm{Alg}_\Bbbk}(\Bbbk G, Z)$.

§1.2.7. Graded algebras and filtered algebras

We now define algebras that are decomposed into pieces labeled by elements of $\mathbb{N}$. One could define similar notions by replacing $\mathbb{N}$ with any monoid N. For instance, using $N = \mathbb{Z}$ is common in the literature, and using $N = \mathbb{Z}_2$ is foundational in "super" structures in physics and in Lie theory.

i. Graded algebras

An algebra $(A,\, m : A \otimes A \to A,\, u : \Bbbk \to A)$ is **($\mathbb{N}$-)graded** if

- $A_{\mathrm{vs}} = \bigoplus_{i \in \mathbb{N}} A_i$, for some subspaces A_i of A_{vs}, for all $i \in \mathbb{N}$,

- the image of $m|_{A_i \otimes A_j}$ lies in A_{i+j}, for all $i, j \in \mathbb{N}$, and

- the image of u lies in A_0.

In this case, A_i is the **homogeneous part of A of degree** i. For graded algebras A and A', an algebra map $\phi : A \to A'$ is **graded** if $\phi(A_i) \subset A'_i$, for all $i \in \mathbb{N}$.

The underlying vector space of a graded algebra is a **graded vector space**, that is, a vector space $V = \bigoplus_{i \in \mathbb{N}} V_i$, for some subspaces V_i of V.

Example 1.9. Free algebras, tensor algebras, polynomial algebras, symmetric algebras, exterior algebras, and path algebras are all examples of graded algebras. For instance, $T(V)_i = V^{\otimes i}$ and $(\Bbbk Q)_i = \Bbbk Q_i$, for $i \in \mathbb{N}$. Moreover, for a nonzero scalar $q \in \Bbbk^\times$, we have that the following algebras are graded:

$$\Bbbk_q[v_1, \ldots, v_n] := \Bbbk\langle v_1, \ldots, v_n\rangle / (v_i v_j - q v_j v_i)_{1 \leq i < j \leq n},$$
$$\Lambda_q(v_1, \ldots, v_n) := \Bbbk\langle v_1, \ldots, v_n\rangle / (v_i v_j + q v_j v_i)_{1 \leq i \leq j \leq n}.$$

Here, $\Bbbk_q[v_1, \ldots, v_n]$ is called a **quantum (q-)polynomial algebra**, and $\Lambda_q(v_1, \ldots, v_n)$ is called a **quantum (q-)exterior algebra**. See Exercise 1.8.

ii. Filtered algebras

An algebra $(A,\ m : A \otimes A \to A,\ u : \Bbbk \to A)$ is **($\mathbb{N}$-)filtered** if

- $A_{\mathrm{vs}} = \bigcup_{i \in \mathbb{N}} A_i$, for subspaces A_i of A_{vs} with $A_i \subset A_{i+1}$, for all $i \in \mathbb{N}$,

- the image of $m|_{A_i \otimes A_j}$ lies in A_{i+j}, for all $i, j \in \mathbb{N}$, and

- the image of u lies in A_0.

In this case, A_i is the **filtered part of A of degree** i. For filtered algebras A and A', an algebra map $\phi : A \to A'$ is **filtered** if $\phi(A_i) \subset A'_i$ for all $i \in \mathbb{N}$.

Example 1.10. The n-th **Weyl algebra** $A_n(\Bbbk)$ is a filtered algebra generated by variables $v_1, \ldots, v_n, w_1, \ldots, w_n$, subject to relations:

$$w_i v_j - v_j w_i - \delta_{i,j} 1, \quad v_i v_j - v_j v_i, \quad w_i w_j - w_j w_i \quad \text{for } 1 \leq i, j \leq n.$$

For instance, the first Weyl algebra is $A_1(\Bbbk) = \Bbbk\langle v, w\rangle / (wv - vw - 1)$.

iii. Associated graded algebras

Graded algebras $\bigoplus_{j \in \mathbb{N}} A_j$ are filtered with degree i filtered part: $A_0 \oplus \cdots \oplus A_i$. Conversely, for a filtered algebra $A = \bigcup_{i \in \mathbb{N}} A_i$, we can build the graded algebra:

$$\mathrm{gr}(A) = \bigoplus_{i \in \mathbb{N}} \mathrm{gr}(A)_i, \quad \text{for} \quad \mathrm{gr}(A)_i := A_i / A_{i-1}, \quad \text{where } A_{-1} = 0,$$

with multiplication $(a_i + A_{i-1})(a_j + A_{j-1}) := a_i a_j + A_{i+j-1}$ for $a_i \in A_i$ and $a_j \in A_j$, and with unit $1_{\mathrm{gr}(A)} = 1_A + A_{-1}$. This is called the **associated graded algebra**, and check that this is indeed an algebra. See also Exercise 1.9.

§1.3. Representations and modules

A *representation* of an algebraic structure S in terms of another algebraic structure U is a structure-preserving map

$$\rho : S \to \mathrm{End}(U)$$

from S to the structure $\mathrm{End}(U)$ consisting of certain endomorphisms of U. Here, S *acts* on U, and U inherits the structure of an S-*module*. We also say that S *captures symmetries* of U (see Walton [2019]). Often, U is a set, an abelian group, or a vector space. We focus on actions of algebras and of groups on vector spaces here.

§1.3.1. Representations

Fix an algebra $A := (A, m, u)$. A **representation of** A is a vector space V equipped with an algebra map

$$\rho := \rho_V : A \to \mathrm{End}_\Bbbk(V), \quad a \mapsto [\rho(a) : V \to V].$$

The **dimension** or **degree** of (V, ρ) is defined by $\deg(\rho_V) := \dim_\Bbbk V$.

If $\dim_\Bbbk V = n$, then $\mathrm{End}_\Bbbk(V) \cong \mathrm{Mat}_n(\Bbbk)$ (see §1.2.1). So, a n-dimensional representation of A is realized as an algebra map, $A \to \mathrm{Mat}_n(\Bbbk)$. The elements of A are then identified as matrices, and the operations of A (addition, multiplication, scalar multiplication) are encoded as the corresponding matrix operations.

A representation of A is called **trivial** if it is a 1-dimensional representation of the form: $A \to \Bbbk$, $a \mapsto 1_\Bbbk$, for all $a \in A$.

A representation (V, ρ) of A is called **faithful** if ρ is injective; that is, if $\rho(a)(v) = 0_V$ for all $v \in V$, then $a = 0_A$.

Faithfulness ensures that the action of A on V does not factor through the action of a proper quotient algebra A/I on V. Indeed, if ρ is not faithful, then $\ker(\rho) \neq 0$ and $\overline{\rho} : A/\ker(\rho) \to \mathrm{End}(V)$, $a + \ker(\rho) \mapsto \rho(a)$, is a well-defined representation of $A/\ker(\rho)$ using V. Conversely, if I is a nonzero ideal of A and $\overline{\rho} : A/I \to \mathrm{End}(V)$, $a + I \mapsto \rho(a)$, is a representation of A/I, then ρ is not faithful.

Now take $V = A_{\mathrm{vs}}$ with $\rho(a)(b) = m(a \otimes b) =: ab$, for all $a, b \in A$. This yields the **regular representation of** A, denoted by ρ_{reg} or A_{reg}, of degree equal to $\dim_\Bbbk A_{\mathrm{vs}}$. This is an example of a faithful representation because $a 1_A = a$ for all $a \in A$.

For instance, let $A := \Bbbk[v]/(v^2 - 1_A)$, which has vector space basis 1_A and v. The regular representation of A is the algebra map,

$$\Bbbk[v]/(v^2 - 1_A) \to \mathrm{Mat}_2(\Bbbk), \quad 1_A \mapsto \begin{pmatrix} 1 & 0 \\ 0 & 1 \end{pmatrix}, \quad v \mapsto \begin{pmatrix} 0 & 1 \\ 1 & 0 \end{pmatrix}.$$

Given two representations (V, ρ) and (V', ρ') of A, a linear map $\phi : V \to V'$ is a **representation morphism** if $\rho'(a) \circ \phi = \phi \circ \rho(a)$, for all $a \in A$.

The terminology for the various types of morphisms (namely, **mono-**, **epi-**, **iso-**, **endo-**, **auto-**) from the vector space setting also applies in this setting. In particular, an isomorphism of representations is commonly referred to as an **equivalence**.

Take a representation $(V, \rho : A \to \mathrm{End}_{\Bbbk}(V))$ of A. Take a subspace W of V, with embedding $\iota : W \to V$ and with projection $\pi : V \to V/W$.

Here, W is a **subrepresentation** of (V, ρ) if $\mathrm{im}(\rho(a) \circ \iota) \subset W$, for all $a \in A$.

The quotient space V/W of V is a **quotient representation** of (V, ρ) if, for all $a \in A$, we have that $\ker(\pi \circ \rho(a)) \supset W$. Indeed, in this case, the composition $\psi(a) := \pi \circ \rho(a) : V \to V/W$ induces a well-defined map $\overline{\psi(a)} : V/W \to V/W$, for $\overline{\psi(a)}(v + W) = \psi(a)(v)$. Namely, for $v + W = v' + W$ in V/W we get that $v - v' \in W$. Then, $\psi(a)(v - v') = 0$, and $\overline{\psi(a)}(v + W) = \overline{\psi(a)}(v' + W)$.

Consider the example above for $A = \Bbbk[v]/(v^2 - 1_A)$ and $V = A_{\mathrm{reg}} = \Bbbk 1_A \oplus \Bbbk v$. Check that $W = \Bbbk(1_A + v)$ is both a subspace and quotient space of V, and W naturally becomes a subrepresentation and a quotient representation of A_{reg}.

See Exercises 1.10 and 1.11 for practice; see also §1.3.4 for Exercise 1.11(a).

§1.3.2. Modules

Fix an algebra $A := (A, m, u)$. A **left A-module** is a vector space V with a linear map,

$$\triangleright := \triangleright_V : A \otimes V \to V, \quad a \otimes v \mapsto a \triangleright v \quad (\textbf{action map}),$$

such that the following diagrams commute.

$$
\begin{array}{ccc}
A \otimes (A \otimes V) \cong (A \otimes A) \otimes V & \xrightarrow{\;m \otimes \mathrm{id}\;} & A \otimes V \\
{\scriptstyle \mathrm{id} \otimes \triangleright} \downarrow & & \downarrow {\scriptstyle \triangleright} \\
A \otimes V & \xrightarrow{\qquad\qquad \triangleright \qquad\qquad} & V
\end{array}
\qquad
\begin{array}{ccc}
V \cong \Bbbk \otimes V & \xrightarrow{\;u \otimes \mathrm{id}\;} & A \otimes V \\
& {\scriptstyle \mathrm{id}} \searrow & \downarrow {\scriptstyle \triangleright} \\
& & V
\end{array}
$$

Such modules are sometimes denoted by $_A V$. The **dimension** of $(V, \triangleright)$ is $\dim_{\Bbbk} V$.

Likewise, a **right A-module** is a vector space V equipped with a linear map $\triangleleft := \triangleleft_V : V \otimes A \to V$ such that

$$\triangleleft \circ (\mathrm{id}_V \otimes m) = \triangleleft \circ (\triangleleft \otimes \mathrm{id}_A), \qquad \triangleleft \circ (\mathrm{id}_V \otimes u) = \mathrm{id}_V.$$

We will focus on left modules below; the details for right modules hold similarly.

There is a bijection between left A-modules $\{(V, \triangleright)\}$ and right A^{op}-modules $\{(V, \triangleleft)\}$, given by sending V to V, and by sending $a \triangleright v$ to $v \triangleleft a$, for $a \in A_{\mathrm{vs}}$ and $v \in V$.

A left A-module is called **trivial** if it is equal to $\Bbbk$ and equipped with the linear map, $\triangleright : A \otimes \Bbbk \to \Bbbk$, where $a \otimes \lambda \mapsto \lambda$, for all $a \in A$ and $\lambda \in \Bbbk$.

A left A-module $(V, \triangleright)$ is **faithful** if, for any nonzero ideal I of A, the action map $\triangleright$ does not descend to linear map $(A/I) \otimes V \to V$ making V a left (A/I)-module.

The **left regular module** over A is the vector space $V = A_{\text{vs}}$, with $\triangleright$ given by the multiplication map of A; this is denoted by $_A(A_{\text{reg}})$.

For instance, let $A := \Bbbk[v]/(v^2 - 1_A)$. Then, the left regular A-module is the vector space $V := \Bbbk 1_A \oplus \Bbbk v$ with:

$$1_A \triangleright 1_A = 1_A, \quad v \triangleright 1_A = v, \quad 1_A \triangleright v = v, \quad v \triangleright v = 1_A.$$

Given two left A-modules $(V, \triangleright)$ and $(V', \triangleright')$, a linear map $\phi : V \to V'$ is a **left module morphism** or **(left) module map** if the following diagram commutes.

$$
\begin{array}{ccc}
A \otimes V & \xrightarrow{\ \triangleright\ } & V \\
{\scriptstyle \text{id} \otimes \phi} \downarrow & & \downarrow {\scriptstyle \phi} \\
A \otimes V' & \xrightarrow{\ \triangleright'\ } & V'
\end{array}
$$

The terminology for the various types of morphisms (namely, **mono-**, **epi-**, **iso-**, **endo-**, **auto-**) from the vector space setting also applies in this setting.

Take a left A-module $(V, \triangleright)$, with a subspace W of V, an embedding $\iota : W \to V$, and a projection $\pi : V \to V/W$. Then:

- W is a **left A-submodule** of $(V, \triangleright)$ if $\text{im}(\triangleright (\text{id}_A \otimes \iota)) \subset W$;

- A quotient space V/W of V is a **left A-quotient module** of $(V, \triangleright)$ if $\ker(\pi \circ \triangleright) \supset A \otimes W$.

Now assume that A is ($\mathbb{N}$-)graded, where $A = \bigoplus_{i \in \mathbb{N}} A_i$; see §1.2.7. Then, a **graded left A-module** is a direct sum of vector spaces $V = \bigoplus_{j \in \mathbb{N}} V_j$ with left A-action map, $\triangleright : A \otimes V \to V$, where $\text{im}(\triangleright|_{A_i \otimes V_j}) \subset V_{i+j}$ for all $i, j \in \mathbb{N}$.

In fact, there is a bijection between the set of representations of A and the set of left A-modules; see Exercise 1.12. So, actions of A on V are captured by representations, or equivalently, by left modules. Therefore, all of the notions about modules can be transported to representations, and vice versa. For instance, we can define graded representations of graded algebras.

Remark 1.11. One may opt to use representations when studying actions of A on a fixed vector space V (e.g., to study an induced action of a quotient algebra A/I on V). Yet modules are useful when studying actions of a fixed algebra A on vector spaces V (e.g., to study an induced A-action on a quotient space V/W).

§1.3.3. Bimodules

Next, we consider a structure that is both a left and a right module. Take algebras A, B_1, and B_2. A (B_1, B_2)-**bimodule** is a vector space V equipped with linear maps,

$$\triangleright : B_1 \otimes V \to V \qquad \text{and} \qquad \triangleleft : V \otimes B_2 \to V,$$

where $(V, \triangleright)$ is a left B_1-module, $(V, \triangleleft)$ is a right B_2-module, and the following diagram commutes.

$$
\begin{array}{ccc}
B_1 \otimes (V \otimes B_2) \cong (B_1 \otimes V) \otimes B_2 \xrightarrow{\;\triangleright \otimes \, \mathrm{id}\;} V \otimes B_2 \\
\mathrm{id} \otimes \triangleleft \downarrow \qquad\qquad\qquad\qquad\qquad\qquad \downarrow \triangleleft \\
B_1 \otimes V \xrightarrow{\qquad\qquad\qquad \triangleright \qquad\qquad\qquad} V
\end{array}
$$

Such bimodules are sometimes denoted by ${}_{B_1}V_{B_2}$. We refer to the (A, A)-bimodule ${}_A V_A$ as an A-**bimodule**.

Remark 1.12. If C is a commutative algebra, then a left C-module $(V, \triangleright)$ is also a right C-module, where $v \triangleleft c := c \triangleright v$. Indeed,

$$v \triangleleft (cc') \;=\; (cc') \triangleright v \stackrel{C \text{ com}}{=\!=} (c'c) \triangleright v \;=\; c' \triangleright (c \triangleright v) \;=\; c' \triangleright (v \triangleleft c) \;=\; (v \triangleleft c) \triangleleft c'.$$

Furthermore, $(V, \triangleright, \triangleleft)$ is a C-bimodule in this case.

The **regular bimodule** over A is the vector space $V = A_{\mathrm{vs}}$, with $\triangleright$ and $\triangleleft$ given by the multiplication map of A, denoted by ${}_A(A_{\mathrm{reg}})_A$.

Given (B_1, B_2)-bimodules $(V, \triangleright, \triangleleft)$ and $(V', \triangleright', \triangleleft')$, a linear map $\phi : V \to V'$ is a **bimodule morphism** or **bimodule map** if it is simultaneously a left B_1-module map and a right B_2-module map.

Likewise, one can define **subbimodules, quotient bimodules**, the **trivial bimodule** $\Bbbk$, and **graded bimodules** over graded algebras.

§1.3.4. Over groups

Let $(G, \star, e)$ be a group, with operation $\star : G \times G \to G$, and with identity element e identified with a function $\vec{e} : \{\cdot\} \to G$ as in §1.1.1. Here, we consider actions of G on a vector space $V := (V, +, *)$, which historically predates actions of algebras on V discussed in the previous sections. Yet representations and modules over a group wind up 'being the same' as those over the corresponding group algebra.

Take $\mathrm{GL}(V)$ to be the **general linear group** on V, which is the group of automorphisms of V under composition. Then, a **representation of** G is a vector space V equipped with a group homomorphism,

$$\pi := \pi_V : G \to \mathrm{GL}(V), \quad g \mapsto [\pi(g) : V \to V].$$

Moreover, a **left G-module** is a vector space V equipped with a function,

$$\rhd : G \times V \to V, \quad (g, v) \mapsto g \rhd v,$$

such that the following compatibility conditions hold:

$$\rhd \circ (\star \times \mathrm{id}_V) = \rhd \circ (\mathrm{id}_G \times \rhd), \qquad \rhd \circ (\vec{e} \times \mathrm{id}_V) = \mathrm{id}_V,$$

$$\rhd \circ (\mathrm{id}_G \times +) = + \circ (\rhd_{V \times V}), \qquad \rhd \circ (\mathrm{id}_G \times *) = * \circ (\mathrm{id}_G \times \rhd).$$

Here, $\rhd_{V \times V} : G \times V \times V \to V \times V$ is the **diagonal action**: $(g, v, w) \mapsto (g \rhd v, \, g \rhd w)$.

The sets of representations of G, of left G-modules, of representations of $\Bbbk G$, and of left $\Bbbk G$-modules are all in bijective correspondence. See Exercise 1.13.

A representation of (V, π) of G is **faithful** if π is injective, i.e., if $\pi(g)(v) = v$ for all $v \in V$, then $g = e$.

Remark 1.13. Faithfulness for representations of G is different than for representations of $\Bbbk G$. If $\rho : \Bbbk G \to \mathrm{End}(V)$ is faithful as in §1.3.1, then the group representation $\pi = \rho|_G : G \to \mathrm{End}(V)^\times = \mathrm{GL}(V)$ is faithful; to see this, take the contrapositive. Yet if $\pi : G \to \mathrm{GL}(V)$ is faithful, then its $\Bbbk$-extension to $\rho : \Bbbk G \to \mathrm{End}(V)$ given by $\rho(\sum_{g \in G} \lambda_g \, g) := \sum_{g \in G} \lambda_g \, \pi(g)$ need not be faithful (see Exercise 1.11(a)).

See Exercise 1.14 for notions of groups and rings acting on other structures.

§1.4. Operations on algebras and modules

Now we present various recurring operations used to build modules and algebras, including (direct) sums, tensor products, free products, Homs, and duals. Each of these constructions applies to morphisms of modules and algebras as well.

§1.4.1. Direct products, sums, and direct sums

Here, we study direct products, sums, and direct sums of modules and algebras. We leave it to the reader to verify the results in this part; see Exercise 1.15.

The next result defines **direct products and (direct) sums of left modules**. Likewise, one can define such constructions for right modules and bimodules.

Proposition 1.14. *Fix a nonzero algebra (A, m, u). Then, the following statements hold.*

(a) *Let $(V_1, \rhd_1), \dots, (V_r, \rhd_r)$ be left A-modules. Then, the direct product of vector spaces $V_1 \times \cdots \times V_r$ forms a left A-module with action map:*

$$\rhd_{V_1 \times \cdots \times V_r} : \quad A \otimes (V_1 \times \cdots \times V_r) \to V_1 \times \cdots \times V_r,$$
$$a \otimes (v_1, \dots, v_r) \mapsto ((a \rhd_1 v_1), \dots, (a \rhd_r v_r)).$$

(b) *Let $(V_1, \triangleright), \ldots, (V_r, \triangleright)$ be left A-submodules of a left A-module $(V, \triangleright)$. Then, the direct sum $V_1 \oplus \cdots \oplus V_r$ is a left A-module using action map:*

$$\triangleright_{V_1 \oplus \cdots \oplus V_r} : \quad A \otimes (V_1 \oplus \cdots \oplus V_r) \to V_1 \oplus \cdots \oplus V_r,$$
$$a \otimes (v_1 + \cdots + v_r) \mapsto (a \triangleright v_1) + \cdots + (a \triangleright v_r).$$

The sum of vector spaces $V_1 + \cdots + V_r$ is also a left A-module using a similar action.

(c) *The canonical embeddings and projections between a vector space V_j and the direct product (or direct sum) of vector spaces $\{V_i\}_{i=1}^r$ given in Lemma 1.2 can be upgraded to left A-module maps.* □

We call a nonzero left A-module V **decomposable** if $V = V_1 \oplus V_2$, for some nonzero left A-submodules V_1 and V_2 of V. Else, we say that a nonzero left A-module V is **indecomposable**. A necessary and sufficient condition for V to be indecomposable is given in Proposition 1.22 later.

Indecomposable modules (resp., indecomposable representations, defined analogously) serve as one type of 'building block' in module theory (resp., representation theory). Other types of building blocks are the *simple* modules (resp., *irreducible* representations) discussed later in §1.5.

Moreover, we can discuss the generation of modules with the notions above. We say that a left (resp., right) A-module V is **finitely generated** if there exists a surjective left (resp., right) A-module map from $(A_{\mathrm{reg}})^{\oplus n}$ to V, for some $n \in \mathbb{N}$.

Next, we turn our attention to defining **direct products and (direct) sums of algebras**, which is achieved by way of the result below.

Proposition 1.15. *Let $(A_1, m_1, u_1), \ldots, (A_r, m_r, u_r)$ be algebras.*

(a) *Then, the direct product of vector spaces $A_1 \times \cdots \times A_r$ is an algebra, where*

$$m_{A_1 \times \cdots \times A_r} : \quad (A_1 \times \cdots \times A_r) \otimes (A_1 \times \cdots \times A_r) \to A_1 \times \cdots \times A_r,$$
$$(a_1, \ldots, a_r) \otimes (a_1', \ldots, a_r') \mapsto \big(m_1(a_1 \otimes a_1'), \ldots, m_r(a_r \otimes a_r')\big),$$
$$u_{A_1 \times \cdots \times A_r} : \quad \Bbbk \to A_1 \times \cdots \times A_r,$$
$$1_{\Bbbk} \mapsto (u_1(1_{\Bbbk}), \ldots, u_r(1_{\Bbbk})) =: (1_{A_1}, \ldots, 1_{A_r}).$$

The components (A_i, m_i, u_i) are nonunital subalgebras of $A_1 \times \cdots \times A_r$.

(b) *If $A_1, \ldots, A_r$ are subspaces of an algebra (A, m_A, u_A), such that*

$$(m_A)|_{A_i \otimes A_j} = 0 \quad \text{for } i \neq j, \tag{1.16}$$

then the vector space $A_1 + \cdots + A_r$ is an algebra, with multiplication m_A and unit u_A. Likewise, $(A_1 \oplus \cdots \oplus A_r, m_A, u_A)$ is an algebra if (1.16) holds.

(c) *The embeddings and projections between a vector space A_j and the direct product $A_1 \times \cdots \times A_r$ (resp., the direct sum $A_1 \oplus \cdots \oplus A_r$) in Lemma 1.2 can be upgraded to not-necessarily-unital algebra maps (resp., when (1.16) holds).* $\square$

We call a nonzero algebra A **indecomposable** if $A \not\cong A_1 \times A_2$, for some nonzero algebras A_1 and A_2; else, A is **decomposable**. Note that direct products are used for the definition of (in)decomposability of algebras, as opposed to direct sum as used for (in)decomposable modules above; see Remark 1.1.

Proposition 1.17. *A nonzero algebra A is indecomposable if and only if the only central idempotents of A (that is, elements $e \in Z(A)$ with $e^2 = e$) are 0_A and 1_A.* $\square$

It can be shown with the result above that, for a finite quiver Q, a path algebra $\Bbbk Q$ is an indecomposable algebra if and only if the quiver Q is connected.

§1.4.2. Tensor products and free products

Now we combine modules and algebras via a tensor product of vector spaces. Let A, B_1, B_2 be algebras, and let V, W be vector spaces. We write $_{B_1}V$ if V is a left B_1-module with action map $\triangleright : B_1 \otimes V \to V$, and write V_{B_2} if V is a right B_2-module with action map $\triangleleft : V \otimes B_2 \to V$. We also write $_{B_1}V_{B_2}$ if V is a (B_1, B_2)-bimodule with action maps $\triangleright$ and $\triangleleft$.

i. Tensor product of modules $_{B_1}V$ and W_{B_2}

For modules $_{B_1}V$ and W_{B_2}, the vector space $V \otimes W$ is a (B_1, B_2)-bimodule, where

$$b_1 \triangleright (v \otimes w) := (b_1 \triangleright v) \otimes w \quad \text{and} \quad (v \otimes w) \triangleleft b_2 := v \otimes (w \triangleleft b_2), \tag{1.18}$$

for each $b_1 \in B_1, b_2 \in B_2, v \in V$, and $w \in W$. See Exercise 1.16.

ii. Tensor product modules $_C V$ and $_C W$

For modules $_C V$ and $_C W$ over a commutative algebra C, we obtain that the tensor product $V \otimes W$ is a C-bimodule; see Remark 1.12 and §1.4.2i. But this fails when C is not commutative. We will discuss a way of making $V \otimes W$ a left C-module when C is not necessarily commutative in a future volume; there, we will give C the structure of a *bialgebra*.

iii. Tensor product $\otimes_A$ of modules V_A and $_A W$

We now form a tensor product of modules V_A and $_A W$. To do so, take a vector space Z. Then, a bilinear map $\beta : V \times W \to Z$ is called A-**balanced** if

$$\beta(v \triangleleft a, w) = \beta(v, a \triangleright w),$$

for all $a \in A, v \in V$, and $w \in W$.

The **tensor product V and W over A** is defined as the unique vector space $V \otimes_A W$ attached to $V \times W$, via an A-balanced map $\alpha : V \times W \to V \otimes_A W$, that satisfies the commutative diagram below.

$$V \times W \xrightarrow{\ \alpha\ } V \otimes_A W$$

$$\forall A\text{-balanced map } \beta \qquad \exists! \text{ linear map } \gamma$$

$$Z \text{ vs}$$

Here, we have a bijection:

$$\{\beta : V \times W \to Z,\ A\text{-balanced maps}\} \overset{1\text{-}1}{\longleftrightarrow} \{\gamma : V \otimes_A W \to Z,\ \text{linear maps}\}.$$

Concretely, $V \otimes_A W$ is realized as a quotient space of $V \otimes W$ as follows:

$$V \otimes_A W \cong \frac{V \otimes W}{\operatorname{span}_{\Bbbk}\langle (v \triangleleft a) \otimes w - v \otimes (a \triangleright w)\rangle_{a \in A,\, v \in V,\, w \in W}}.$$

Note that $({}_{B_1}V_A) \otimes_A ({}_A W_{B_2})$ is a (B_1, B_2)-bimodule. See Exercise 1.16.

In particular, if C is commutative, then $(V_C) \otimes_C ({}_C W)$ is a C-bimodule (see Remark 1.12). See also Exercises 1.17 and 1.18 for practice.

Similar to §1.1.4, we can form a linear map $f \otimes_A f' : V \otimes_A V' \to W \otimes_A W'$, for a right A-module map $f : V \to W$, and a left A-module map $f' : V' \to W'$. See Exercises 1.19 and 1.20.

iv. Algebras constructed using $(- \otimes_A -)$

An **A-bimodule algebra** is a triple (D, m, u) consisting of an A-bimodule D, and A-bimodule maps, $m : D \otimes_A D \to D$ and $u : A_{\mathrm{reg}} \to D$, satisfying associativity and unitality axioms:

$$m(m \otimes_A \operatorname{id}_D) = m(\operatorname{id}_D \otimes_A m) \qquad \text{and} \qquad m(u \otimes_A \operatorname{id}_D) = \operatorname{id}_D = m(\operatorname{id}_D \otimes_A u).$$

Indeed, $(D \otimes_A D) \otimes_A D \cong D \otimes_A (D \otimes_A D)$ and $D \otimes_A A_{\mathrm{reg}} \cong D \cong A_{\mathrm{reg}} \otimes_A D$ as discussed in Exercise 1.18.

A **morphism** between A-bimodule algebras is an A-bimodule map that is simultaneously an algebra map.

Now for an A-bimodule V, the **bimodule tensor algebra** is defined to be the unique A-bimodule algebra $T_A(V)$ attached to V, which comes equipped with an A-bimodule embedding,

$$\alpha : V \to T_A(V)_{A\text{-bimod}} \sqsubset T_A(V),$$

such that for any A-bimodule algebra Z, the commutative diagram below holds.

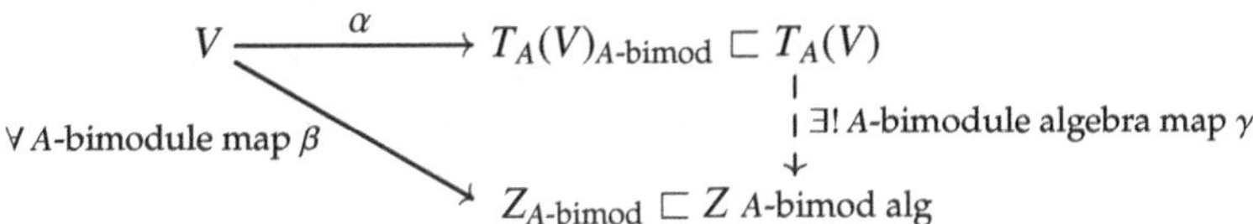

We obtain that $\mathrm{Hom}_{A\text{-bimod}}(V, Z_{A\text{-bimod}}) \cong \mathrm{Hom}_{A\text{-bimod alg}}(T_A(V), Z)$ as sets. Here, the symbol $\sqsubset$ denotes taking the underlying A-bimodule.

When A equals the ground field $\Bbbk$, we recover the tensor algebra $T(V)$ constructed in §1.2.2. Indeed, as an A-bimodule, we have that

$$T_A(V) = A_{\mathrm{reg}} \oplus V \oplus (V \otimes_A V) \oplus (V \otimes_A V \otimes_A V) \oplus \cdots .$$

Moreover, path algebras $\Bbbk Q$ arise as bimodule tensor algebras; see Exercise 1.21.

v. Tensor product of algebras

Consider the following result; the proof is Exercise 1.22. See also Exercise 1.23.

Proposition 1.19. *Let (A, m_A, u_A) and (B, m_B, u_B) be algebras. Then, the tensor product of underlying vector spaces $A \otimes B$ is an algebra, with*

$$m_{A\otimes B} := (m_A \otimes m_B)(\mathrm{id}_A \otimes \mathrm{flip} \otimes \mathrm{id}_B) \quad and \quad u_{A\otimes B} := u_A \otimes u_B.$$

Here, the linear map, $\mathrm{flip} : B \otimes A \to A \otimes B$, is given by $b \otimes a \mapsto a \otimes b$. $\square$

vi. Free product of algebras

One can also combine two algebras A and B universally in the following manner.

The **free product of A and B** is the unique algebra $A \circledast B$ equipped with injective algebra maps $\alpha_A : A \to A \circledast B$ and $\alpha_B : B \to A \circledast B$, such that for any algebra Z equipped with algebra maps $\beta_A : A \to Z$ and $\beta_B : B \to Z$, we have that the following diagram commutes. See Exercise 1.24 for practice.

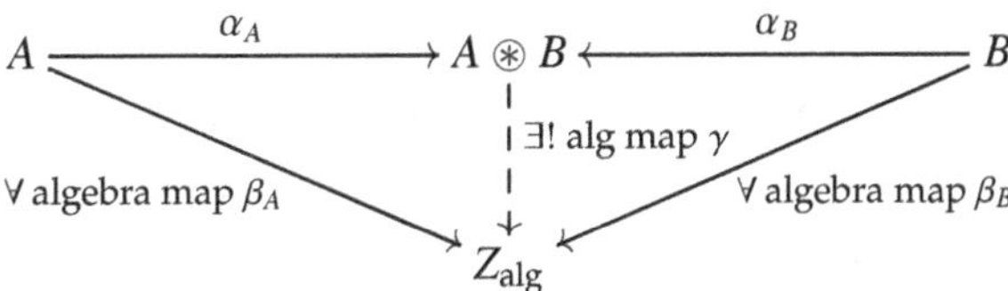

§1.4.3. Homs and duals

We discuss when a Hom space of modules (resp., algebras) has the structure of a module (resp., an algebra). We adopt the notation $_AV$, V_A, $_{B_1}V_{B_2}$ from §1.4.2.

Given an algebra A and left A-modules $_AV$ and $_AW$, recall from §1.1.4iv that the collection of linear maps from the underlying vector space V to the underlying vector space W is itself a vector space, denoted by $\mathrm{Hom}_{\Bbbk}(V, W)$. In fact, the collection of left A-module maps from $_AV$ to $_AW$ is a subspace of $\mathrm{Hom}_{\Bbbk}(V, W)$; we denote this subspace by $\mathrm{Hom}_{A\text{-mod}}(V, W)$.

Likewise, for right A-modules V_A and W_A, we let $\mathrm{Hom}_{\mathrm{mod}\text{-}A}(V, W)$ be the vector space of right A-module maps from V_A to W_A. Moreover, for algebras B_1 and B_2, we take $\mathrm{Hom}_{(B_1, B_2)\text{-bimod}}(V, W)$ to be the vector space of (B_1, B_2)-bimodule maps from $_{B_1}V_{B_2}$ to $_{B_1}W_{B_2}$.

i. Homs of (bi)modules

Consider the next result, the proof of which is part of Exercise 1.25.

Proposition 1.20. *For algebras A, B_1, B_2, we have that the following statements hold.*

(a) *If V is an (A, B_1)-bimodule and W is a (A, B_2)-bimodule, then the vector space $\mathrm{Hom}_{A\text{-mod}}(V, W)$ is a (B_1, B_2)-bimodule.*

(b) *If V is a (B_1, A)-bimodule and W is a (B_2, A)-bimodule, then the vector space $\mathrm{Hom}_{\mathrm{mod}\text{-}A}(V, W)$ is a (B_2, B_1)-bimodule.* $\qquad\square$

Now applying Proposition 1.20 to the linear dual $V^* = \mathrm{Hom}_{\Bbbk}(V, \Bbbk)$ of a vector space V, we immediately obtain the result below.

Corollary 1.21. *If V is a right (resp., left) B_1-module and $\Bbbk$ is a trivial right (resp., left) B_2-module, then V^* is a (B_1, B_2)-bimodule (resp., a (B_2, B_1)-bimodule).* $\qquad\square$

ii. Algebras constructed using $\mathrm{Hom}_{A\text{-mod}}(-, -)$

Let A be an algebra, and let V be a left A-module. Then,

$$\mathrm{End}_{A\text{-mod}}(V) := \mathrm{Hom}_{A\text{-mod}}(V, V)$$

is an algebra, with composition as multiplication, and with $1_{\mathrm{End}_{A\text{-mod}}(V)} = \mathrm{id}_V$. We call $\mathrm{End}_{A\text{-mod}}(V)$ a **left A-module endomorphism algebra**.

Note that when $A = \Bbbk$, we recover the endomorphism algebra, $\mathrm{End}_{\Bbbk}(V)$, from §1.2.1. See Exercise 1.26.

This endomorphism algebra can be used to detect when V is indecomposable; the proof of the result below is part of Exercise 1.15.

Proposition 1.22. *Let A be a nonzero algebra, let V be a left A-module, and consider the algebra $E := \mathrm{End}_{A\text{-mod}}(V)$. Then, V is indecomposable if and only if the only idempotents of E (that is, elements $e \in E$ with $e^2 = e$) are 0_E and 1_E.* $\qquad\square$

iii. On Homs of algebras

Given two algebras A and B, one may wish to put an algebra structure on the Hom space of underlying vector spaces, $\mathrm{Hom}_{\Bbbk}(A, B)$. But recall from §1.1.4v that the direction of a morphism gets reversed when applying the operation $\mathrm{Hom}_{\Bbbk}(-, B)$. So, it is more natural to replace the algebra,

$$(A,\ m_A : A \otimes A \to A,\ u_A : \Bbbk \to A),$$

in the first slot of $\mathrm{Hom}_{\Bbbk}(A, B)$ with an algebraic structure of the form,

$$(A,\ \Delta_A : A \to A \otimes A,\ \varepsilon_A : A \to \Bbbk),$$

and require that Δ_A and ε_A satisfy a reversed version of the associativity and unitality axioms (namely, *coassociativity* and *counitality*, respectively). Such a structure $(A, \Delta_A, \varepsilon_A)$ is called a *coalgebra*, and the resulting algebraic structure on the Hom space, $\mathrm{Hom}_{\Bbbk}(A, B)$, is a *convolution algebra*. Details will be discussed in a future volume.

Example 1.23. We consider the special case of when the algebra in the first slot of a Hom space is the ground field $\Bbbk$ (which is a coalgebra with $\Delta_{\Bbbk} : \Bbbk \xrightarrow{\sim} \Bbbk \otimes \Bbbk$ and $\varepsilon_{\Bbbk} = \mathrm{id}_{\Bbbk}$). We also assume that the algebra in the second slot, denoted by (A, m_A, u_A), is arbitrary. Then, the vector space $\mathrm{Hom}_{\Bbbk}(\Bbbk, A)$ is an algebra, where for $f, f' \in \mathrm{Hom}_{\Bbbk}(\Bbbk, A)$, the multiplication and unit are given as follows:

$$m_{\mathrm{Hom}_{\Bbbk}(\Bbbk,A)}(f \otimes f') : \Bbbk \xrightarrow{\sim} \Bbbk \otimes \Bbbk \xrightarrow{f \otimes f'} A \otimes A \xrightarrow{m_A} A, \qquad u_{\mathrm{Hom}_{\Bbbk}(\Bbbk,A)}(1_{\Bbbk}) := u_A.$$

We leave it to the reader to verify that this is indeed an associative, unital algebra.

§1.4.4. Restriction and (co)induction

Fix an algebra map $\phi : A \to B$. Then, there is a way to build modules over B from modules over A via ϕ, and vice versa. These operations will be called *restriction*, *induction*, and *coinduction*, as discussed below.

Let $(V, \triangleright_V)$ be a left B-module. Then, V is also a left A-module, with action

$$A \otimes V \to V, \qquad a \otimes v \mapsto \phi(a) \triangleright_V v.$$

This left A-module is called the **restriction of $_BV$ to A along** $\phi : A \to B$, and is denoted by $\mathrm{Res}_A^B(V)$. Exercise 1.27 asks the reader to verify that

$$\mathrm{Res}_A^B(V) \cong \mathrm{Hom}_{B\text{-mod}}(B, V), \quad \text{as left } A\text{-modules.} \tag{1.24}$$

Here, B is a left (B, A)-bimodule with $\triangleright_B = m_B$ and $b \triangleleft_B a = m_B\,(b \otimes \phi(a))$.

Now let V be a left A-module. We can build a left B-module from V with:

$$\mathrm{Ind}_A^B(V) := B \otimes_A V. \tag{1.25}$$

Exercise 1.27 asks the reader to verify that $\mathrm{Ind}_A^B(V)$ is a left B-module. This construction is called the **induction of** $_A V$ **to** B **along** $\phi : A \to B$.

Keep V as a left A-module. We can also build a left B-module from V by taking:

$$\mathrm{Coind}_A^B(V) := \mathrm{Hom}_{A\text{-mod}}(B, V). \tag{1.26}$$

Exercise 1.27 asks the reader to verify that $\mathrm{Coind}_A^B(V)$ is a left B-module. This construction is called the **coinduction of** $_A V$ **to** B **along** $\phi : A \to B$.

The construction of algebras built from restriction, induction, or coinduction is also explored as part of the open-ended Exercise 1.28.

§1.5. Simple algebras

As remarked in §1.4.1, indecomposable modules (resp., representations) are viewed as building blocks in module theory (resp., representation theory). Here, we study a notion finer than indecomposability, namely *simplicity*, and we also examine the corresponding notion for algebras. The reader may wish to view Figure 1.2 in §1.8 for a preview of the results for simple algebras; some details are deferred to Exercises 1.33 and 1.35.

§1.5.1. Simple modules

Given an algebra A, we say that a nonzero left A-module V is **simple** if the only nonzero left A-submodule of V is V itself. Simple right A-modules and simple A-bimodules can be defined similarly. Likewise, a representation of A is said to be **irreducible** if it does not have a proper subrepresentation; see Exercise 1.29.

Remark 1.27. (a) A simple module is indecomposable. Indeed, if V is a decomposable left A-module, then there exist nonzero left A-submodules V_1, V_2 of V such that $V \cong V_1 \oplus V_2$. So, V_1 is a proper submodule of V, and V is not simple.

(b) But indecomposable modules need not be simple. For instance, take the polynomial algebra $\Bbbk[v]$ from §1.2.3. Then, by Proposition 1.22 and Exercise 1.26(a), the regular left $\Bbbk[v]$-module $_{\Bbbk[v]}(\Bbbk[v])_{\mathrm{reg}}$ is indecomposable. But $_{\Bbbk[v]}(\Bbbk[v])_{\mathrm{reg}}$ has proper submodules, e.g., $_{\Bbbk[v]}(v)$. So, $_{\Bbbk[v]}(\Bbbk[v])_{\mathrm{reg}}$ is not simple.

We will see later in Proposition 1.48 in §1.6 that indecomposable modules are simple when the algebra A is *semisimple*. For now, we will consider *simple* algebras;

we will also provide a structure result for such algebras in the finite-dimensional case. To proceed, we present some useful results below, the first of which is due to Schur [1905].

Proposition 1.28 (Schur's Lemma). *Let A be an algebra, let V and W be simple left A-modules, and take $\phi \in \mathrm{Hom}_{A\text{-mod}}(V, W)$. Then, either $\phi = 0$ or ϕ is an isomorphism of left A-modules. In particular, $\mathrm{End}_{A\text{-mod}}(V)$ is a division algebra.*

Proof. Take a nonzero element ϕ of $\mathrm{Hom}_{A\text{-mod}}(V, W)$. Then, $\ker(\phi)$ and $\mathrm{im}(\phi)$ are left A-submodules of V and W, respectively. Since V and W are simple and $\phi \neq 0$, we have that $\ker(\phi) = 0$ and $\mathrm{im}(\phi) = W$. Therefore, ϕ is an isomorphism. For the last statement, recall that $\mathrm{End}_{A\text{-mod}}(V)$ is an algebra (see §1.4.3ii). Since its nonzero elements are invertible, $\mathrm{End}_{A\text{-mod}}(V)$ is then a division algebra. $\qquad\square$

Next, we measure how far an A-module V is from being simple. A **composition series** for a left A-module V is a sequence of left A-submodules

$$0 = V_0 \to V_1 \to V_2 \to \cdots \to V_n \to \cdots \to V,$$

such that each quotient module V_{i+1}/V_i is a simple left A-module for all i. If $V = V_n$ above for some n, then we say that V has **finite length**, and we refer to the minimum such n as the **length** of V.

Length is well-defined due to the following results of Jordan [1989] (reprint of 1870 work) and Hölder [1889] that were established for groups. The proof for modules holds similarly and we will skip this here.

Theorem 1.29 (Jordan-Hölder Theorem). *Take a left A-module V of finite length. If V has two composition series*

$$0 = V_0 \to V_1 \to \cdots \to V_n = V \quad \text{and} \quad 0 = W_0 \to W_1 \to \cdots \to W_m = V,$$

then $n = m$ and there exists a permutation σ of $\{1, \ldots, n\}$ such that $V_{\sigma(i)+1}/V_{\sigma(i)} \cong W_{i+1}/W_i$ as left A-modules for all i. $\qquad\square$

Observe that a left A-module is simple precisely when its length is 1.

Now for a left A-module V, we will study its decomposition into a direct sum of submodules. Given a left A-submodule W of V, we call a left A-submodule X of V a **complement** for W in V if $V \cong W \oplus X$ as left A-modules.

Lemma 1.30. *Given an algebra A, we have that a sum of simple left A-modules is a direct sum. Moreover, for a (direct) sum V of simple left A-modules, we have that any left A-submodule W of V has a complement.*

Proof. Let $\{V_i\}_{i\in I}$ be simple left A-modules, and take the sum $V := \sum_{i\in I} V_i$. Next, using Zorn's lemma, there is a maximal subset J of I such that $V' := \sum_{i\in J} V_i$ is direct. Now for V_j with $j \in I$, we have that $V' \cap V_j$ is either 0 or equal to V_j, since V_j is simple. If $V' \cap V_j = 0$, then $V' + V_j$ is direct, contradicting J being maximal. So, $V' \cap V_j = V_j$, for each $j \in I$, and $\sum_{j\in I} V_j \subset V'$. Hence, $V' = V$ is a direct sum.

Likewise, if W is a left A-submodule of V, then there exists a maximal subset J of I such that $W' := W + \sum_{i\in J} V_i$ is direct. We can also repeat the arguments above to obtain that $W' = V$. Thus, $V = W \oplus X$, for $X := \sum_{i\in J} V_i = \bigoplus_{i\in J} V_i$. $\qquad\square$

So, modules that are a direct sum of simple modules have an internal structure that is well-understood. We also have a general decomposition result when we use the weaker notion of indecomposable modules, due to the work of Krull [1925] and Schmidt [1913]. We refer the reader to Section 1.2.6 of Lorenz [2018] for a proof of this result for representations in the finite-dimensional case, which can be translated to modules via Exercise 1.12.

Theorem 1.31 (Krull-Schmidt Theorem). *If V is a left A-module of finite length, then*

$$V \;\cong\; V_1 \oplus \cdots \oplus V_n,$$

for a unique choice of indecomposable left A-modules V_i, up to isomorphism. $\qquad\square$

Finally, we relate simple modules to ideals of an algebra. A left ideal of an algebra A is **minimal** if the only left ideals of A contained in I are 0 and I itself. A similar notion holds for right ideals and (two-sided) ideals. The result below is then clear by definition.

Lemma 1.32. *Given an algebra A, we have the following statements.*

(a) *A minimal left ideal of A is a simple left A-module.*

(b) *If $_A(A_{\mathrm{reg}})$ is a direct sum of simple left A-modules $\bigoplus_{i\in I} V_i$, then each V_i is a minimal left ideal of A.* $\qquad\square$

§1.5.2. Simple algebras

Recall that ideals (rather than subalgebras) are the substructures of algebras that arise as kernels of algebra maps, and thus, yield quotient algebras. So, ideals are 'ideal' when breaking an algebra down into pieces. We consider when an algebra cannot be broken down in such a manner.

A nonzero algebra A is called **simple** if its only ideals are 0 and itself.

Note that left (resp., right) ideals of A coincide with the left (resp., right) A-submodules of regular A-module. So, an algebra A is simple if and only if the regular bimodule $_A(A_{\mathrm{reg}})_A$ is simple.

Simplicity also implies indecomposability for algebras; cf. Remark 1.27.

Remark 1.33. (a) A simple algebra is indecomposable. Indeed, suppose that A is a decomposable algebra. Then, $A \cong A_1 \times A_2$, for some nonzero algebras A_1 and A_2. Now $(A_1)_{\mathrm{reg}} \times \{0_{A_2}\}$ is a proper ideal of A. Thus, A is not simple.

(b) But indecomposable algebras need not be simple. For instance, the polynomial algebra $\Bbbk[v]$ (from §1.2.3) is indecomposable. One way to see this is to use Figure 1.1 and Exercise 1.15(d), or by using Proposition 1.17. On the other hand, $\Bbbk[v]$ has proper ideals, e.g., (v). So, $\Bbbk[v]$ is not a simple algebra.

Example 1.34. Matrix algebras $\mathrm{Mat}_n(\Bbbk)$ (from §1.2.1) are simple. To see this, recall the elementary matrices $E_{k,\ell}$, and take a nonzero ideal I of $\mathrm{Mat}_n(\Bbbk)$. Then, I contains a matrix $X := (c_{i,j})$, with an entry $c_{p,q} \neq 0$ for some indices p, q. Now,

$$E_{k,\ell} = \frac{1}{c_{p,q}} E_{k,p} \, X \, E_{q,\ell}$$

is contained in I, for each k, ℓ. Thus, $I = \mathrm{Mat}_n(\Bbbk)$.

We will show in §1.5.3 that matrix algebras $\mathrm{Mat}_n(\Bbbk)$ are the only finite-dimensional simple algebras. Moreover, for a division algebra D, we have that a matrix algebra $\mathrm{Mat}_n(D)$ is simple; see Exercise 1.33.

Now consider the following result about commutative, simple algebras.

Proposition 1.35. *Let C be a nonzero commutative algebra. Then, C is simple if and only if C is a field. In this case, C is finite-dimensional if and only if $C \cong \Bbbk$.*

Proof. First, note that fields are commutative and simple. Conversely, suppose that C is a commutative simple algebra. Let x be a nonzero element of C. Then, the nonzero ideal generated by x must be equal to C, since C is simple. Now there exists a nonzero element y of C such that $xy = 1_C$. Since C is commutative, we also have that $yx = 1_C$. Thus, x is invertible, and C is a field.

Further, if C is finite-dimensional, then C is a finite field extension of $\Bbbk$, and thus (using field theory), C is algebraic over $\Bbbk$. Since $\Bbbk$ is assumed to be algebraically closed, we get that $C \cong \Bbbk$. Conversely, $\Bbbk$ is a finite-dimensional simple algebra. $\square$

§1.5.3. Classification of simple algebras

Our next goal is to classify finite-dimensional simple algebras. Consider the important lemma below.

Lemma 1.36. *The only finite-dimensional division algebra D (over $\Bbbk$) is $\Bbbk$.*

Proof. Take an element $x \in D$. Since D is finite-dimensional, there exists some minimal $n \in \mathbb{N}$ such that $1_D = x^0, x = x^1, x^2, \ldots, x^n$ are $\Bbbk$-linearly dependent. In this case, $p(x) := x^n + \lambda_{n-1}x^{n-1} + \cdots + \lambda_1 x + \lambda_0 = 0$, for some scalars $\lambda_i \in \Bbbk$. This polynomial has a root λ in $\Bbbk$ because $\Bbbk$ is assumed to be algebraically closed. Now $p(x) = (x - \lambda)q(x) = 0$, and by the minimality of n, we get that $q(x) \neq 0$. Since division algebras are domains, we have that $x = \lambda \in \Bbbk$. Thus, $D = \Bbbk$. $\qquad\square$

Next, we present a classification result for finite-dimensional simple algebras; see Section 71 of Cartan [1898] (for $\Bbbk = \mathbb{C}$) and Theorem 21 of Wedderburn [1908], which builds on Satz 30 of Molien [1892].

Theorem 1.37 (Cartan-Wedderburn Theorem). *Take a finite-dimensional algebra A. Then, A is simple if and only if $A \cong \mathrm{Mat}_n(\Bbbk)$, for some $n \in \mathbb{N}$.*

Proof. First, if A is isomorphic to $\mathrm{Mat}_n(\Bbbk)$, then A is simple by Example 1.34.

Conversely, assume that A is a finite-dimensional simple algebra. Then, there exists a minimal left ideal I of A since A is finite-dimensional. Now $I = Ax$ for some nonzero $x \in A$. On the other hand, the A-bimodule AxA is equal to A since A is simple. So,

$$_A(A_{\mathrm{reg}}) = \textstyle\sum_{a \in A} I_a,$$

for left ideals $I_a := Axa$ of A. Note that there is a left A-module epimorphism $I \to I_a$ given by $x \mapsto xa$. Since I is a simple left A-module [Lemma 1.32(a)], each I_a is either 0 or isomorphic to I. Hence,

$$_A(A_{\mathrm{reg}}) \cong \textstyle\sum_{a \in S} I,$$

for a subset S of A. Further, this sum is direct by Lemma 1.30, and finite as $_A(A_{\mathrm{reg}})$ is finitely generated by 1_A. Therefore,

$$_A(A_{\mathrm{reg}}) \cong I^{\oplus n} \tag{1.38}$$

as left A-modules, for some $n \in \mathbb{N}$. We then obtain the algebra isomorphisms:

$$A^{\mathrm{op}} \cong \mathrm{End}_{A\text{-mod}}(_A(A_{\mathrm{reg}})) \cong \mathrm{End}_{A\text{-mod}}(I^{\oplus n}) \cong \mathrm{Mat}_n\left(\mathrm{End}_{A\text{-mod}}(I)\right).$$

The first and last isomorphisms hold by Exercise 1.26(a,d). Apply Schur's Lemma [Proposition 1.28] to get that $\mathrm{End}_{A\text{-mod}}(I)$ is a division algebra. We then apply Lemma 1.36 to get that $\mathrm{End}_{A\text{-mod}}(I) \cong \Bbbk$, since A (and thus $\mathrm{End}_{A\text{-mod}}(I)$) is finite-dimensional. Now by Exercise 1.33(a), we have that as algebras:

$$A \cong \mathrm{Mat}_n(\Bbbk)^{\mathrm{op}} \cong \mathrm{Mat}_n(\Bbbk). \qquad\square$$

Remark 1.39. There is no known analogue of the Cartan-Wedderburn Theorem for infinite-dimensional algebras. For instance, the Weyl algebras $A_n(\Bbbk)$ from Example 1.10 are simple by the work of Hirsch [1937], and are not isomorphic to a matrix algebra. In general, classification problems in infinite settings are too tough to pursue without imposing strong hypotheses on the structures of interest.

§1.6. Semisimple algebras

Here, we study *semisimple* algebras, which are generalizations of simple algebras (see §1.5) in the finite-dimensional case. We will present a version of the Cartan-Wedderburn Theorem [Theorem 1.37] for these algebras below. The reader may wish to view Figure 1.2 in §1.8 for a preview of the results for semisimple algebras; some details are deferred to Exercises 1.33 and 1.35.

Given an algebra A, we say that a left A-module V is **semisimple** if it is the sum of simple left A-modules; this sum is direct by Lemma 1.30. Moreover, we say that the algebra A is **semisimple** if the regular module $_A(A_{\mathrm{reg}})$ is semisimple; else, we call A **nonsemisimple**.

There are several characterizations of semisimple algebras in the literature, and the one below is module-theoretic like the definition.

Lemma 1.40. *An algebra A is semisimple if and only if each of its left A-modules is semisimple.*

Proof. The reverse direction is clear. For the forward direction, note that any left A-module M has a set of generators $\{m_i\}_{i\in I}$ that is not necessarily finite. Now there is a surjective A-module homomorphism:

$$A^{\oplus I} \to M, \quad (a_i)_{i\in I} \mapsto \textstyle\sum_{i\in I}(a_i \triangleright m_i).$$

Since the direct sum of semisimple modules is semisimple, and a homomorphic image of a semisimple module is semisimple (think about this), we have that M is semisimple, as desired. $\square$

One useful feature of semisimple algebras is given as follows.

Lemma 1.41. *Let A be a semisimple algebra. Then, any descending chain of left ideals of A must stabilize.*

Proof. We have that $_A(A_{\mathrm{reg}})$ is a direct sum of finitely minimal left ideals $\{I_j\}_{j=1}^r$ of A due to Lemma 1.32(b) and due to $_A(A_{\mathrm{reg}})$ being finitely generated by 1_A. Since, for each j, each chain of left A-submodules of I_j stabilizes due to the minimality of I_j, we then get the result for $_A(A_{\mathrm{reg}})$. $\square$

An algebra A is called **left Artinian** if any descending chain of left ideals of A stabilizes; thus, we have shown that semisimple algebras are left Artinian. On the other hand, an algebra A is **left Noetherian** if any ascending chain of left ideals of A stabilizes. Artinian and Noetherian algebras are very important classes of algebras in mathematics, and we will refer the reader to Goodearl and Warfield [2004] and other references in §1.10 for further information.

§1.6.1. Classification of semisimple algebras

Next, we turn our attention to classifying semisimple algebras. Consider the preliminary result below.

Proposition 1.42. *Let A be an algebra. Then, A is simple and semisimple if and only if $A \cong \mathrm{Mat}_n(D)$ for some $n \in \mathbb{N}$ and some division algebra D.* $\qquad\square$

Proof. For the forward direction, we use the proof of the Cartan-Wedderburn Theorem [Theorem 1.37]. Note that the regular module $_A(A_{\mathrm{reg}})$ is (finitely) generated by 1_A as a left A-module. Next, since A is semisimple, $_A(A_{\mathrm{reg}}) \cong \bigoplus_{j=1}^{n} V_j$, a direct sum of finitely many simple left A-modules, which is a direct sum of finitely many minimal left ideals of A (see Lemma 1.32(b)). One can then argue as in the proof of Theorem 1.37 to obtain (1.38). Then, we use the simplicity of A and follow that argument to get that $A \cong \mathrm{Mat}_n(D)$ as algebras, for a division algebra D; see Exercise 1.33(a). The backward direction holds by Exercise 1.33(b,c). $\qquad\square$

As a consequence, we obtain a family of semisimple algebras.

Corollary 1.43. *Take division algebras $D_1, \ldots, D_r$. Then, $\prod_{i=1}^{r} \mathrm{Mat}_{n_i}(D_i)$ is a semisimple algebra. In particular, $\prod_{i=1}^{r} \mathrm{Mat}_{n_i}(\Bbbk)$ is a finite-dimensional semisimple algebra.*

Proof. Let $A_i := \mathrm{Mat}_{n_i}(D_i)$. Since A_i is semisimple by Proposition 1.42, we get that the regular left module over A_i is a direct sum of minimal left ideals $\{I_{i,j}\}_j$ of A_i. We can embed each $I_{i,j}$ to identify it as an ideal of $\prod_{i=1}^{r} A_i$, and it remains minimal. Therefore, the regular left module over $\prod_{i=1}^{r} A_i$ is now $\bigoplus_{i,j} I_{i,j}$, which implies that $\prod_{i=1}^{r} A_i$ is semisimple via Lemma 1.40. $\qquad\square$

We now present the full classification of semisimple algebras, which is due to Theorem 22 of Wedderburn [1908] in the finite-dimensional case, and the work of Artin [1927] extending Wedderburn's result to the infinite-dimensional case while assuming the Artinian and Noetherian conditions. Works by Noether [1929] and Hopkins [1939] were then applied to remove the chain conditions in the initial hypotheses. This yields the result below.

Theorem 1.44 (Artin-Wedderburn Theorem). *Let A be an algebra. Then, A is semisimple if and only if*

$$A \cong \prod_{i=1}^{r} \mathrm{Mat}_{n_i}(D_i),$$

for some unique choice of $r, n_1, \ldots, n_r \in \mathbb{N}$ and division algebras $D_1, \ldots, D_r$. In this case, A is finite-dimensional if and only if $A \cong \prod_{i=1}^{r} \mathrm{Mat}_{n_i}(\Bbbk)$.

Proof. The algebra $\prod_{i=1}^{r} \mathrm{Mat}_{n_i}(D_i)$ is semisimple by Corollary 1.43. Conversely, if A is semisimple, then $_A(A_{\mathrm{reg}})$ is a direct sum of minimal left ideals $\{I_j\}_{j\in J}$ of A, and $|J| < \infty$ since the regular module is finitely generated by 1_A. In this case,

$$_A(A_{\mathrm{reg}}) \cong \bigoplus_{j=1}^{r}(I_j^{\oplus n_j}),$$

where $\{I_j\}_j$ are pairwise non-isomorphic left A-modules. Now as in the proof of Theorem 1.37, we obtain that

$$A^{\mathrm{op}} \cong \mathrm{End}_{A\text{-mod}}(_A(A_{\mathrm{reg}})) \cong \mathrm{End}_{A\text{-mod}}(\bigoplus_{j=1}^{r}(I_j^{\oplus n_j})) \cong \prod_{j=1}^{r} \mathrm{Mat}_{n_j}\big(\mathrm{End}_{A\text{-mod}}(I_j)\big).$$

The first and last isomorphism follows from Exercise 1.26 and Schur's Lemma [Proposition 1.28]. Now apply Schur's Lemma again and Exercise 1.33(a) to conclude that A is of the desired form. We leave it to the reader to verify the uniqueness statement via the Jordan-Hölder Theorem [Theorem 1.29]. Lastly, we apply Lemma 1.36 to yield the result for the finite-dimensional case. $\square$

We refer to the values $\{r; n_1, \ldots, n_r\}$ in the decomposition of A above as the **Artin-Wedderburn parameters** of A.

Next, note the following consequence of the Cartan-Wedderburn Theorem and the Artin-Wedderburn Theorem [Theorems 1.37 and 1.44].

Corollary 1.45. *Finite-dimensional simple algebras are semisimple.* $\square$

Remark 1.46. (a) However, finite-dimensional semisimple algebras are not necessarily simple; e.g., consider the direct product of fields, $\Bbbk \times \Bbbk$.

(b) Also, infinite-dimensional simple algebras need not be semisimple. Namely, the Weyl algebra $A_n(\Bbbk)$ from Example 1.10 is simple (see Remark 1.39), but is not semisimple. To see why $A := A_n(\Bbbk)$ is not semisimple, consider the infinite chain $A \supset Av_1 \supset Av_1^2 \supset Av_1^3 \supset \cdots$ of left ideals of A. Since this chain does not stabilize, Lemma 1.41 yields the desired result.

Finally, we point out a special case of the Artin-Wedderburn Theorem, namely for the group algebras $\Bbbk G$ from §1.2.6; see §1.3.4 on modules over $\Bbbk G$.

Theorem 1.47 (Maschke's Theorem). *Let G be a group. Then, a group algebra $\Bbbk G$ is semisimple precisely when G is finite.* $\square$

This vital result is due to the work of Maschke [1899], predating the Artin-Wedderburn Theorem. We refer the reader to textbooks on (group) representation theory for the proof; see, e.g., Theorem 8.1 of James and Liebeck [2001] where $\Bbbk = \mathbb{C}$ and G is assumed to be finite, or Theorem 3.4.1 of Lorenz [2018] for the most general case with no assumptions on $\Bbbk$.

§1.6.2. Modules over semisimple algebras

Let us now show that indecomposability is the same as simplicity for modules over semisimple algebras; see Remark 1.27.

To do so, we say that a left A-module V is **completely reducible** if V is isomorphic to a direct sum of simple left A-submodules.

Proposition 1.48. *An algebra A is semisimple if and only if any left A-module is completely reducible. In this case, an indecomposable left A-module must be simple.*

Proof. Suppose that A is semisimple. Then, any left A-module V is of the form $\sum_{i \in I} V_i$, for some left A-submodules V_i generated by one element. Such modules V_i are homomorphic images of the regular left module $_A(A_{\mathrm{reg}})$, where 1_A maps to the generator of V_i. Now $V = \sum_{i \in I} V_i$ is the homomorphic image of a direct sum of simple left A-modules by the semisimplicity of A and by Lemma 1.30. Since homomorphic images of semisimple modules are semisimple, the forward direction holds. Conversely, if the left regular module over A is completely reducible, then A is semisimple by definition.

Lastly, if a left A-module V is not simple, then there exists a proper left A-submodule W of V. Now by Lemma 1.30, there exists a left A-submodule X of V such that $V = W \oplus X$. Thus, V is not indecomposable. $\qquad\square$

Remark 1.49. If A is an algebra that has an indecomposable left A-module that is not simple, then by the result above, A is a nonsemisimple algebra. For instance, we can use this to show that the polynomial algebra $\Bbbk[v]$ is nonsemisimple; see Remark 1.27.

Semisimple algebras have finitely many simple modules, up to isomorphism. This result follows from Lemma 1.32 and the proofs of Theorems 1.37 and 1.44.

Proposition 1.50. *Take a semisimple algebra A with Artin-Wedderburn parameters $\{r; n_1, \ldots, n_r\}$. Then, A has r simple left A-modules, $V_1, \ldots, V_r$, up to isomorphism, with $\dim_{\Bbbk} V_i = n_i$, for $i = 1, \ldots, r$, up to reordering.* $\qquad\square$

Example 1.51. Take $A := \mathrm{Mat}_n(\Bbbk)$. Then, A has precisely one simple left A-module V, up to isomorphism, and V has dimension n. Here, V is isomorphic to the left ideal of matrices, $\{(c_{i,j}) \in \mathrm{Mat}_n(\Bbbk) \mid c_{i,j} = 0 \ \text{for} \ j \neq 1\}$.

See Exercises 1.30, 1.31, and 1.32 for the commutative case and more practice.

§1.7. Separable algebras

Here, we discuss separable algebras, which are algebraic versions of separable field extensions in field theory. We will see that these algebras are the same as finite-dimensional semisimple algebras (due to our assumptions on the ground field). However, separable algebras have an advantage over semisimple algebras in that they can be defined using compositions of morphisms, i.e., with commutative diagrams. Defining algebraic structures via commutative diagrams will play a role throughout the categorical chapters in this book.

Take an algebra (A, m, u), and consider the linear map:

$$\mu : A \otimes A^{\mathrm{op}} \to A, \quad a \otimes b \mapsto m(a \otimes b).$$

Note that μ is an A-bimodule map, with $\rhd_A = m = \lhd_A$, and $\rhd_{A \otimes A^{\mathrm{op}}} = m \otimes \mathrm{id}_A$, and $\lhd_{A \otimes A^{\mathrm{op}}} = \mathrm{id}_A \otimes m$.

We call an algebra (A, m, u) **separable** if μ has a right inverse $\phi : A \to A \otimes A^{\mathrm{op}}$ as A-bimodules, i.e., if the following diagrams commute.

$$
\begin{array}{ccc}
A \xrightarrow{\phi} A \otimes A^{\mathrm{op}} & &
A \otimes A^{\mathrm{op}} \otimes A \xleftarrow{\phi \otimes \mathrm{id}} A \otimes A \xrightarrow{\mathrm{id} \otimes \phi} A \otimes A \otimes A^{\mathrm{op}}
\end{array}
\tag{1.52}
$$

Proposition 1.53. *We have that an algebra A is separable if and only if there exists an element $e^A := e^1 \otimes e^2 \in A \otimes A^{\mathrm{op}}$ such that*

$$e^1 e^2 = 1_A \quad \text{and} \quad a e^1 \otimes e^2 = e^1 \otimes e^2 a, \quad \forall a \in A.$$

Here, we are using sumless notation for elements in $A \otimes A^{\mathrm{op}}$. In this case, the element e is idempotent, and we call it a **separability idempotent** *for A.*

Proof. Take a right inverse ϕ of μ as above, and consider the notation:

$$\phi(1_A) := e^1 \otimes e^2 \in A \otimes A^{\mathrm{op}}.$$

We also identify this element with the morphism below:

$$e^A := \phi \circ u : \Bbbk \to A \otimes A^{\mathrm{op}}, \quad 1_{\Bbbk} \mapsto e^1 \otimes e^2.$$

The first commutative diagram in (1.52) implies the equation $e^1 e^2 = 1_A$. Also, the unit axiom and the second commutative diagram in (1.52) implies that

$$a e^1 \otimes e^2 = a \phi(1_A) \overset{(*)}{=} \phi(a 1_A) = \phi(a) = \phi(1_A a) = \phi(1_A) a = e^1 \otimes e^2 a$$

for all $a \in A$. In particular for step $(*)$, we have that

$$\phi(a 1_A) = \phi m(a \otimes 1_A) = (m \otimes \mathrm{id})(\mathrm{id} \otimes \phi)(a \otimes 1_A) = (m \otimes \mathrm{id})(a \otimes e^1 \otimes e^2) = a\phi(1_A).$$

Moreover, the element $e \in A \otimes A^{\mathrm{op}}$ is idempotent: for a copy $\overline{e^A}$ of e^A, we get:

$$(e^A)^2 = e^A \overline{e^A} = (e^1 \otimes e^2)(\overline{e}^1 \otimes \overline{e}^2) = e^1 \overline{e}^1 \otimes \overline{e}^2 e^2 = \overline{e}^1 \otimes \overline{e}^2 e^2 e^1 = \overline{e}^1 \otimes \overline{e}^2 = e^A.$$

Here, the third equation holds by multiplication in $A \otimes A^{\mathrm{op}}$ (see Proposition 1.19), the fourth equation holds by the second separability axiom, and the fifth equation holds by the first separability axiom applied in A^{op} (see Exercise 1.34(b)).

We leave it to the reader to verify the converse. $\qquad\square$

Next, we provide some examples of separable algebras; see Exercise 1.34(a).

Example 1.54. Recall the matrix algebras $\mathrm{Mat}_n(\Bbbk)$ and the elementary matrices $E_{k,\ell}$ from §1.2.1. Then, $(e^{\mathrm{Mat}_n(\Bbbk)})_k := \sum_{\ell=1}^n E_{\ell,k} \otimes E_{k,\ell}$ is a separability idempotent, for each $k = 1,\ldots,n$. So, separability idempotents are not unique.

Example 1.55. Recall the group algebra $\Bbbk G$ from §1.2.6. If G is finite, then the element $e^{\Bbbk G} := \frac{1}{|G|} \sum_{g \in G} g \otimes g^{-1}$ is a separability idempotent for $\Bbbk G$.

Moreover, the name for separable algebras stems from the example below.

Example 1.56. Let $\mathbb{F}$ be an arbitrary field and consider a finite field extension K over $\mathbb{F}$. Then, we have that the field extension $K/\mathbb{F}$ is separable if and only if the $\mathbb{F}$-algebra K is separable. See Lemma 10.7b from Pierce [1982].

We will now discuss why there is no difference between finite-dimensional semisimple algebras studied in §1.6 and separable algebras studied here.

Proposition 1.57. *An algebra A is separable if and only if A is semisimple and finite-dimensional.*

Proof. For the forward direction, see Corollary 10.3 and Corollary 10.4b of Pierce [1982]. (Note that this direction holds over an arbitrary field.)

For the reverse direction, we can apply the Artin-Wedderburn theorem to get that A is isomorphic to a finite direct product of matrix algebras over $\Bbbk$. Now $\mathrm{Mat}_n(\Bbbk)$ is separable as shown in Example 1.54, and a finite direct product of separable algebras is separable by Exercise 1.34(b). $\qquad\square$

We will explore generalizations of algebras, including separable algebras, in the categorical setting later in Chapter 4. There, the convention for a separable algebra is slightly modified due to the following remark.

Remark 1.58. Notice that the opposite multiplication in $A \otimes A^{\mathrm{op}}$ is only used when showing that e^A is an idempotent element. However, the idempotent condition is not used later in the categorical version of a separable algebra. As a result, the map μ is later replaced with m, while still requiring (1.52).

§1.8. Summary

Recall that our ground field $\Bbbk$ is algebraically closed and has characteristic 0 throughout the book. The main structure of interest in this chapter is an algebra over a field $\Bbbk$, which is a combination of a unital ring and a $\Bbbk$-vector space. We discussed ways that the structure of an algebra gets translated to a collection of linear maps (or matrices) via representations and modules. We also presented various constructions of algebras and modules, namely direct products, sums, direct sums, tensor products, Homs, and duals.

We then investigated classification results for various types of algebras, i.e., those that are simple, semisimple, and separable, along with (counter-)examples. This is captured by Figure 1.2 below. Exercise 1.35 asks the reader to examine this figure in detail and to derive more (counter-)examples that fit into the diagram.

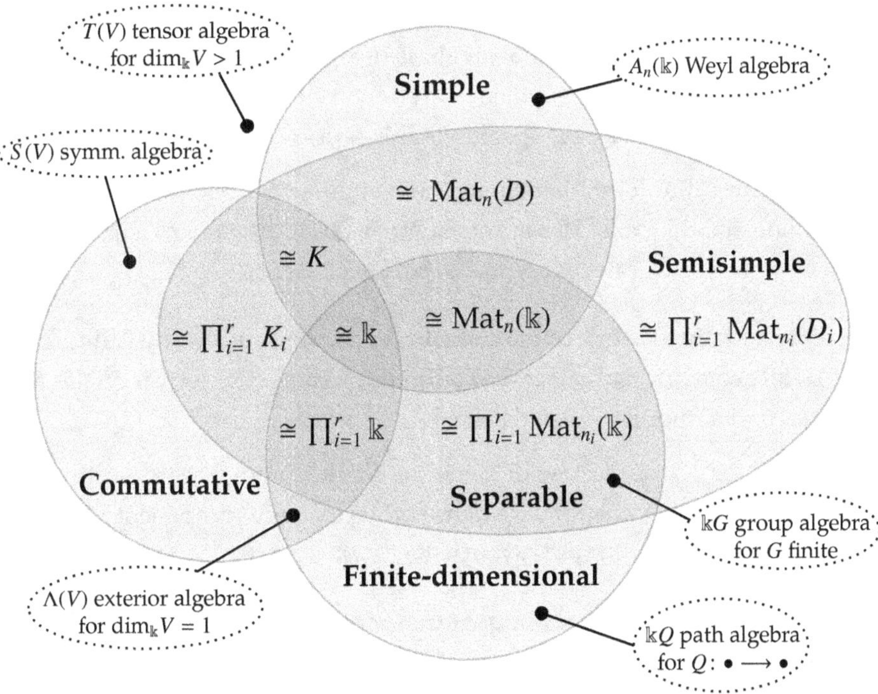

Figure 1.2: Classification results for algebras over $\Bbbk$, with (counter-)examples. Here, n, n_i, r are positive integers; K, K_i are fields over $\Bbbk$; and D, D_i are division algebras over $\Bbbk$.

§1.9. Modern applications

We now illustrate how various notions that were introduced in this chapter on algebras over a field are used in modern mathematics. A full understanding of the resources here is not expected. Instead, we aim to put the chapter's material into context by offering videos and content to casually explore.

The **Artin–Wedderburn Theorem** has been used in several fields, such as:

* **Cryptography**, see, e.g., Kuz'min et al. [2015];

* **Languages and automata**, see, e.g., Almeida and Rodaro [2016];

* **Linear codes**, see, e.g., Olteanu and Van Gelder [2015];

* **Network synchronization**, see, e.g., Zhang and Motter [2020];

* **Noncommutative geometry**, see, e.g., Marcolli and van Suijlekom [2014];

* **Semidefinite programming**, see, e.g., Vallentin [2009], Burgdorf et al. [2013];

* **Symbolic dynamics**, see, e.g., Kwapisz [2000].

The importance of the **polynomial algebras in algebraic geometry** is highlighted in the following video. An excellent introductory text on this field, Smith et al. [2000], is also authored by the speaker of this series.

Karen Smith's 2022 Joint Mathematics Meetings AMS Colloquium Lecture on "Understanding and Measuring Singularities in Algebraic Geometry"
https://youtu.be/k6sk9_6EzuQ

The **representation theory of path algebras** has been studied by Chindris et al. [2015] with geometric and categorical techniques, and a clear lecture on this article by the second author is available below.

Ryan Kinser's 2012 Worldwide Center of Mathematics research lecture on "Module varieties with dense orbits in every component"
https://youtu.be/BnVXT64JSx0

The role of the **Weyl algebras in quantum mechanics** is discussed in Section 3 of the expository article by Walton [2019], and a lecture on this material can be found at the link below (starting at 15:08).

C. Walton's 2021 Joint Mathematics Meetings NAM Claytor-Woodard Lecture on "An Invitation to Noncommutative Algebra"
https://youtu.be/G2ZX0Zq0BxM

Weyl algebras are also examples of **algebras of differential operators**; see Coutinho [1995] for a user-friendly introduction to this area of research.

§1.10. References for further exploration

- A great historical account of the development of algebras over a field before 1927 is provided by LaDuke [1983].

- The manuscript by Dickson [1923] is compelling as it established numerous notions about algebras that are still used 100 years later in ring theory, in representation theory, and in algebraic number theory.

- The textbook by Aluffi [2009] provides an insightful discussion of algebraic structures with a view towards *Category Theory*. It discusses why the categories of groups and of rings are not well-behaved, and why the categories of abelian groups, of vector spaces, and of modules are preferable, for instance.

- The textbook by Assem et al. [2006] is recommended to learn about path algebras $\Bbbk Q$ (see §1.2.5), their quotient algebras, and their module theory. *Representations of quivers Q* are also discussed.

- The textbook by James and Liebeck [2001] is a friendly resource to learn about the representation theory of groups G, and equivalently, of group algebras $\Bbbk G$. It assumes that G is finite, and that $\Bbbk = \mathbb{R}$ or $\mathbb{C}$.

- A holistic investigation of representation theory, including that of algebras, of groups, and of other algebraic structures such as *Lie algebras* and *Hopf algebras*, is provided in the textbook by Lorenz [2018].

- The textbooks by Pierce [1982] and Goodearl and Warfield [2004] are excellent introductory resources for the classical theory of rings, of (associative) algebras, and of their modules.

- The Database of Ring Theory (https://ringtheory.herokuapp.com/) is a helpful repository for examples of rings that satisfy certain properties. It also maintains a list of useful theorems and websites in ring theory.

§1.11. Exercises

1.1 [Open-ended] Recall the notion of structure versus property in §1.1.1.

   (a) List various properties of groups.

   (b) Describe structures that one could impose on a certain group G.

   (c) Repeat parts (a) and (b) for rings, vector spaces, and algebras, after reading §1.1.2, §1.1.3, and §1.1.5, respectively.

1.2   Recall the chat about universal property in §1.1.4iii. Let X be a gadget, and let $\mathrm{Univ}(X)$ and $\overline{\mathrm{Univ}(X)}$ be universal structures attached to X, say of Form I. Prove that there is a structure isomorphism between $\mathrm{Univ}(X)$ and $\overline{\mathrm{Univ}(X)}$.

Hint. Use commutative diagrams as described in §1.1.4iii, the feature of uniqueness in the definition of a universal property, and the notion of isomorphism that requires the existence of mutually inverse morphisms.

1.3   Recall direct sums and tensor products of vector spaces from §1.1.4i,iii. Let U, V, W be vector spaces, and consider $\Bbbk$ as a vector space over itself. Prove the isomorphisms below using the universal property of $\otimes := \otimes_\Bbbk$.

(a) $\Bbbk \otimes V \cong V$ and $V \otimes \Bbbk \cong V$.

(b) $(U \otimes V) \otimes W \cong U \otimes (V \otimes W)$.

(c) $U \otimes (V \oplus W) \cong (U \otimes V) \oplus (U \otimes W)$.

An aside. An analogue to part (c) holds if we replace $V \oplus W$ with $\bigoplus_{i \in I} V_i$.

1.4   Recall tensor products, Hom spaces, and duals of vector spaces from §1.1.4iii,iv. Let V and W be vector spaces.

(a) Show that the following linear map is an embedding:

$$W \otimes V^* \to \mathrm{Hom}_\Bbbk(V, W), \quad w \otimes f \mapsto [v \mapsto f(v)w].$$

(b) Prove that the linear map in part (a) is an isomorphism when V and W are finite-dimensional.

(c) Verify that $W^* \otimes V^* \cong (V \otimes W)^*$ when V and W are finite-dimensional.

(d) Consider the **double dual** V^{**} of V. Show that the linear map

$$V \to V^{**}, \quad v \mapsto [f \mapsto f(v)].$$

is an embedding, which is an isomorphism when $\dim_\Bbbk V < \infty$.

1.5   Prove that the notions of an algebra given in Definitions 1.6 and 1.7 in §1.1.5 are equivalent. Namely, show that there is a bijection between the sets of algebras in Definitions 1.6 and of algebras in Definitions 1.7.

1.6   Let $S(V)$ be a symmetric algebra of a vector space V from §1.2.3.

(a) Show that if V has basis $\{v_1, \dots, v_n\}$, then $S(V)$ is isomorphic to the polynomial algebra $\Bbbk[v_1, \dots, v_n]$, as algebras.

(b) If $\dim_\Bbbk V < \infty$, is $S(V)$ finite-dimensional? If so, what is $\dim_\Bbbk S(V)$?

1.7 Consider the exterior algebra $\Lambda(V)$ of a vector space V from §1.2.4.

   (a) Show that if V has basis $\{v_1, \ldots, v_n\}$, then as algebras,

   $$\Lambda(V) \cong \Bbbk\langle v_1, \ldots, v_n\rangle/(v_i v_j + v_j v_i)_{1 \le i \le j \le n}.$$

   (b) If $\dim_\Bbbk V < \infty$, is $\Lambda(V)$ finite-dimensional? If so, what is $\dim_\Bbbk \Lambda(V)$?

1.8 Recall Example 1.9 from §1.2.7. Verify that free algebras, $(q\text{-})$polynomial algebras, $(q\text{-})$exterior algebras, and path algebras are all graded algebras. For each of these algebras A, determine if the homogeneous parts A_i are finite-dimensional. If so, compute the closed form of the generating function,

$$H_A(t) := \sum_{i \in \mathbb{N}} (\dim_\Bbbk A_i)\, t^i.$$

That is, express $H_A(t)$ as a fraction. This is called the **Hilbert series** of A.

1.9 Recall Example 1.10 on the Weyl algebras $A_n(\Bbbk)$ from §1.2.7. Prove that $\mathrm{gr}(A_1(\Bbbk)) \cong \Bbbk[v, w]$ as graded algebras. This shows that $A_1(\Bbbk)$ is a **filtered deformation** of $\Bbbk[v, w]$.

1.10 Pertaining to §1.3.1, write down the precise definition of two representations (V, ρ) and (V', ρ') over an algebra A being equivalent.

1.11 Recall the discussion of representations from §1.3.1. Take the symmetric group on three letters, $S_3 = \{e, (12), (13), (23), (123), (132)\}$, which can be generated as a group by (12) and (123), and take its group algebra $\Bbbk S_3$ from §1.2.6. Now consider the linear maps below.

$$\rho_1 : \Bbbk S_3 \to \mathrm{Mat}_2(\Bbbk), \quad (12) \mapsto \begin{pmatrix} -1 & 2 \\ 0 & 1 \end{pmatrix} \quad (123) \mapsto \begin{pmatrix} 1 & 0 \\ 0 & 1 \end{pmatrix}$$

$$\rho_2 : \Bbbk S_3 \to \mathrm{Mat}_2(\Bbbk), \quad (12) \mapsto \begin{pmatrix} -1 & 1 \\ 0 & 1 \end{pmatrix} \quad (123) \mapsto \begin{pmatrix} 0 & -1 \\ 1 & -1 \end{pmatrix}$$

$$\rho_3 : \Bbbk S_3 \to \mathrm{Mat}_2(\Bbbk), \quad (12) \mapsto \begin{pmatrix} -1 & -1 \\ 0 & 1 \end{pmatrix} \quad (123) \mapsto \begin{pmatrix} -1 & -1 \\ 1 & 0 \end{pmatrix}$$

Extend these maps to representations of $\Bbbk S_3$ using the operation of S_3.

   (a) Which of these representations ρ are faithful? And which of the corresponding group representations $\rho|_{S_3} : S_3 \to \mathrm{GL}_2(\Bbbk)$ are faithful?

   (b) Which pairs of representations are equivalent? See Exercise 1.10.

1.12 Given an algebra A, show that there is a bijection between the set of representations of A (see §1.3.1) and the set of left A-modules (see §1.3.2).

1.13 Building on Exercise 1.12 and §1.3.4: Given a group G, show that there are bijective correspondences between the following sets.

(a) Left G-modules.

(b) Representations of G.

(c) Representations of $\Bbbk G$.

(d) Left $\Bbbk G$-modules.

Hint. Use the universal property of $\Bbbk G$ to relate (b) and (c).

1.14 Recall the discussion of left modules over a group G in §1.3.4.

(a) One can weaken the definition of a left G-module by imposing that G acts on an abelian group M from the left. Write down such a definition. This is also referred to as a **left G-module** in the literature.

(b) One can further weaken the definition of a left G-module in part (a) by imposing that G acts on a set X from the left. Write down such a definition. This is often referred to as a **left G-set**.

(c) Now take a ring R, and deduce of the definitions of a **left R-module M** and a **right R-module M**, for M an abelian group.

(d) Show that M is an abelian group (resp., is a $\Bbbk$-vector space) if and only if M is a left $\mathbb{Z}$-module (resp., is a left $\Bbbk$-module) as in part (c).

1.15 Consider the products and sums of modules and algebras from §1.4.1.

(a) Verify Proposition 1.14 on direct products and sums of modules.

(b) Verify Proposition 1.15 on direct products and sums of algebras.

(c) Verify Proposition 1.17 on indecomposable algebras.

(d) Take a finite quiver Q. Show that a path algebra $\Bbbk Q$ is an indecomposable algebra if and only if the quiver Q is connected.

(e) Verify Proposition 1.22 on indecomposable modules from §1.4.3ii.

(f) When is the regular left $\Bbbk Q$-module $(\Bbbk Q)_{\mathrm{reg}}$ indecomposable?

1.16 Recall the tensor products of (bi)modules from §1.4.2i,iii.

(a) Verify that the tensor product $\otimes$ of modules $_{B_1}V$ and W_{B_2} with action maps in (1.18) is a (B_1, B_2)-bimodule.

(b) Show that $(_{B_1}V_A) \otimes_A (_A W_{B_2})$ is a (B_1, B_2)-bimodule.

1.17 Consider the notion of a left module over a ring R in Exercise 1.14(c).

(a) Write down the universal property of the tensor product over R, namely $M \otimes_R N$, of a right R-module M and a left R-module N.

(b) When R is commutative, upgrade part (a) to derive the universal property of the tensor product over R of two R-modules M and N.

(c) Simplify the following tensor products over the rings $\mathbb{Z}$ and $\mathbb{Q}$, where the action maps below are all given by multiplication.

(i) $\mathbb{Q} \otimes_{\mathbb{Z}} \mathbb{Z}$. (ii) $\mathbb{Q} \otimes_{\mathbb{Z}} \mathbb{Q}$. (iii) $\mathbb{Q} \otimes_{\mathbb{Q}} \mathbb{Q}$. (iv) $\mathbb{Q} \otimes_{\mathbb{Z}} \mathbb{Q}/\mathbb{Z}$.

1.18 Recall the tensor products of (bi)modules from §1.4.2iii. For an algebra A, let U, V, W be A-bimodules, and take A_{reg} to be the regular A-bimodule.

(a) Verify that $A_{\text{reg}} \otimes_A V \cong V$ and $V \otimes_A A_{\text{reg}} \cong V$ as A-bimodules.

(b) Verify that $(U \otimes_A V) \otimes_A W \cong U \otimes_A (V \otimes_A W)$ as A-bimodules.

1.19 Recall the tensor product construction from §1.4.2iii. Take the 2-dimensional commutative $\Bbbk$-algebra: $A := \Bbbk[x]/(x^2)$. Here, we show that if $f : V \to W$ is an injective A-module morphism, then for an A-module X, we may not have that the linear map $\text{id}_X \otimes_A f : X \otimes_A V \to X \otimes_A W$ is injective.

(a) Take the 1-dimensional A-module $V = \Bbbk v$ with $1_A \triangleright v = v$ and $x \triangleright v = 0$. Show that the $\Bbbk$-linear map $f : V \to A_{\text{reg}}$ defined by $f(v) = x$ is an injective A-module morphism.

(b) Show that $V \otimes_A V \cong V$ as A-modules.

(c) Verify that the A-module morphism $\text{id}_V \otimes_A f : V \otimes_A V \to V \otimes_A A_{\text{reg}}$ is not injective.

1.20 In contrast to the previous exercise, the operation $(X \otimes_A -)$ from §1.4.2iii preserves surjective morphisms. Verify this as follows.

(a) Take a surjective left A-module morphism $f : V \to W$, and a right A-module X, and prove that $\text{id}_X \otimes_A f : X \otimes_A V \to X \otimes_A W$ is a surjective $\Bbbk$-linear map.

(b) Likewise, for any left A-module X, show that $(- \otimes_A X)$ preserves surjective $\Bbbk$-linear maps.

1.21 Take a path algebra $\Bbbk Q$ of a quiver $Q = (Q_0, Q_1, s, t)$ from §1.2.5. Show that $\Bbbk Q$ is isomorphic to a bimodule tensor algebra $T_A(V)$ from §1.4.2iv, for a choice of an algebra A and an A-bimodule V.

1.22 For the tensor product of algebras in §1.4.2v, verify Proposition 1.19. Use commutative diagrams for fun.

1.23 Regarding §1.4.2v and §1.2.1, show that $\mathrm{Mat}_n(\Bbbk) \otimes \mathrm{Mat}_m(\Bbbk) \cong \mathrm{Mat}_{nm}(\Bbbk)$ as algebras. Also show that $\mathrm{Mat}_n(\Bbbk) \otimes A \cong \mathrm{Mat}_n(A)$, for any algebra A.

1.24 Recall the tensor product and the free product of algebras in §1.4.2v,vi, along with free/tensor algebras, polynomial/symmetric algebras, and presentations of algebras from §§1.2.2, 1.2.3.

(a) Take the following finitely presented algebras:

$$A := \Bbbk\langle v_1, \ldots, v_n \rangle / (f_1, \ldots, f_r), \quad B := \Bbbk\langle w_1, \ldots, w_m \rangle / (g_1, \ldots, g_s).$$

Express $A \circledast B$ and $A \otimes B$ as quotient algebras of $\Bbbk\langle v_1, \ldots, v_n, w_1, \ldots, w_m \rangle$.

(b) For finite-dimensional vector spaces V and W, verify the following algebra isomorphism statements:

(i) $S(V) \otimes S(W) \cong S(V \oplus W)$;　　(iii) $T(V) \otimes T(W) \not\cong T(V \oplus W)$;

(ii) $S(V) \circledast S(W) \not\cong S(V \oplus W)$;　　(iv) $T(V) \circledast T(W) \cong T(V \oplus W)$.

1.25 Recall the Hom space of (bi)modules discussed in §1.4.3i.

(a) Verify Proposition 1.20 and Corollary 1.21.

(b) Use part (a) to write down and prove a bimodule upgrade of the Tensor-Hom adjunction presented in §1.1.4iv.

1.26 Recall the left A-module Hom space and endomorphism algebra from §1.4.3ii. Let $V, V_1, \ldots, V_r$ be left A-modules, and verify the items below.

(a) $A^{\mathrm{op}} \cong \mathrm{End}_{A\text{-mod}}(A_{\mathrm{reg}})$ and $A \cong \mathrm{End}_{\mathrm{mod}\text{-}A}(A_{\mathrm{reg}})$ as algebras.

(b) Proposition 1.17 is a special case of Proposition 1.22.

(c) $\mathrm{End}_{A\text{-mod}}(\bigoplus_{i=1}^r V_i) \cong \prod_{i,j=1}^r \mathrm{Hom}_{A\text{-mod}}(V_i, V_j)$ as vector spaces.

(d) $\mathrm{End}_{A\text{-mod}}(V^{\oplus n}) \cong \mathrm{Mat}_n(\mathrm{End}_{A\text{-mod}}(V))$ as algebras.

1.27 Recall the restriction and (co)induction constructions in §1.4.4.

(a) Verify the module isomorphism (1.24).

(b) Show that the vector spaces in (1.25) and (1.26) are left B-modules.

(c) Now interpret Tensor-Hom adjunction of Exercise 1.25(b) in terms of restricted and (co)induced modules. The resulting vector space isomorphisms are known as **Frobenius reciprocity**.

1.28 [Open-ended] Recall the various constructions of algebras using the tensor product and Hom operations in §§1.4.2 and 1.4.3.

(a) Are there additional constructions of algebras using the tensor product and Hom operations beyond what is presented in §§1.4.2, 1.4.3?

(b) Explore the constructions of algebras in §§1.4.2 and 1.4.3, along with any results from part (a), for the path algebras $\Bbbk Q$ in §1.2.5. Do the constructions depend on operations of the underlying quivers Q?

(c) Explore the constructions of algebras in §§1.4.2 and 1.4.3, along with any results from part (a), for the group algebras $\Bbbk G$ in §1.2.6. Do the constructions depend on operations of the groups G?

1.29 To accompany the notion of a simple module over an algebra A as in §1.5.1, we examine irreducible representations of a group G here.

(a) Use the notion of a left A-submodule in §1.3.2 to write down the definition of a left G-submodule of a left G-module $(V, \triangleright)$ as in §1.3.4.

(b) Use the correspondence between left G-modules and representations of G in Exercise 1.13 to write down the definition of a subrepresentation of a representation $\pi_V : G \to \mathrm{GL}(V)$ as in §1.3.4.

(c) Given that simple A-modules as in §1.5.1 correspond to irreducible representations of A, write down the definition of an irreducible representation of G.

(d) Define a completely reducible representation of G (see §1.6.2).

(e) Take the representations $\pi_1, \pi_2\ \pi_3$ of S_3 given by ρ_1, ρ_2, ρ_3 in Exercise 1.11, respectively, without extending linearly to $\Bbbk S_3$.

   (i) Which of π_1, π_2, π_3 are irreducible representations of S_3?

   (ii) Which of π_1, π_2, π_3 are completely reducible?

1.30 Recall the material about modules over semisimple algebras from §1.6.2.

(a) Show that a simple module over a commutative semisimple algebra must be 1-dimensional (as $\Bbbk$-vector space).

(b) Verify that an irreducible representation of an abelian group (as in Exercise 1.29) must be 1-dimensional.

1.31 Recall the material about (modules over) semisimple algebras from §1.6.

(a) Show that the 2-dimensional algebra $\Bbbk[x]/(x^2 - 1)$ is semisimple.

(b) Show that the 2-dimensional algebra $\Bbbk[x]/(x^2)$ is nonsemisimple.

1.32 Here, we examine the representation theory of groups (resp., of group algebras) outside of the setting for Maschke's Theorem [Theorem 1.47] (resp., for Artin-Wedderburn's Theorem [Theorem 1.44]).

(a) Consider the additive group $\mathbb{Z}$, and show that it has infinitely many inequivalent irreducible representations over the field $\mathbb{C}$.

(b) Show that the assignment

$$\pi : \mathbb{Z} \to \mathrm{GL}_2(\mathbb{C}), \quad n \mapsto \begin{pmatrix} 1 & n \\ 0 & 1 \end{pmatrix}$$

is a representation of $\mathbb{Z}$ that is not completely reducible.

(c) Now consider the finite field $\mathbb{F}_2$ of order 2, and the cyclic group $C_2 = \langle g \rangle$ of order 2. Show that the only irreducible representation C_2 over $\mathbb{F}_2$ is the trivial representation (where e, g maps to $1_{\mathbb{F}_2}$).

(d) Construct a representation $\pi : C_2 \to \mathrm{GL}_2(\mathbb{F}_2)$ that is not completely reducible.

1.33 Let D be a division algebra, and consider the matrix algebra $\mathrm{Mat}_n(D)$.

(a) Show that $\mathrm{Mat}_n(D)^{\mathrm{op}} \cong \mathrm{Mat}_n(D)$ as algebras.

(b) Verify that $\mathrm{Mat}_n(D)$ is simple (see §1.5.2).

(c) Verify that $\mathrm{Mat}_n(D)$ is semisimple (see §1.6).

1.34 Recall the discussion of separable algebras in §1.7. Verify the following:

(a) The elements $(e^{\mathrm{Mat}_n(\Bbbk)})_k$ for $\mathrm{Mat}_n(\Bbbk)$, and $e^{\Bbbk G}$ for $\Bbbk G$, in Examples 1.54 and 1.55, respectively, are separability idempotents.

(b) If A and B are separable algebras, then so are $A^{\mathrm{op}}, A \times B$, and $A \otimes B$.

1.35 [Open-ended] Consider Figure 1.2 in §1.8

(a) Verify the inclusions and (counter-)examples in Figure 1.2.

(b) Derive more (counter-)examples for this figure by using the examples of algebras over $\Bbbk$ discussed in §1.2.

(c) Discover further (counter-)examples for the figure by searching for examples of algebras over $\Bbbk$ in the literature, not included in §1.2.

· CHAPTER 2 ·

CATEGORIES

History

A *category* is a system of objects and structure-preserving maps between them, which satisfy certain predictable axioms. Categories, along with functors between categories and equivalences of categories, were introduced in the landmark work by Eilenberg and MacLane [1945] for applications in algebraic topology. The field took a substantial leap forward due to the work of Buchsbaum [1955] and Grothendieck [1957] in their introduction of abelian categories, especially in algebraic geometry in the latter work. Kan [1958] then developed the notion of adjunction for functors to provide a framework for ubiquitous concepts in homotopy theory and other parts of mathematics. The work of Lawvere [1963] fundamentally launched category theory as a subject of independent interest.

Overview

An introduction to categories is covered in §2.1; many examples are provided. Universal constructions and abelian categories are then discussed in §2.2. How to move from one category to another via *functors*, and how two categories are considered the same, are presented in §§2.3, 2.4. Key relationships between functors (*adjunction, representability*) are the focus of §§2.5, 2.6. Next, 'building blocks' in categories (*indecomposability, simplicity, semisimplicity*) are highlighted in §§2.7, 2.8, and finiteness conditions are introduced in §2.9. The chapter ends with summarizing diagrams in §2.10, modern applications in §2.11, references in §2.12, and several exercises.

Standing hypotheses.   Linear structures are over an algebraically closed field $\Bbbk$ of characteristic 0, and algebras over $\Bbbk$ are associative and unital.

69

§2.1. Categories

We define categories and present many examples, some building on material from Chapter 1. We also define products of categories here.

§2.1.1. Categories

A **category** C consists of the following data.

(a) A collection of **objects** $\mathrm{Ob}(C)$ of C. Here, we write $X \in C$ for $X \in \mathrm{Ob}(C)$.

(b) For every pair of objects $X, Y \in C$, a collection of **morphisms** from X to Y, denoted by $\mathrm{Hom}_C(X, Y)$. Here, we write $f : X \to Y$ for $f \in \mathrm{Hom}_C(X, Y)$. (The collection of all morphisms in C is denoted by $\mathrm{Hom}(C)$.)

(c) For any object $X \in C$, an **identity morphism** id_X in $\mathrm{Hom}_C(X, X)$.

(d) For each pair of morphisms $f \in \mathrm{Hom}_C(W, X)$ and $g \in \mathrm{Hom}_C(X, Y)$, a **composition** $gf := g \circ f \in \mathrm{Hom}_C(W, Y)$. Here, we call f and g **composable**.

This data must satisfy the axioms below.

- **(associativity)** For $f \in \mathrm{Hom}_C(W, X)$, $g \in \mathrm{Hom}_C(X, Y)$, $h \in \mathrm{Hom}_C(Y, Z)$, we have that $(hg)f = h(gf)$ as morphisms from W to Z in C.

- **(unitality)** For $f \in \mathrm{Hom}_C(W, X)$ and $g \in \mathrm{Hom}_C(X, Y)$, we have that $\mathrm{id}_X f = f$ in $\mathrm{Hom}_C(W, X)$, and that $g \, \mathrm{id}_X = g$ in $\mathrm{Hom}_C(X, Y)$.

Collection versus set, and the importance of morphisms. Notice that we use the term *collection* instead of *set* when describing the objects and morphisms of a category C. If $\mathrm{Hom}(C)$ is a set, then we refer to C as **small**; else, C is **large**. Also, when C is small, $\mathrm{Ob}(C)$ is a set because objects are in bijective correspondence with the identity morphisms in C. So, the morphisms of a category are more intriguing than its objects. For instance, we could have that $\mathrm{Hom}_C(X, Y) = \varnothing$, for some $X, Y \in C$.

A notion weaker than smallness is to require that $\mathrm{Hom}_C(X, Y)$ is a set for each pair $X, Y \in C$. In this case, C is **locally small**. Most of the categories here will be locally small, so we will often forgo set-theoretic subtleties.

If $f \in \mathrm{Hom}_C(X, Y)$, then X (resp., Y) is the **domain** (resp., **codomain**) of f.

To compare objects X and Y in a category C, let us consider certain types of morphisms in C.

We say that $g : X \to Y \in C$ is a **mono**, or is **monic**, if for any $f, f' \in \mathrm{Hom}_C(W, X)$ with $gf = gf'$, we get that $f = f'$. In other words, monos are **left cancellative**. If $g : X \to Y$ is monic, then we refer to $X := (X, g)$ as a **subobject** of Y.

We say that $g : X \to Y \in C$ is an **epi**, or is **epic**, if for any $h, h' \in \mathrm{Hom}_C(Y, Z)$ with $hg = h'g$, we get that $h = h'$. In other words, epis are **right cancellative**. If $g : X \to Y$ is epic, then we refer to $Y := (Y, g)$ as a **quotient object** of X.

We also call $g : X \to Y \in C$ an **iso** if there exists a morphism $g' \in \mathrm{Hom}_C(Y, X)$ if $g'g = \mathrm{id}_X$ and $gg' = \mathrm{id}_Y$. In this case, we write g^{-1} for g', and write $X \cong Y$.

One might expect that a morphism is an iso precisely when it is both monic and epic, but this is not true in general. (It is true in *abelian categories* as we will see later in §2.2.2.) See Exercises 2.1 and 2.2 for more details.

Mono/ epi vs. monomorphism/ epimorphism.   Here, we do not use *monomorphism* (resp., *epimorphism*) to refer to a morphism in a category that is a *mono* (resp., an *epi*). Indeed, the former terminology (e.g., in Chapter 1 material) does not always match with the categorical notions here. For instance, there is a non-surjective ring homomorphism that is epic in the category of rings defined later; see Exercise 2.2(c).

Next, we turn our attention to subcategories. Given a category C, we have that a **subcategory** D of C consists of the following data.

(a) A subcollection $\mathrm{Ob}(D)$ of $\mathrm{Ob}(C)$.

(b) A subcollection $\mathrm{Hom}(D)$ of $\mathrm{Hom}(C)$.

We also require that the following conditions hold.

- If $X \in D$, then $\mathrm{id}_X \in \mathrm{Hom}(D)$.

- If $f \in \mathrm{Hom}(D)$, then the domain and codomain of f are objects of D.

- If $f, g \in \mathrm{Hom}(D)$ with $\mathrm{codomain}(f) = \mathrm{domain}(g)$, then $gf \in \mathrm{Hom}(D)$.

If, further, $\mathrm{Hom}_C(X, Y) = \mathrm{Hom}_D(X, Y)$ for any objects $X, Y \in D$, then we say that D is a **full subcategory** of C.

Sometimes it is useful to reverse the directions of morphisms in a category. Given a category C, its **opposite category** C^{op} is a category defined by:

(a) $\mathrm{Ob}(C^{\mathrm{op}}) = \mathrm{Ob}(C)$;

(b) There exists $f^{\mathrm{op}} \in \mathrm{Hom}_{C^{\mathrm{op}}}(X, Y)$ if and only if there exists $f \in \mathrm{Hom}_C(Y, X)$.

Exercise 2.3 is on verifying that C^{op} is indeed a category, given the data above. In particular, monos (resp., epis) in C are epis (resp., monos) in C^{op}.

§2.1.2. Examples of categories

We provide many examples of categories in terms of their collections of objects and morphisms, but this is by no means an exhaustive list (see Exercise 2.5). Some non-algebraic terminology below is outside the scope of this book, and will not be defined here. Recall that we assume that all vector spaces and algebras are over the (algebraically closed) ground field $\Bbbk$ (of characteristic 0) below.

i. Algebraic categories

These examples build on many constructions presented in Chapter 1.

- Monoid: monoid and monoid homomorphisms. (Think about what the definition of a monoid homomorphism should be.)

- Group: groups and group homomorphisms.

- Ab: abelian groups and group homomorphisms. (Recall the discussion about structure versus property in §1.1.1.)

- A **groupoid** $\mathcal{G}$ is a category in which all morphisms are isomorphisms. This is a generalization of a group because if $\mathcal{G}$ has one object, then the morphisms are identified with elements of a group. See Exercise 2.4.

- Ring: unital rings and unital ring homomorphisms.

- Rng: not-necessarily-unital rings and ring homomorphisms.

- ComRing: commutative unital rings and unital ring homomorphisms.

- Vec: vector spaces and linear maps.

- FdVec: finite-dimensional vector spaces and linear maps.

- Alg: algebras and algebra homomorphisms.

- ComAlg: commutative algebras and algebra homomorphisms.

- FdAlg: finite-dimensional algebras and algebra homomorphisms.

- FgAlg: finitely generated algebras and algebra homomorphisms.

- $\mathbb{N}$-GrAlg: $\mathbb{N}$-graded algebras and $\mathbb{N}$-graded algebra homomorphisms.

- N-GrAlg: N-graded algebras and their homomorphisms, for any monoid N.

- Vec_N: N-graded vector spaces and N-graded linear maps, for a monoid N.

- $\mathrm{Rep}(A)$: representations of an algebra A and representation morphisms.

- A-Mod (resp., Mod-A): left (resp., right) modules over algebra A and module morphisms. In particular,

$$\text{Hom}_{A\text{-Mod}}(X, Y) \text{ here } \text{ is } \text{Hom}_{A\text{-mod}}(X, Y) \text{ in Chapter 1.}$$

- Rep(G), G-Mod, Mod-G, for a group G, are defined likewise.

- (A, B)-Bimod: (A, B)-bimodules over algebras A and B, and bimodule morphisms. Likewise, we can define the category A-Bimod.

- FdRep(A), A-FdMod, FdMod-A, (A, B)-FdBimod, and A-FdBimod are the finite-dimensional version of the representation and (bi)module categories above.

- Bim: objects are algebras, and morphisms from an algebra A to an algebra B are given by isoclasses of (A, B)-bimodules. Compositions are given by tensor product over algebras, and the identity morphisms are the regular bimodules.

The category Ab is an example of a subcategory of Group. In fact, Ab is a full subcategory of Group because if we take a morphism between two abelian groups in Group, then this morphism also belongs to Ab.

But Ring is a non-full subcategory of Rng. Indeed, a morphism between two unital rings in Rng may not be unital, and thus, may not belong to Ring.

ii. Logical and categorical categories

- $\varnothing$: the **empty category**. It consists of no objects and no morphisms.

- Set: sets and functions. This is a large category, as there is no set of all sets.

- FinSet: finite sets and functions.

- Set$_*$: sets with a fixed base point and base-point-preserving functions.

- Rel: objects are sets $X, Y, Z, \ldots$; morphisms are subsets $R_{X \times Y} \subset X \times Y$; and

$$S_{Y \times Z} \circ R_{X \times Y} := \{(x, z) \in X \times Z \mid \exists\, y \in Y \text{ with } (x, y) \in R_{X \times Y}, (y, z) \in S_{Y \times Z}\},$$

and $\text{id}_X := \{(x, x) \in X \times X \mid x \in X\}$. See Section 0.1.3 of Heunen and Vicary [2019].

- Cat: small categories and *functors* (to be defined in §2.3.1).

iii. Geometric and topological categories

- Aff: affine varieties and regular maps.

- Mfld: manifolds and smooth maps. One can also get subcategories by adjusting the smoothness (or differentiability) class as desired.

- nCob: objects are manifolds of dimension $n - 1$ and morphisms are cobordisms of dimension n.

- Top: topological spaces and continuous maps.

- Top$_*$: topological spaces with a fixed base point and morphisms are base-point-preserving continuous maps.

iv. Analytic categories

- Meas: measure spaces and measurable functions.

- Hilb: Hilbert spaces and bounded linear maps.

- FdHilb: finite-dimensional Hilbert spaces and bounded linear maps.

- Ban: Banach spaces and bounded linear maps. There are variations by imposing further hypotheses on the collection of morphisms.

v. Combinatorial categories

- Poset: partially ordered set and order-preserving functions.

- Graph: graphs and incidence-preserving functions sending vertices to vertices and edges to edges.

- DirGraph (or Quiv): Likewise, we can define a category of directed graphs (also known as quivers), by preserving the orientation of arrows.

vi. Amusing (non-)categories

Not all collections of objects and corresponding morphisms form categories. For instance, Exercise 2.6 asks if the following collections of objects and morphisms form categories.

- 80sMusic: Let the objects be persons, and let there be a morphism from Person A to Person B if A and B both like a certain track from the 1980s.

- SharePw: Pick your favorite streaming service. Then, let the objects be persons, and let there be a morphism from Person A to Person B if A shared a service password for that streaming service with B.

- SameTaste: Pick your favorite food. Let the objects be persons, and let there be a morphism from Person A to Person B if A and B both enjoy this food.

§2.1.3. Products of categories

Here, we will construct new categories from old ones via a product. This will be needed later to discuss when a category is *(in)decomposable* (see §2.2.2ii).

For categories C and C', the **product category** $C \times C'$ is defined by the data below.

(a) $\mathrm{Ob}(C \times C') = \{(X, X') \mid X \in C,\ X' \in C'\}$.

(b) $\mathrm{Hom}_{C \times C'}((X, X'), (Y, Y')) = \{(g, g') \mid g \in \mathrm{Hom}_C(X, Y),\ g' \in \mathrm{Hom}_{C'}(X', Y')\}$.

We have that $\mathrm{id}_{(X, X')} = (\mathrm{id}_X, \mathrm{id}_{X'})$, for all $X \in C$ and $X' \in C'$. For $f \in \mathrm{Hom}_C(W, X)$, $f' \in \mathrm{Hom}_{C'}(W', X')$, $g \in \mathrm{Hom}_C(X, Y)$, $g' \in \mathrm{Hom}_{C'}(X', Y')$, we also have that

$$(g, g') \circ (f, f') = (gf, g'f') \in \mathrm{Hom}_{C \times C'}((W, W'), (Y, Y')).$$

§2.2. Universal constructions and abelian categories

In this part, we discuss some useful objects and morphisms in a category C that are constructed via a universal property. (See the chat about universal property in §1.1.4iii.) As usual for universal constructions, the objects below need not exist. But if they do exist, they are unique up to iso; compare to Exercise 1.2. Then, we introduce the notion of an abelian category, which is a category in which each of the universal constructions discussed here exists. The reader may wish to view Figure 2.1 in §2.10 for a preview of the terminology in this section.

§2.2.1. Universal objects and morphisms in categories

Here, we will introduce the following universal constructions: (i) initial, terminal, and zero objects; (ii) coproducts and products of objects; (iii) pushouts and pullbacks of morphisms; (iv) coequalizers and equalizers of morphisms; (v) zero morphisms; and (vi) cokernels, kernels, and images of morphisms. To proceed, take C to be an arbitrary category, and consider the following conventions.

Universal object vs. universal morphism? Or both? Universal constructions in a category consist of an object $\mathrm{Univ}(X)$, equipped with a (collection of) morphism(s) α_X (or α'_X). When working with these constructions, sometimes one only considers the object $\mathrm{Univ}(X)$, and at other times one may just work with the morphism(s) α_X (or α'_X), but both go by the same name.

For instance, we will construct a categorical version of a kernel of a morphism $f : X \to Y$ below. This will consist of an object $\ker(f)$ equipped with a certain morphism given by $\alpha'_f : \ker(f) \to X$. By "the kernel of f", we could mean the object $\ker(f)$, or the morphism α'_f, or both.

---

i. Initial, terminal, and zero objects

An object $I \in \mathcal{C}$ is called **initial** if for every object $X \in \mathcal{C}$ there exists a unique morphism $\vec{0}_X \colon I \to X$. Dually, an object $T \in \mathcal{C}$ is **terminal** if for every object $X \in \mathcal{C}$ there exists a unique morphism $_X\vec{0} \colon X \to T$.

$$I \xdashrightarrow{\exists! \, \vec{0}_X} X \qquad\qquad X \xdashrightarrow{\exists! \, _X\vec{0}} T$$

If an object is both initial and terminal, then we call it a **zero** object and denote this by 0. Here are a few examples; see Exercise 2.7.

- Set: The initial object is $\varnothing$, and the terminal object is the one-point set.

- Group, Ab: The (initial, terminal, and) zero object is the trivial group, $\{e\}$.

- Ring: The initial object is $\mathbb{Z}$, and the terminal object is the zero ring.

- Vec: The (initial, terminal, and) zero object is the zero vector space.

- A-Mod: The (initial, terminal, and) zero object is the zero module.

- Alg: The initial object is $\Bbbk$, and the terminal object is the zero algebra.

- Top: The initial object is the empty space, and the terminal object is the one-point space.

ii. Coproducts and products of objects

Now we discuss ways of combining objects in categories via (co)products.

The **coproduct** of two objects $X, Y \in \mathcal{C}$ is an object $X \sqcup Y$ in $\mathcal{C}$, equipped with two morphisms $\alpha_X \colon X \to X \sqcup Y$ and $\alpha_Y \colon Y \to X \sqcup Y$ in $\mathcal{C}$, such that for any object C equipped with morphisms $\beta_X \colon X \to C$ and $\beta_Y \colon Y \to C$ in $\mathcal{C}$, we have a unique morphism $\gamma \colon X \sqcup Y \to C$ with $\beta_X = \gamma \alpha_X$ and $\beta_Y = \gamma \alpha_Y$.

The **product** of two objects $X, Y \in \mathcal{C}$ is an object $X \sqcap Y$ in $\mathcal{C}$, equipped with two morphisms $\alpha'_X \colon X \sqcap Y \to X$ and $\alpha'_Y \colon X \sqcap Y \to Y$ in $\mathcal{C}$, such that for any object P equipped with morphisms $\beta'_X \colon P \to X$ and $\beta'_Y \colon P \to Y$ in $\mathcal{C}$, we have a unique morphism $\gamma' \colon P \to X \sqcap Y$ with $\beta'_X = \alpha'_X \gamma'$ and $\beta'_Y = \alpha'_Y \gamma'$.

These constructions are visualized via the commutative diagrams below.

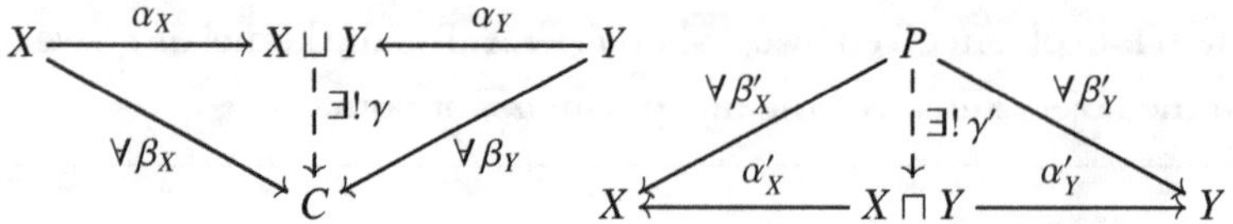

When Y and P are equal to X, and $\beta'_X = \beta'_Y = \mathrm{id}_X$, the resulting map γ' from the universal property of products is the **diagonal morphism** of X in $\mathcal{C}$, denoted by

$$\mathrm{diag}_X : X \to X \sqcap X.$$

Here are a few examples of coproducts and products of objects in categories.

- Set: The coproduct is disjoint union $\uplus$, and the product is cartesian product.

- Group: The coproduct is free product, and the product is direct product.

- Ab: The coproduct is direct sum, and the product is direct product.

- Ring: The coproduct is free product (similar to that for algebras in §1.4.2), and the product is direct product.

- Vec: The coproduct is direct sum, and the product is direct product.

- A-Mod: The coproduct is direct sum, and the product is direct product.

- Alg: The coproduct is the free product discussed in §1.4.2, and the product is direct product.

- Top: The coproduct is disjoint union with disjoint union topology, and the product is cartesian product with product topology.

Note that the coproduct (or product) of a collection of objects $\{X_i\}_{i \in I}$ can be formed in the same manner as above; we denote these constructions by $\coprod_{i \in I} X_i$ (or by $\prod_{i \in I} X_i$, resp.). See Exercise 2.8 for practice with coproducts and products.

iii. Pushouts and pullbacks of morphisms

Here, we discuss how one can produce an object from a pair of morphisms in a category, via the pushout and pullback constructions. This will be a generalization of the coproduct and product of objects discussed in §2.2.1ii.

The **pushout** (or **fiber coproduct**) of $f : Z \to X$ and $g : Z \to Y$ is an object

$$X \sqcup_Z Y := X \sqcup_{Z,f,g} Y$$

equipped with two morphisms $\alpha_X : X \to X \sqcup_Z Y$ and $\alpha_Y : Y \to X \sqcup_Z Y$, where $\alpha_X f = \alpha_Y g$, such that for every object C with morphisms $\beta_X : X \to C$ and $\beta_Y : Y \to C$ where $\beta_X f = \beta_Y g$, there exists a unique morphism $\gamma : X \sqcup_Z Y \to C$ with $\beta_X = \gamma \alpha_X$ and $\beta_Y = \gamma \alpha_Y$.

The **pullback** (or **fiber product**) of $f : X \to Z$ and $g : Y \to Z$ is an object

$$X \sqcap_Z Y := X \sqcap_{Z,f,g} Y$$

equipped with two morphisms $\alpha'_X : X \sqcap_Z Y \to X$ and $\alpha'_Y : X \sqcap_Z Y \to Y$, where $f\alpha'_X = g\alpha'_Y$, such that for every object P with morphisms $\beta'_X : P \to X$ and $\beta'_Y : P \to Y$ where $f\beta'_X = g\beta'_Y$, there exists a unique morphism $\gamma' : P \to X \sqcap_Z Y$ with $\beta'_X = \alpha'_X \gamma'$ and $\beta'_Y = \alpha'_Y \gamma'$.

The constructions are visualized via the commutative diagrams below.

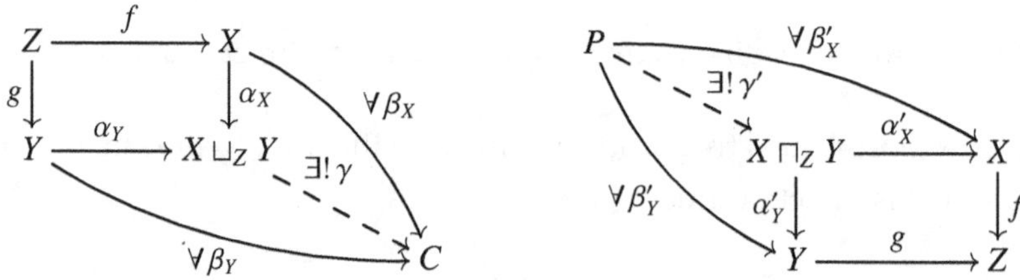

For example, in Set, we have that for objects $X, Y, Z \in$ Set with functions f and g above, the following results.

- $X \sqcup_Z Y$ is the quotient set, $(X \uplus Y)/\sim$, with $f(z) \sim g(z)$ for each $z \in Z$, equipped with set maps α_X, α_Y from X, Y respectively. That is, $(X \uplus Y)/\sim$ is the collection of equivalence classes $\{[w]\}_{w \in X \uplus Y}$, where

$$[w] = \begin{cases} \{f(z), g(z)\}, & \text{if } w = f(z) \text{ or } g(z), \text{ for some } z \in Z, \\ \{w\}, & \text{otherwise.} \end{cases}$$

- $X \sqcap_Z Y$ is the set $\{(x, y) \in X \times Y \mid f(x) = g(y) \text{ in } Z\}$, equipped with set maps α'_X and α'_Y to X and Y, respectively.

See Exercise 2.9. See also Exercise 2.10 for practice.

iv. Coequalizers and equalizers of morphisms

Now we explain how one forms the coequalizer, and the equalizer, of a pair of morphisms. In particular, for two functions f and g on a set X, we will generalize the subset of elements x of X for which $f(x) = g(x)$. To do so, fix a pair of **parallel morphisms** $f, g : X \to Y$ in $\mathcal{C}$; these are, by definition, morphisms f and g with the same domain and codomain.

The **coequalizer** of $f, g : X \to Y$ is an object $\mathrm{coeq}(f, g)$ equipped with a morphism $\alpha : Y \to \mathrm{coeq}(f, g)$ where $\alpha f = \alpha g$, such that for any object C equipped with a morphism $\beta : Y \to C$ where $\beta f = \beta g$, there exists a unique morphism $\gamma : \mathrm{coeq}(f, g) \to C$ with $\beta = \gamma\alpha$.

The **equalizer** of $f, g : X \to Y$ is an object $\mathrm{eq}(f, g)$ equipped with a morphism $\alpha' : \mathrm{eq}(f, g) \to X$ where $f\alpha' = g\alpha'$, such that for any object K equipped with a morphism $\beta' : K \to X$ where $f\beta' = g\beta'$, there exists a unique morphism $\gamma' : K \to \mathrm{eq}(f, g)$ with $\beta' = \alpha'\gamma'$.

These constructions are visualized via the commutative diagrams below.

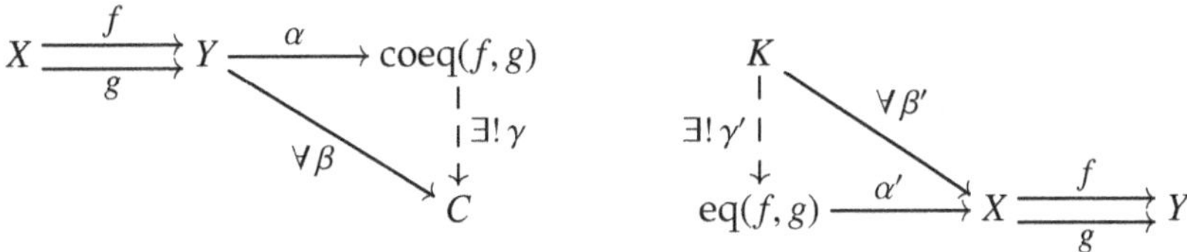

For example, in Set, we have the following results for functions $f, g : X \to Y$.

- $\mathrm{coeq}(f, g)$ is the quotient set, $Y/{\sim}$, with $f(x) \sim g(x)$ for each $x \in X$, equipped with a set map α from Y. That is, $Y/{\sim} = \{[y]\}_{y \in Y}$, where

$$[y] = \begin{cases} \{f(x), g(x)\}, & \text{if } y = f(x) \text{ or } g(x), \text{ for some } x \in X, \\ \{y\}, & \text{otherwise.} \end{cases}$$

- $\mathrm{eq}(f, g)$ is the set $\{x \in X \mid f(x) = g(x)\}$, equipped with a set map α' to X.

We also have that in Vec, the tensor product of modules over a $\Bbbk$-algebra (as in §1.4.2iii) arises as a coequalizer of two $\Bbbk$-linear maps. See Exercises 2.11 and 2.12 for practice. In particular, the latter exercise explains why any category in which pushouts and coproducts (resp., pullbacks and products) exist also has coequalizers (resp., equalizers).

v. Zero morphisms

To generalize kernels and cokernels from the algebraic setting to the categorical one, we first need the notion of a zero morphism in a category.

A morphism $g : X \to Y$ is **constant** if $gf = gf'$ for any pair of morphisms $f, f' : W \to X$, and g is **coconstant** if $hg = h'g$ for any morphisms $h, h' : Y \to Z$. A **zero morphism** is a morphism that is constant and coconstant.

Next, note that if C has a zero object 0, then $_0\vec{0} = \vec{0}_0 = \mathrm{id}_0$.

Also, for any objects X and Y, we get unique morphisms $\vec{0}_Y : 0 \to Y$ and $_X\vec{0} : X \to 0$. The composition,

$$\vec{0} := \vec{0}_{X,Y} := \vec{0}_Y \circ {_X\vec{0}} : X \to 0 \to Y,$$

is then a zero morphism. This is independent of the choice of zero object. That is, if $0'$ is another zero object of C, then $\vec{0}'_Y \circ {_X\vec{0}'} : X \to 0' \to Y$ satisfies the following:

$$\vec{0}'_{X,Y} = \vec{0}'_Y \circ {_X\vec{0}'} = \vec{0}_Y \circ {_{0'}\vec{0}} \circ {_X\vec{0}'} = \vec{0}_Y \circ \vec{0}'_0 \circ {_X\vec{0}'} = \vec{0}_Y \circ {_X\vec{0}} = \vec{0}_{X,Y}.$$

We say that C **has zero morphisms** if for any triple of objects $X, Y, Z \in C$ and for all morphisms $f : X \to Y$ and $g : Y \to Z$, there exist morphisms $\tilde{0}_{*,**} : * \to **$ that make the diagram below commute.

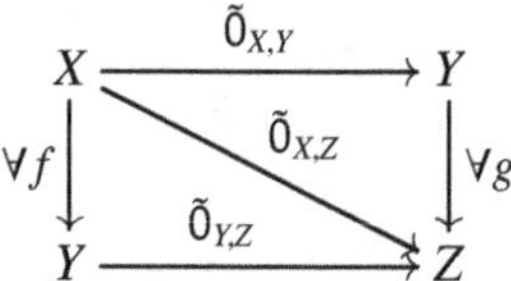

When C has a zero object, then C has zero morphisms by taking $\tilde{0}_{*,**} = \vec{0}_{*,**}$.

For emphasis, note that for all morphisms f in C: $f \circ \vec{0} = \vec{0}$ and $\vec{0} \circ f = \vec{0}$.

vi.   Cokernels, kernels, and images of morphisms

Now we introduce categorical generalizations of the cokernels and kernels of morphisms discussed in the algebraic setting of §1.1. Assume that a category C has a zero object. Thus, the zero morphism $\vec{0}_{X,Y}$ exists, for each $X, Y \in C$ here.

The **cokernel** of a morphism $f : X \to Y$ is an object coker(f) equipped with a morphism $\alpha : Y \to \mathrm{coker}(f)$, where $\alpha f = \vec{0}_{X,\mathrm{coker}(f)}$, such that for any object C equipped with a morphism $\beta : Y \to C$ where $\beta f = \vec{0}_{X,C}$, we have a unique morphism $\gamma : \mathrm{coker}(f) \to C$ with $\beta = \gamma\alpha$ and $\vec{0}_{X,C} = \gamma \circ \vec{0}_{X,\mathrm{coker}(f)}$.

The **kernel** of a morphism $f : X \to Y$ is an object ker(f) equipped with a morphism $\alpha' : \mathrm{ker}(f) \to X$, where $f\alpha' = \vec{0}_{\mathrm{ker}(f),Y}$, such that for any object K equipped with a morphism $\beta' : K \to X$ where $f\beta' = \vec{0}_{K,Y}$, we have a unique morphism $\gamma' : K \to \mathrm{ker}(f)$ with $\beta' = \alpha'\gamma'$ and $\vec{0}_{K,Y} = \vec{0}_{\mathrm{ker}(f),Y} \circ \gamma'$.

These constructions are visualized via the commutative diagrams below.

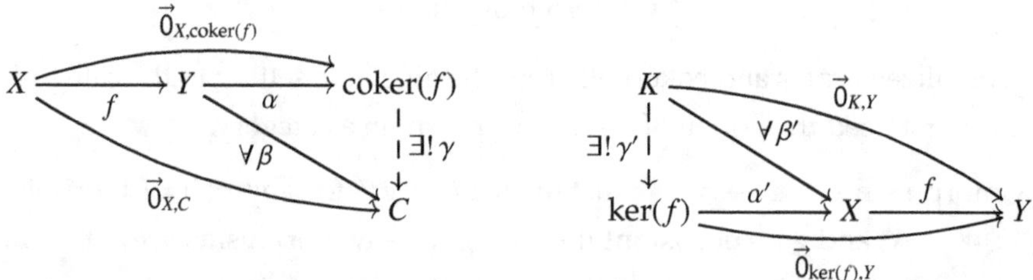

See Exercise 2.13 for practice. In particular, we have that kernels are monic, and that cokernels are epic.

If $f : X \to Y$ is a mono (so that X is a subobject of Y), then we write

$$Y/X := \mathrm{coker}(X \to Y)$$

and refer to this as a **quotient object** of Y by X, if it exists.

On the other hand, if a mono (resp., epi) is the kernel (resp., cokernel) of a morphism, we say it is a **normal mono** (resp., **normal epi**).

The **image** of a morphism $f : X \to Y$ is an object im(f) equipped with a mono $\alpha : \mathrm{im}(f) \to Y$ and a morphism $\tilde{\alpha} : X \to \mathrm{im}(f)$ where $f = \alpha\tilde{\alpha}$, such that for any object Z equipped with a mono $\beta : Z \to Y$ and morphism $\tilde{\beta} : X \to Z$, where $f = \beta\tilde{\beta}$, we have a unique morphism $\gamma : \mathrm{im}(f) \to Z$ with $\tilde{\beta} = \gamma\tilde{\alpha}$ and $\alpha = \beta\gamma$. This is visualized via the commutative diagram below. See also Exercise 2.14.

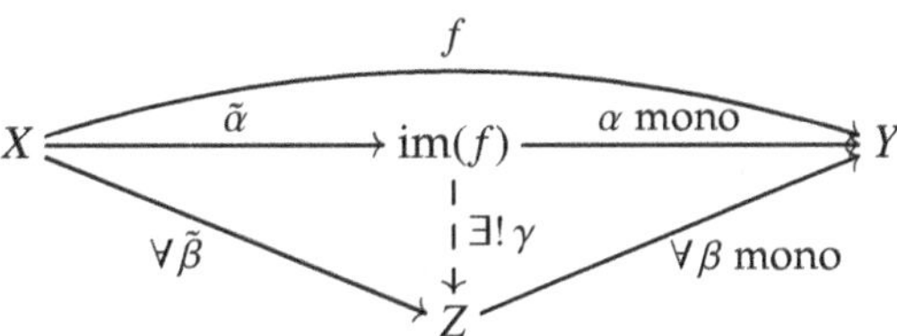

We recover the kernels, cokernels, and images of morphisms from §1.1 in the algebraic setting, especially for categories with a zero object. See Exercise 2.15.

Remark 2.1. Each of the universal constructions above is realized as either a categorical *colimit* or a *limit*. This will be discussed in §2.3.6.

§2.2.2. Abelian categories

Finally, we examine categories in which all of the universal constructions in §2.2.1 exist: abelian categories. Thus, abelian categories are quite important; there are also numerous examples of such categories as discussed below. The reader may wish to view Figure 2.1 in §2.10 for a summary of terminology.

i. Preadditive and linear categories

First, a category C is said to be **preadditive** if for each $X, Y \in C$, we have that $\mathrm{Hom}_C(X, Y)$ is an abelian group with:

- group operation +,

- additive identity (which is $\vec{0} := \vec{0}_{X,Y}$ when C has a zero object), and

- additive inverse of $f : X \to Y$, denoted by $-f : X \to Y$.

We also require composition of morphisms to distribute over +: that is, for morphisms $f, f' : X \to Y$ and $g, g' : Y \to Z$ in C, we have that

$$g \circ (f + f') = (g \circ f) + (g \circ f'), \qquad (g + g') \circ f = (g \circ f) + (g' \circ f).$$

Here, C is also called an Ab-**category** or is **enriched over** Ab (see §3.11 later). An example of an Ab-category is Ab itself.

Moreover, C is called **($\Bbbk$-)linear** if $\mathrm{Hom}_C(X, Y)$ is a $\Bbbk$-vector space for each $X, Y \in C$, and composition distributes over addition and over scalar multiplication, i.e.:

$$g \circ (\lambda f) = \lambda(g \circ f),$$

for all $f : X \to Y$, $g : Y \to Z \in C$ and $\lambda \in \Bbbk$. Linear categories are also said to be **enriched over** Vec (see §3.11 later). An example of such a category includes Vec itself; see §1.1.4iv.

ii. Additive categories

Now recall from Remark 1.1 that the direct sum (or coproduct) of two vector spaces V and W is isomorphic to the direct product (or product) of V and W. This fact is generalized in the result below.

Proposition 2.2. *Take an Ab-category C with objects X and Y in C. Then, the following statements are equivalent.*

(a) *There exists a coproduct $(X \sqcup Y,\ \alpha_X : X \to X \sqcup Y,\ \alpha_Y : Y \to X \sqcup Y)$ in C.*

(b) *There exists a product $(X \sqcap Y,\ \alpha'_X : X \sqcap Y \to X,\ \alpha'_Y : X \sqcap Y \to Y)$ in C.*

(c) *There exists an object $X \square Y \in C$, with morphisms $p_X : X \to X \square Y$, $p_Y : Y \to X \square Y$, $p'_X : X \square Y \to X$, and $p'_Y : X \square Y \to Y$ in C such that*

$$p'_X\, p_X = \mathrm{id}_X, \qquad p'_Y\, p_Y = \mathrm{id}_Y, \qquad p'_Y\, p_X = \vec{0}_{X,Y}, \qquad p'_X\, p_Y = \vec{0}_{Y,X},$$

$$p_X\, p'_X + p_Y\, p'_Y = \mathrm{id}_{X \square Y}.$$

In this case, $X \sqcup Y \cong X \sqcap Y \cong X \square Y$.

The tuple in part (c) above is called the **(binary) biproduct** of X and Y in C

Proof. We will sketch why (a) is equivalent to (c), and leave the rest of the proof as Exercise 2.16. To show that (a) implies (c), take $X \square Y = X \sqcup Y$, along with $p_X = \alpha_X$ and $p_Y = \alpha_Y$. Now using the universal property of the coproduct construction, define the desired maps p'_X and p'_Y below.

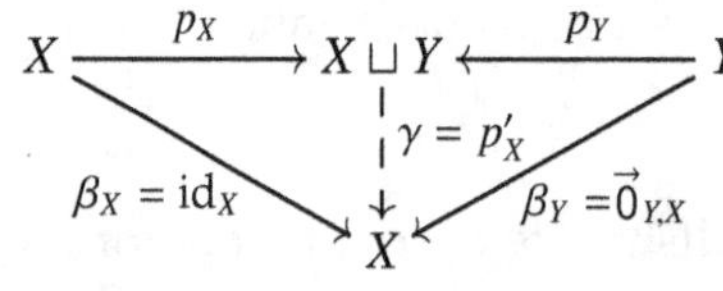

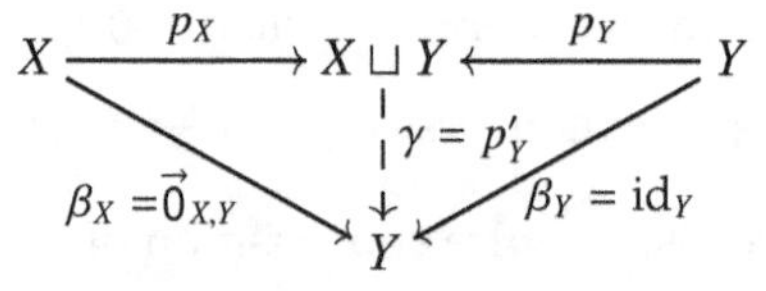

Indeed, the first four conditions of (c) hold by the diagrams above. One can show that $p_X\, p'_X + p_Y\, p'_Y = \mathrm{id}_{X \square Y}$, which we leave to the reader.

To show that (c) implies (a), take $X \sqcup Y = X \square Y$, along with $\alpha_X = p_X$ and $\alpha_Y = p_Y$. Now for any morphisms $\beta_X : X \to C$ and $\beta_Y : Y \to C$, define the morphism

$$\gamma := \beta_X p'_X + \beta_Y p'_Y : X \square Y \to C.$$

Then, $\beta_X = \gamma\, \alpha_X$ and $\beta_Y = \gamma\, \alpha_Y$, and one can show γ is the unique morphism with this property. We leave this, and the rest of the proof, to the reader. $\square$

Now a preadditive category $\mathcal{C}$ is said to be **additive** if there exists

- a zero object in $\mathcal{C}$, and

- the binary biproduct between any two objects in $\mathcal{C}$.

Inductively, additive categories contain biproducts of finitely many objects. See also Exercise 2.17.

Example 2.3. One example of an additive category is the **zero category** consisting of only the zero object 0 and the zero morphism $\vec{0}_{0,0} = \mathrm{id}_0$.

Next, we turn our attention to product categories. If $\mathcal{C}$ and $\mathcal{C}'$ are additive, then it is straightforward to see that so is the product category $\mathcal{C} \times \mathcal{C}'$.

An additive category is said to be **decomposable** if it is equivalent to the product of two nonzero categories; else, it is **indecomposable**.

iii. Abelian categories

This brings us to the main concept of the section. An additive category $\mathcal{C}$ is **abelian** if satisfies the conditions below.

- Every morphism in $\mathcal{C}$ has a cokernel and a kernel in $\mathcal{C}$.

- All monos and epis in $\mathcal{C}$ are normal.

One nice feature about abelian categories is given as follows.

Proposition 2.4. *Isos are precisely monic epis (or, epic monos) in abelian categories $\mathcal{C}$.*

Proof. Let $f : W \to X$ be a monic epi in $\mathcal{C}$. Since f is monic, by normality we have that $f = \alpha'_g : \ker(g) \to X$ for some morphism $g : X \to Y$ in $\mathcal{C}$. Thus, $gf = \vec{0}_{\ker(g),Y} = \vec{0}_{X,Y} \circ f$. Since f is epic, we then get that $g = \vec{0}_{X,Y}$. Therefore,

$$f = \alpha'_{\vec{0}_{X,Y}} : \ker(\vec{0}_{X,Y}) \to X.$$

Now we have the commutative diagram below from the definition of a kernel.

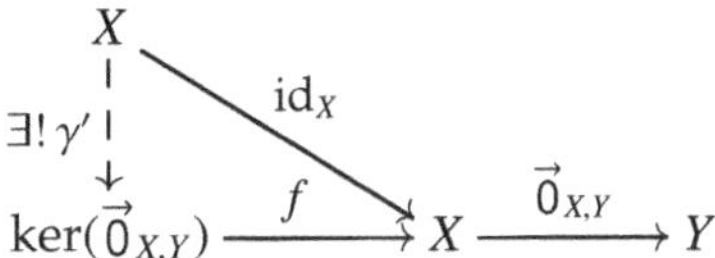

So, $f\gamma' = \mathrm{id}_X$. This implies that $f\gamma'f = f$. Since f is monic, $\gamma'f = \mathrm{id}_{\ker(\vec{0}_{X,Y})}$. So, f is an iso with inverse γ'. The converse direction is covered in Exercise 2.1(b). $\square$

The result above can be used to rule out examples of abelian categories. For instance, Exercise 2.2(c) shows that Ring is not an abelian category.

But there are many nice examples of abelian categories, including:

$$\text{Ab}, \quad \text{Vec}, \quad \text{Rep}(A), \quad A\text{-Mod}, \quad \text{Mod-}A, \quad (A, B)\text{-Bimod}.$$

In particular, this is why we work with the definition of a $\Bbbk$-algebra in §1.1.5 that builds on the structure of a vector space, instead of building on a unital ring structure. We leave it to the reader to explore whether the rest of the categories in §2.1.2 are abelian; see Remark 2.17 later.

Next, we show that abelian categories contain vital universal objects.

Proposition 2.5. *In an abelian category $\mathcal{C}$, pushouts and pullbacks exist.*

Proof. We will sketch the proof of the statement about pushouts; and leave to the rest to Exercise 2.18. Moreover, we leave it to the reader to dualize arguments to get the proof of the statement for pullbacks.

Let $f : Z \to X$ and $g : Z \to Y$ be morphisms, for which we aim to construct the pushout. Consider the biproduct:

$$(X \square Y, \ p_X : X \to X \square Y, \ p_Y : Y \to X \square Y, \ p'_X : X \square Y \to X, \ p'_Y : X \square Y \to Y).$$

Since $(X \square Y, \ p'_X : X \square Y \to X, \ p'_Y : X \square Y \to Y)$ is a product (see Proposition 2.2), there exists a unique morphism $q : Z \to X \square Y$ that makes the diagram below commute.

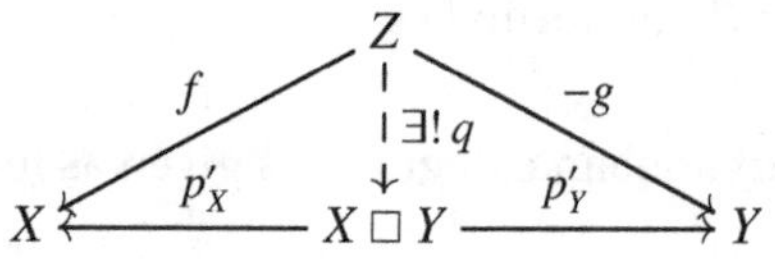

Next, take the cokernel of q, given by $\alpha_q : X \square Y \to \operatorname{coker}(q)$. Define

$$\alpha_X := \alpha_q \circ p_X : X \to \operatorname{coker}(q) \quad \text{and} \quad \alpha_Y := \alpha_q \circ p_Y : Y \to \operatorname{coker}(q).$$

Then, we get the following computation (see Proposition 2.2):

$$\alpha_X f - \alpha_Y g \ = \ \alpha_X p'_X q + \alpha_Y p'_Y q \ = \ \alpha_q(p_X p'_X + p_Y p'_Y)q \ = \ \alpha_q q \ = \ \vec{0}_{Z,\operatorname{coker}(q)}.$$

So, the diagram below commutes, and $(\operatorname{coker}(q), \alpha_X, \alpha_Y)$ is a candidate for the pushout of the morphisms f and g.

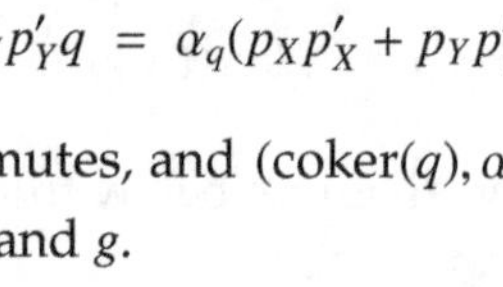

Now for every object C with morphisms $\beta_X : X \to C$ and $\beta_Y : Y \to C$, where $\beta_X f = \beta_Y g$, one can construct a morphism $\gamma : \operatorname{coker}(q) \to C$, with $\beta_X = \gamma \alpha_X$ and $\beta_Y = \gamma \alpha_Y$. One can also show that the choice of γ is unique. Thus, the pushout of f and g exists in $\mathcal{C}$. $\qquad\square$

Finally, each of the universal constructions in §2.2.1 exists in an abelian category C, due to the reasoning below (corresponding to subsection numbers).

(i) An initial object and a terminal object exist because C has a zero object.

(ii) Coproducts and products exist due to Proposition 2.2.

(iii) Pushouts and pullbacks exist due to Proposition 2.5.

(iv) Coequalizers and equalizers exist due to (ii,iii) and Exercise 2.12.

(v) C has zero morphisms since it has a zero object.

(vi) Cokernels and kernels exist by definition.

§2.3. Functors and natural transformations

Now we set up a framework to answer the question of whether two categories are the 'same'; this will be made clear in §2.4. For now, we study how to move from one category to another. Let $C, C', D,$ and $\mathcal{E}$ be categories throughout.

§2.3.1. Functors

A **functor** $F : C \to D$ from C to D consists of the following data.

(a) An object $F(X)$ in D, for each $X \in C$.

(b) A morphism $F(g) : F(X) \to F(Y)$ in D, for each $g \in \mathrm{Hom}_C(X, Y)$.

This data must satisfy the axioms below.

- (**respects identity morphisms**) We have that $F(\mathrm{id}_X) = \mathrm{id}_{F(X)}$ for all $X \in C$.

- (**respects composition**) For all $g \in \mathrm{Hom}_C(X, Y)$ and $h \in \mathrm{Hom}_C(Y, Z)$, we have that $F(h \circ g) = F(h) \circ F(g)$ in $\mathrm{Hom}_D(F(X), F(Z))$.

Sometimes, $F : C \to D$ above is referred to as a **covariant functor** as it preserves the directions of the morphisms in part (b).

We can also reverse the direction of morphisms when transporting categories as follows. A **contravariant functor** $F : C \to D$ consists of the following data.

(a) An object $F(X)$ in D, for each $X \in C$.

(b) A morphism $F(g) : F(Y) \to F(X)$ in D, for each $g \in \mathrm{Hom}_C(X, Y)$.

This data must satisfy the axioms below.

- (**respects identity morphisms**) We have that $F(\mathrm{id}_X) = \mathrm{id}_{F(X)}$ for all $X \in C$.

- (**respects composition contravariantly**) For each pair $f \in \mathrm{Hom}_C(W, X)$ and $g \in \mathrm{Hom}_C(X, Y)$, we have that $F(g \circ f) = F(f) \circ F(g)$ in $\mathrm{Hom}_D(F(Y), F(W))$.

In any case, C (resp., D) is the **domain** (resp., **codomain**) of $F : C \to D$. Also, if $C = D$, then we refer to F as an **endofunctor**.

An example of an endofunctor on any category C is the **identity functor** Id_C given by $\mathrm{Id}_C(X) = X$ and $\mathrm{Id}_C(f) = f$, for all $X \in C$ and $f \in \mathrm{Hom}(C)$.

Given functors $F : C \to D$ and $G : D \to \mathcal{E}$, we can **compose** functors object-wise and morphism-wise to yield a functor $GF := G \circ F : C \to \mathcal{E}$.

Next, we will discuss "injectivity" and "surjectivity" for functors $F : C \to D$. To do so on morphisms, consider the functions below (which are between sets as we assume C and D are locally small):

$$F_{X,Y} : \mathrm{Hom}_C(X, Y) \to \mathrm{Hom}_D(F(X), F(Y)), \ f \mapsto F(f).$$

- F is called **faithful** if $F_{X,Y}$ is injective, for each $X, Y \in C$.
- F is called **full** if $F_{X,Y}$ is surjective, for each $X, Y \in C$.
- F is called **fully faithful** if $F_{X,Y}$ is bijective, for each $X, Y \in C$.

See Exercise 2.19 for practice.

Now we consider the following notions for the injectivity and surjectivity of objects of a functor $F : C \to D$.

- F is said to be an **embedding** if F is faithful and if F is injective on objects.
- F is said to be **essentially surjective** if, for each object $Y \in D$, there exists an object $X \in C$ such that $Y \cong F(X)$.

The image of a functor $F : C \to D$ is not necessarily a subcategory of D, but we can form the **essential image** of F: the full subcategory $\mathrm{Im}^{\mathrm{ess}}(F)$ of D on objects $Y \cong F(X)$ of D, for some $X \in C$. In particular, F is essentially surjective precisely when its essential image is D. See Exercise 2.20, after reading §2.4.2, for practice.

The notions below pertain to a functor $F : C \to D$ preserving certain categorical properties from §2.2.2.

- F is **additive** if C, D are preadditive and $F_{X,Y}$ is a group map, for each $X, Y \in C$.
- F is **linear** if C, D are linear and $F_{X,Y}$ is a linear map, for each $X, Y \in C$.

In fact, we have a convenient characterization of additive functors between additive categories. The proof of the result below is Exercise 2.21.

Lemma 2.6. *Suppose that $(C, \square^C)$ and $(D, \square^D)$ are additive categories. Then, a functor $F : C \to D$ is additive if and only if F preserves biproducts, i.e., for all $X, Y \in C$ we have:*

$$F(X \square^C Y) \cong F(X) \square^D F(Y). \qquad \square$$

§2.3.2. Examples of functors

In this part, we present interesting examples of functors using some of the categories from §2.1.2. See Exercise 2.22 for practice.

Forgetful functors. These are functors $\mathrm{Forg} : C \to D$, where each object of D is obtained from an object in C by *forgetting certain structure*. Consider the following.

- $\mathrm{Forg} : \mathsf{Ring} \to \mathsf{Ab}$ (forgets multiplication)

- $\mathrm{Forg} : \mathsf{Vec} \to \mathsf{Ab}$ (forgets scalar multiplication)

- $\mathrm{Forg} : \mathsf{Top} \to \mathsf{Set}$ (forgets topology)

- $\mathrm{Forg} : \mathsf{Poset} \to \mathsf{Set}$ (forgets partial ordering)

- $\mathrm{Forg} : \mathsf{Quiv} \to \mathsf{Graph}$ (forgets direction of arrows / sources and targets)

Forgetful functors are usually faithful, but could be not full nor essentially surjective. For instance, $\mathrm{Forg} : \mathsf{Ring} \to \mathsf{Ab}$ is not full as the trivial group homomorphism $\mathbb{Z}/n\mathbb{Z} \to \mathbb{Z}$ cannot be upgraded to a ring homomorphism. On the other hand, $\mathrm{Forg} : \mathsf{Ring} \to \mathsf{Ab}$ is not essentially surjective as the quotient group $\mathbb{Q}/\mathbb{Z}$ cannot admit the structure of a ring.

Inclusions. These are functors $\mathrm{Inc} : C \to D$, where $\mathrm{Ob}(C)$ is a subcollection of $\mathrm{Ob}(D)$. Here, C is a subcategory of D. The examples below *impose a certain property* on objects in D to get objects in C. (One can also impose properties on morphisms.)

- $\mathrm{Inc} : \mathsf{Ab} \to \mathsf{Group}$

- $\mathrm{Inc} : \mathsf{ComAlg} \to \mathsf{Alg}$

- $\mathrm{Inc} : \mathsf{FinSet} \to \mathsf{Set}$

Inclusions are always faithful. If $\mathrm{Inc} : C \to D$ is a full (so, a fully faithful) inclusion, then C is a full subcategory of D.

Free functors. These are functors $\mathrm{Free} : C \to D$, where for $X \in C$, the object $\mathrm{Free}(X)$ is the *free object* in D built from X. For instance, if objects in D have generators, then $\mathrm{Free}(X)$ has no relations amongst its generators. So, free functors are typically not essentially surjective. Consider the examples below.

- $\mathrm{Free} : \mathsf{Set} \to \mathsf{Group}$ (free group)

- $\mathrm{Free} : \mathsf{Vec} \to \mathsf{Alg}$ (tensor algebra, see §1.2.2)

Free objects and free functors will be defined formally later in Example 2.29 in §2.5, and more examples of free functors will be discussed there.

Algebraic functors. Consider the following examples of functors derived from various constructions in Chapter 1. See §§1.1.4iii-v, 1.4.2iii, 1.4.3i, 1.4.4.

- $V \otimes_\Bbbk - : \mathsf{Vec} \to \mathsf{Vec}$ and $- \otimes_\Bbbk W : \mathsf{Vec} \to \mathsf{Vec}$,   for $V, W \in \mathsf{Vec}$.

- $\mathrm{Hom}_\Bbbk(V, -) : \mathsf{Vec} \to \mathsf{Vec}$ and $\mathrm{Hom}_\Bbbk(-, W) : \mathsf{Vec} \to \mathsf{Vec}$,   for $V, W \in \mathsf{Vec}$.

- $V \otimes_A - : (A, B_2)\text{-}\mathsf{Bimod} \to (B_1, B_2)\text{-}\mathsf{Bimod}$,   for $V \in (B_1, A)\text{-}\mathsf{Bimod}$.

- $\mathrm{Hom}_{\mathsf{Mod}\text{-}A}(-, W) : (B_1, A)\text{-}\mathsf{Bimod} \to (B_2, B_1)\text{-}\mathsf{Bimod}$, $W \in (B_2, A)\text{-}\mathsf{Bimod}$.

- $\mathrm{Res}^B_A(-) : B\text{-}\mathsf{Mod} \to A\text{-}\mathsf{Mod}$,   for $A \to B \in \mathrm{Hom}(\mathsf{Alg})$.

- $\mathrm{Ind}^B_A(-) : A\text{-}\mathsf{Mod} \to B\text{-}\mathsf{Mod}$,   for $A \to B \in \mathrm{Hom}(\mathsf{Alg})$.

Functors between mathematical fields. Finally, we list some examples of functors that connect different fields in mathematics. Note that some of the terminology below is outside the scope of this book, and will not be defined.

- $\mathcal{O} : \mathsf{Aff} \to \mathsf{ComAlg}$ (Form the *coordinate algebra* of an affine variety, used in Algebraic Geometry.)

- $\mathcal{L} : \mathsf{Ban} \to \mathsf{Alg}$ (Form the *function algebra* of a Banach space, used in Functional Analysis. Here, $\mathcal{L}(X)$ is a *Banach algebra*.)

- $\pi_1 : \mathsf{Top}_* \to \mathsf{Group}$ (Form the *fundamental group* of a topological space with base point, used in Algebraic Topology.)

- $Z : n\mathsf{Cob} \to \mathsf{FinHilb}$ (or $\mathsf{Vec}_\mathbb{C}$)   (This functor is referred to as a *Topological Quantum Field Theory* in Quantum Physics.)

- $p : \mathsf{FinSet} \to \mathsf{Vec}$ (or Set)   (This is a *species* in Enumerative Combinatorics.)

§2.3.3. Bifunctors and multifunctors

Like bilinear maps for vector spaces (see §1.1.4iii), we consider a way of moving from two categories to a third category, while preserving structure.

Recall the product category $\mathcal{C} \times \mathcal{C}'$ from §2.1.3. Now a functor of the form $F : \mathcal{C} \times \mathcal{C}' \to \mathcal{D}$ is called a **bifunctor**. Here:

- $F(X, -) : \mathcal{C}' \to \mathcal{D}$ is a functor, for a fixed object $X \in \mathcal{C}$, and

- $F(-, X') : \mathcal{C} \to \mathcal{D}$ is a functor, for a fixed object $X' \in \mathcal{C}'$.

Here are a few examples of bifunctors on some categories from §2.1.2i.

$$- \otimes_{\Bbbk} - \,:\, \mathsf{Vec} \times \mathsf{Vec} \to \mathsf{Vec}$$

$$- \otimes_A - \,:\, (B_1, A)\text{-Bimod} \times (A, B_2)\text{-Bimod} \to (B_1, B_2)\text{-Bimod}$$

$$\mathrm{Hom}_{A\text{-Mod}}(-, -) \,:\, ((A, B_1)\text{-Bimod})^{\mathrm{op}} \times (A, B_2)\text{-Bimod} \to (B_1, B_2)\text{-Bimod}$$

See §§1.1.4iii,v for the first example, and §1.4.2iii for the second example. We use Proposition 1.20 in §1.4.3i for the third example; the opposite category is used in the first slot because $\mathrm{Hom}_{A\text{-Mod}}(-, W)$ is contravariant for each $W \in (A, B_2)\text{-Bimod}$.

A **multifunctor** is a functor of the form: $F : \mathcal{C}_1 \times \mathcal{C}_2 \times \cdots \times \mathcal{C}_n \to \mathcal{D}$. Here,

$$F(X_1, \ldots, X_{i-1}, -, X_{i+1}, \ldots, X_n) : \mathcal{C}_i \to \mathcal{D}$$

is a functor for each $i = 1, \ldots, n$, for fixed objects $X_j \in \mathcal{C}_j$ with $j \neq i$.

§2.3.4. Natural transformations and natural isomorphisms

Next, we discuss how to transport one functor to another one with the same domain and codomain categories.

Given two functors $F, F' : \mathcal{C} \to \mathcal{D}$, a **natural transformation**,

$$\phi : F \Rightarrow F',$$

by definition consists of morphisms,

$$\phi_X : F(X) \to F'(X) \ \text{ in } \mathcal{D}, \quad \text{for each } X \in \mathcal{C}.$$

We also require that, for each $f : X \to Y$ in $\mathcal{C}$, the diagram below commutes in $\mathcal{D}$.

$$
\begin{array}{ccc}
F(X) & \xrightarrow{\ F(f)\ } & F(Y) \\
{\scriptstyle \phi_X}\downarrow & & \downarrow{\scriptstyle \phi_Y} \\
F'(X) & \xrightarrow{\ F'(f)\ } & F'(Y)
\end{array}
$$

Here, the morphism ϕ_X is called the **component of ϕ at X**, and the commutative diagram above is referred to as the **naturality of ϕ at f.**

If, further, the component ϕ_X is an iso in $\mathcal{D}$ for every $X \in \mathcal{C}$, then we say that ϕ is a **natural isomorphism**, and we write $\phi : F \overset{\sim}{\Rightarrow} F'$, or just $F \cong F'$.

Pictorially, a natural transformation ϕ from $F : \mathcal{C} \to \mathcal{D}$ to $F' : \mathcal{C} \to \mathcal{D}$ is visualized as the diagram below.

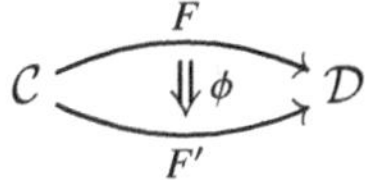

The collection of natural transformations $\phi : F \Rightarrow F'$ is denoted by

$$\mathrm{Nat}(F, F') := \mathrm{Nat}_{C,D}(F, F').$$

This collection may not be a set, even when C and D are locally small. But we will discuss a situation when $\mathrm{Nat}(F, F')$ is a set later in §2.6.2.

Moreover, we denote the collection of natural isomorphisms $\phi : F \overset{\sim}{\Rightarrow} F'$ by

$$\mathrm{NatIsom}(F, F') := \mathrm{NatIsom}_{C,D}(F, F').$$

One example of a natural isomorphism is the **identity natural isomorphism**,

$$\mathrm{ID}_F : F \overset{\sim}{\Rightarrow} F,$$

of a functor $F : C \to D$ to itself. Its components are defined by the morphisms,

$$(\mathrm{ID}_F)_X := \mathrm{id}_{F(X)} : F(X) \to F(X),$$

for all $X \in C$.

More sophisticated examples of natural transformations are given as follows.

Example 2.7. There is a natural transformation j from the identity functor on Vec to the double dual functor Vec (cf. Exercise 1.4(e)).

$$
\mathrm{Vec} \underset{(-)^{**}}{\overset{\mathrm{Id}}{\rightrightarrows}} \mathrm{Vec} \qquad \Downarrow j
$$

The components are given by

$$j_V : V \to V^{**}, \quad v \mapsto [f \mapsto f(v)].$$

If we replace Vec with FdVec, then j is a natural isomorphism. See Exercise 2.23.

Example 2.8. The associativity of the tensor product $\otimes := \otimes_{\Bbbk}$ of vector spaces, $(U \otimes V) \otimes W \cong U \otimes (V \otimes W)$ for each $U, V, W \in \mathrm{Vec}$ (cf. Exercise 1.3(b)), can be upgraded to a natural isomorphism. See Exercise 2.24.

$$
\mathrm{Vec} \times \mathrm{Vec} \times \mathrm{Vec} \underset{\otimes \circ (\mathrm{Id} \times \otimes)}{\overset{\otimes \circ (\otimes \times \mathrm{Id})}{\rightrightarrows}} \mathrm{Vec} \qquad \sim\!\Downarrow a
$$

With the material above, we can construct categories of functors. (See Exercise 2.25 after reading §2.3.5.) Namely, the **functor category** $\mathrm{Fun}(C, D)$ is defined with objects being functors $F : C \to D$ from C to D, and with morphisms given by

$$\mathrm{Hom}_{\mathrm{Fun}(C,D)}(F, F') := \mathrm{Nat}_{C,D}(F, F').$$

In the case, when $C = D$, we denote $\mathrm{Fun}(C, C)$ by $\mathrm{End}(C)$, and refer to it as an **endofunctor category**.

§2.3.5. Compositions of natural transformations

Since the components of natural transformations are encoded by commutative rectangles, we can compose them in two ways – vertically or horizontally – to yield operations on natural transformations.

Given natural transformations ϕ from $F : \mathcal{C} \to \mathcal{D}$ to $F' : \mathcal{C} \to \mathcal{D}$, and ϕ' from $F' : \mathcal{C} \to \mathcal{D}$ to $F'' : \mathcal{C} \to \mathcal{D}$, their **vertical composition** $\phi' \circ^{\text{ver}} \phi$ is the natural transformation pictured below.

$$\mathcal{C} \underset{F''}{\overset{F}{\Longleftrightarrow}} \Downarrow \phi' \circ^{\text{ver}} \phi \quad \mathcal{D} \quad := \quad \mathcal{C} \xrightarrow{\quad} \mathcal{D}$$

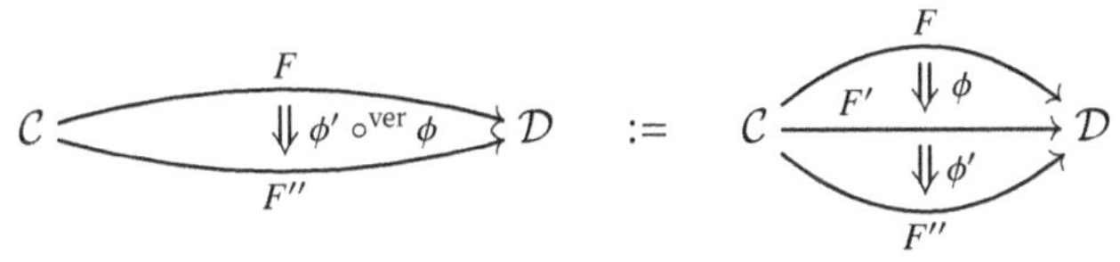

Namely, the components are given $(\phi' \circ^{\text{ver}} \phi)_X := \phi'_X \circ \phi_X : F(X) \to F''(X)$, for $X \in \mathcal{C}$.

On the other hand, take natural transformations ϕ from $F : \mathcal{C} \to \mathcal{D}$ to $F' : \mathcal{C} \to \mathcal{D}$, and ψ from $G : \mathcal{D} \to \mathcal{E}$ to $G' : \mathcal{D} \to \mathcal{E}$. Then, their **horizontal composition** $\psi \circ^{\text{hor}} \phi$ is the natural transformation pictured below.

$$\mathcal{C} \underset{G'F'}{\overset{GF}{\Longleftrightarrow}} \Downarrow \psi \circ^{\text{hor}} \phi \quad \mathcal{E} \quad := \quad \mathcal{C} \underset{F'}{\overset{F}{\Longleftrightarrow}} \Downarrow \phi \quad \mathcal{D} \underset{G'}{\overset{G}{\Longleftrightarrow}} \Downarrow \psi \quad \mathcal{E}$$

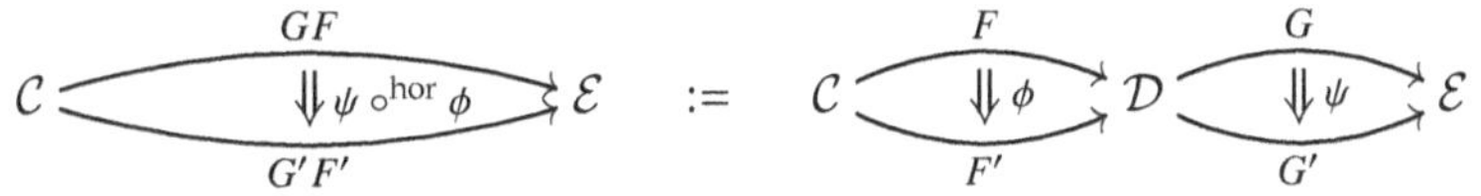

For $X \in \mathcal{C}$, the component at X is given by:

$$(\psi \circ^{\text{hor}} \phi)_X \ := \ \psi_{F'(X)} \circ G(\phi_X) \ = \ G'(\phi_X) \circ \psi_{F(X)} : GF(X) \to G'F'(X).$$

The equality holds by the naturality of ψ at $\phi_X : F(X) \to F'(X)$.

Next, we discuss a special case of horizontal composition– how one can combine a natural transformation with a functor. Take a natural transformation ϕ from $F : \mathcal{C} \to \mathcal{D}$ to $F' : \mathcal{C} \to \mathcal{D}$, with functors $F_0 : \mathcal{B} \to \mathcal{C}$ and $F_1 : \mathcal{D} \to \mathcal{E}$. Then, we can define natural transformations:

$$\phi * F_0 \ := \ \phi \circ^{\text{hor}} \text{ID}_{F_0}, \qquad F_1 * \phi \ := \ \text{ID}_{F_1} \circ^{\text{hor}} \phi.$$

Their components are given by $(\phi * F_0)_W := \phi_{F_0(W)} : FF_0(W) \to F'F_0(W)$ for $W \in \mathcal{B}$, and $(F_1 * \phi)_X := F_1(\phi_X) : F_1F(X) \to F_1F'(X)$ for $X \in \mathcal{C}$, respectively. This is called **whiskering**, and is as pictured below.

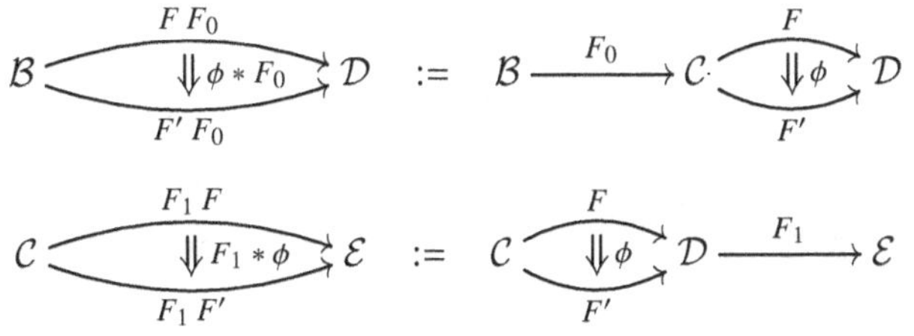

$$\mathcal{B} \underset{F'F_0}{\overset{FF_0}{\Longleftrightarrow}} \Downarrow \phi * F_0 \quad \mathcal{D} \quad := \quad \mathcal{B} \xrightarrow{F_0} \mathcal{C} \underset{F'}{\overset{F}{\Longleftrightarrow}} \Downarrow \phi \quad \mathcal{D}$$

$$\mathcal{C} \underset{F_1F'}{\overset{F_1F}{\Longleftrightarrow}} \Downarrow F_1 * \phi \quad \mathcal{E} \quad := \quad \mathcal{C} \underset{F'}{\overset{F}{\Longleftrightarrow}} \Downarrow \phi \quad \mathcal{D} \xrightarrow{F_1} \mathcal{E}$$

Next, we have that vertical composition and horizontal composition satisfy the compatibility condition below; the proof is left to Exercise 2.26.

$$(\psi' \circ^{\mathrm{ver}} \psi) \circ^{\mathrm{hor}} (\phi' \circ^{\mathrm{ver}} \phi) \;=\; (\psi' \circ^{\mathrm{hor}} \phi') \circ^{\mathrm{ver}} (\psi \circ^{\mathrm{hor}} \phi) \tag{2.9}$$

This is called the **interchange law**, and is pictured as follows.

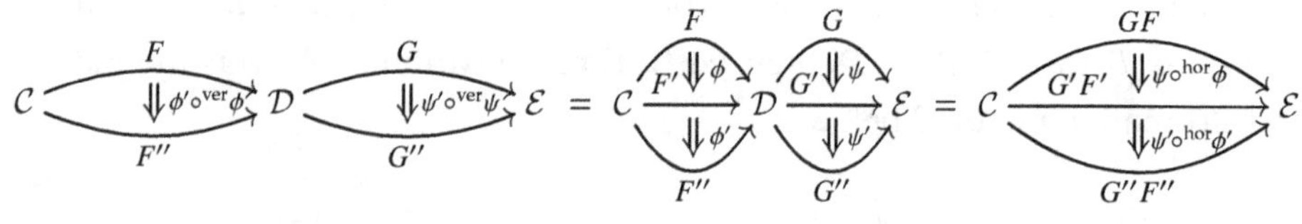

$$
\mathcal{C} \underset{F''}{\overset{F}{\Longrightarrow}} \mathcal{D} \;\Downarrow \phi' \circ^{\mathrm{ver}} \phi \quad
\mathcal{D} \underset{G''}{\overset{G}{\Longrightarrow}} \mathcal{E} \;\Downarrow \psi' \circ^{\mathrm{ver}} \psi
\;=\;
\mathcal{C} \xrightarrow{\ F'\ \Downarrow\phi\ } \mathcal{D} \;\Downarrow\phi'\; \underset{F''}{\ } \xrightarrow{\ G'\ \Downarrow\psi\ } \mathcal{E} \;\Downarrow\psi'\; \underset{G''}{\ }
\;=\;
\mathcal{C} \underset{G''F''}{\overset{GF}{\longrightarrow}} \mathcal{E} \quad G'F' \;\Downarrow\psi\circ^{\mathrm{hor}}\phi \quad \Downarrow\psi'\circ^{\mathrm{hor}}\phi'
$$

§2.3.6. Colimits and limits of functors

Here, we study how each of the universal constructions in §2.2.1 fits into a common framework. To begin, take categories $\mathcal{J}$ and $\mathcal{C}$; here, $\mathcal{J}$ will be considered as an **index category**. Next, take the **diagonal functor** Δ defined below:

$$\Delta : \mathcal{C} \to \mathrm{Fun}(\mathcal{J}, \mathcal{C}), \quad D \mapsto [\Delta(D) : \mathcal{J} \to \mathcal{C}, \; J \mapsto D, \; f \mapsto \mathrm{id}_D],$$

for all $D \in \mathcal{C}$, $J \in \mathcal{J}$, and $f \in \mathrm{Hom}(\mathcal{J})$. Now we consider the constructions below.

A **colimit** of a functor $F : \mathcal{J} \to \mathcal{C}$ consists of the following data.

(a) An object, $\mathrm{colim}_{\mathcal{J}} F$, in $\mathcal{C}$.

(b) A natural transformation, $\alpha : F \Rightarrow \Delta(\mathrm{colim}_{\mathcal{J}} F)$.

This data must satisfy the universality axiom below.

- For each pair $(C, \beta : F \Rightarrow \Delta(C))$, with C an object in $\mathcal{C}$ and $\beta_C : F \Rightarrow \Delta(C)$ a natural transformation, there exists a unique morphism $\gamma_C : \mathrm{colim}_{\mathcal{J}} F \to C$ in $\mathcal{C}$ such that $\beta_C = \Delta(\gamma_C) \circ^{\mathrm{ver}} \alpha$ as natural transformations.

Example 2.10. Many of the universal constructions in §2.2.1 arise as colimits.

(a) An initial object I in $\mathcal{C}$ is a colimit of the functor $F : \varnothing \to \mathcal{C}$. Here, $\mathrm{colim}_{\varnothing} F := I$ and, for any object $X \in \mathcal{C}$, we have that $\gamma_X := \vec{0}_X : I \to X$.

(b) A pushout $(X \sqcup_Z Y, f : Z \to X, g : Z \to Y)$ in $\mathcal{C}$ arises as a colimit as follows. Take $\mathcal{J}$ to be a category with objects J_0, J_1, J_2, and morphisms $J_0 \to J_1, J_0 \to J_2$. Now for the functor $F : \mathcal{J} \to \mathcal{C}$, $\mathrm{colim}_{\mathcal{J}} F := X \sqcup_Z Y$, and for an object C, the morphism $\gamma_C : X \sqcup_Z Y \to C$ is the morphism γ in the pushout construction.

(c) Likewise, either directly or by using Exercises 2.10(a), 2.12(a), and 2.13(c), we have that coproducts, coequalizers, and cokernels arise as colimits.

The details are left to Exercise 2.27.

Dually, a **limit** of a functor $F : \mathcal{J} \to \mathcal{C}$ consists of the following data.

(a) An object, $\lim_{\mathcal{J}} F$, in $\mathcal{C}$.

(b) A natural transformation, $\alpha' : \Delta(\lim_{\mathcal{J}} F) \Rightarrow F$.

This data must satisfy the universality axiom below.

- For each pair $(L, \beta'_L : \Delta(L) \Rightarrow F)$, with L an object in $\mathcal{C}$ and $\beta'_L : \Delta(L) \Rightarrow F$ a natural transformation, there exists a unique morphism $\gamma'_L : L \to \lim_{\mathcal{J}} F$ in $\mathcal{C}$ such that $\beta'_L = \alpha' \circ^{\text{ver}} \Delta(\gamma'_L)$ as natural transformations.

Example 2.11. Dual to Example 2.10, we get that terminal objects, pullbacks, products, equalizers, and kernels arise as limits. The details are left to Exercise 2.27.

§2.4. Isomorphisms and equivalence of categories

Now that we have a framework to move from one category to another, i.e., with functors from §2.3, we will proceed with addressing the question of whether two categories are the same. Let $\mathcal{C}$ and $\mathcal{D}$ be categories throughout.

§2.4.1. Isomorphism of categories

Analogous to two groups being considered the same if there is an isomorphism (e.g., a group homomorphism with an inverse map) between them, we have the notion of sameness for categories below.

We say that $\mathcal{C}$ and $\mathcal{D}$ are **isomorphic** as categories if there are functors $F : \mathcal{C} \to \mathcal{D}$ and $G : \mathcal{D} \to \mathcal{C}$ such that $GF = \text{Id}_{\mathcal{C}}$ and $FG = \text{Id}_{\mathcal{D}}$ as functors. In this case, we write $\mathcal{C} \cong \mathcal{D}$. Isomorphism here is an equivalence relation for categories [Exercise 2.28].

Let us discuss an example and a non-example of isomorphic categories.

Example 2.12. Recall §1.3.4 and consider the categories G-Mod and Rep(G), for a group G. We have that these categories are isomorphic as follows. Define the functors F and F' below, first on objects as such:

- $F : G\text{-Mod} \to \text{Rep}(G)$ sends $(V, \triangleright : G \times V \to V)$ to the G-representation,

$$G \to GL(V), \quad g \mapsto [V \to V, \ v \mapsto g \triangleright v];$$

- $F' : \text{Rep}(G) \to G\text{-Mod}$ sends $(V, \rho : G \to GL(V))$ to the left G-module V with action map,

$$G \times V \to V, \quad (g, v) \mapsto \rho(g)(v).$$

We leave it to the reader to write down the definition of F and F' on morphisms, along with showing that G-Mod $\cong$ Rep(G). See Exercises 2.29 and 2.30.

Example 2.13. Take the category of finite-dimensional $\Bbbk$-vector spaces FdVec. Let $\mathcal{S}$ be the full subcategory of FdVec on the objects $\{\Bbbk^n\}_{n\in\mathbb{N}}$. Since every finite-dimensional $\Bbbk$-vector space is isomorphic to $\Bbbk^n$ for some $n \in \mathbb{N}$, one may expect that FdVec and $\mathcal{S}$ are the "same", say via the functors below:

$$F : \mathsf{FdVec} \to \mathcal{S},\ V \mapsto \Bbbk^{\dim_\Bbbk V} \qquad \text{and} \qquad G : \mathcal{S} \to \mathsf{FdVec},\ \Bbbk^n \mapsto \Bbbk^n.$$

If by 'same' we mean that there exists an isomorphism of categories, then

$$V \;=\; GF(V) \;=\; \Bbbk^{\dim_\Bbbk V},$$

for any $V \in \mathsf{FdVec}$, which is not true. So, we need to replace these vector space equalities with vector space isomorphisms to get that FdVec is the same as $\mathcal{S}$ as categories. That is, we need to weaken the notion of a category isomorphism. See Exercise 2.32 for practice.

The example above illustrates a general phenomenon. The **skeleton** of a category $\mathcal{C}$ is the full subcategory $\mathsf{Skel}(\mathcal{C})$ of $\mathcal{C}$ on objects consisting of exactly one isoclass representative for each isoclass of objects. In particular, there is no iso between distinct objects in $\mathsf{Skel}(\mathcal{C})$.

Even though $\mathcal{C}$ and $\mathsf{Skel}(\mathcal{C})$ carry the 'same' categorical information, they are not isomorphic as categories. But these categories are *equivalent* in the sense that we will define in the next section.

§2.4.2. Equivalence of categories

As discussed above, category isomorphism is too strong of a notion of sameness to have a rich category theory. So, we will consider a weaker notion below.

First, we consider an analogy. Recall that we do not require two groups to be precisely *equal* to be considered the same in group theory; instead, we use the weaker notion of group *isomorphism*. Else, classification problems would be impossible– e.g., the classification of groups of order four would consist of a zoo of examples (e.g., $\mathbb{Z}/4\mathbb{Z}$, $5\mathbb{Z}/20\mathbb{Z}$, symmetries of a rectangle, $3\mathbb{Z}/6\mathbb{Z} \times 4\mathbb{Z}/8\mathbb{Z}$, etc.). It is better to use our two models of groups of order four, namely $C_2 \times C_2$ and C_4, to illustrate how all such groups behave, up to isomorphism.

Now we say that $\mathcal{C}$ and $\mathcal{D}$ are **equivalent** as categories if there exist functors $F : \mathcal{C} \to \mathcal{D}$ and $G : \mathcal{D} \to \mathcal{C}$ such that $GF \cong \mathrm{Id}_\mathcal{C}$ and $FG \cong \mathrm{Id}_\mathcal{D}$ as functors. (Namely, we replace equality of functors for a category isomorphism with natural isomorphism of functors here.) Here, we write $\mathcal{C} \simeq \mathcal{D}$, and refer to F and G as **quasi-inverses** of each other.

Equivalence here is an equivalence relation amongst categories [Exercise 2.28].

If C is isomorphic to $\mathcal{D}$, then C is equivalent to $\mathcal{D}$. But the converse does not hold. For instance, in Example 2.13, we saw that FdVec is not isomorphic to its skeleton $\mathcal{S}$, yet we have that FdVec $\simeq \mathcal{S}$ (see Exercise 2.33).

Notice that the definition of equivalence above is analogous to stating that a homomorphism between groups is a group isomorphism if there exists an inverse group homomorphism. On the other hand, a group isomorphism can equivalently be defined as a group homomorphism that is injective and surjective as a function. Along the latter lines, consider the following definition.

We say that C and $\mathcal{D}$ are **equivalent** as categories if there exists functor $F : C \to \mathcal{D}$ that is fully faithful and essentially surjective. Of course, we need to reconcile this with the notion of equivalence above; this is done as follows.

Theorem 2.14. *Given categories C and $\mathcal{D}$, the following statements are equivalent.*

(a) $C \simeq \mathcal{D}$ *via functors $F : C \to \mathcal{D}$ and $G : \mathcal{D} \to C$, where $GF \cong \mathrm{Id}_C$ and $FG \cong \mathrm{Id}_\mathcal{D}$.*

(b) $C \simeq \mathcal{D}$ *via a fully faithful, essentially surjective functor $F : C \to \mathcal{D}$.*

Proof. We will sketch the proof here, and leave it to the reader to fill in the details (especially by using commutative diagrams); see Exercise 2.34.

Suppose that there exist functors $F : C \to \mathcal{D}$ and $G : \mathcal{D} \to C$ such that we have natural isomorphisms:

$$\phi : \mathrm{Id}_C \overset{\Rightarrow}{\Rightarrow} GF \quad \text{and} \quad \psi : FG \overset{\Rightarrow}{\Rightarrow} \mathrm{Id}_\mathcal{D}. \tag{2.15}$$

We claim that $F : C \to \mathcal{D}$ is fully faithful and essentially surjective. By using the component ψ_Y, we have that $F(G(Y)) \cong Y$, for each $Y \in \mathcal{D}$. This shows that F is essentially surjective. Next, by using ϕ, we have for each $f : X \to X'$ in C that

$$\phi_{X'} f = ((GF)(f)) \phi_X.$$

Therefore, $f = \phi_{X'}^{-1} ((GF)(f)) \phi_X$, and this shows that F is faithful. To get that F is full, we first use ψ to conclude that G is faithful by repeating the argument for F being faithful. Then, take a morphism $h : F(X) \to F(X')$ in $\mathcal{D}$. Also, take the morphism $g := \phi_{X'}^{-1} (G(h)) \phi_X$ from X to X' in C. We obtain that $(GF)(g) = G(h)$. Since G is faithful, we then obtain that F is full.

Conversely, suppose that $F : C \to \mathcal{D}$ is a fully faithful, essentially surjective functor. Our goal is to construct a functor $G : \mathcal{D} \to C$ such that (2.15) holds. For each object $Y \in \mathcal{D}$, there exists an object $Z_Y \in C$ such that $F(Z_Y) \cong Y$ in $\mathcal{D}$ since F is essentially surjective. Now use the following labels:

$$Z_Y =: G(Y) \in C, \qquad \psi_Y : F(Z_Y) = F(G(Y)) \overset{\sim}{\to} Y \in \mathcal{D}.$$

Moreover, since F is fully faithful, we have that for any morphism $g : Y \to Y'$ in $\mathcal{D}$, we get a unique morphism,

$$G(g) : G(Y) \to G(Y') \in C.$$

Now it remains to show that:

(1) the assignments, $Y \mapsto G(Y)$ and $g \mapsto G(g)$, make G into a functor;

(2) there exists a natural isomorphism $\psi : FG \xRightarrow{\sim} \mathrm{Id}_{\mathcal{D}}$;

(3) there exists a natural isomorphism $\phi : \mathrm{Id}_{\mathcal{C}} \xRightarrow{\sim} GF$.

Faithfulness of F is used to establish (1). The morphisms ψ_Y form the components of ψ in (2). For (3), we have by the fully faithfulness of F that it suffices to define a natural isomorphism $v : F \xRightarrow{\sim} FGF$. Taking the components of v to be $v_X := \psi^{-1}_{F(X)} : F(X) \to (FGF)(X)$, for all $X \in \mathcal{C}$, achieves the goal. $\qquad\square$

In any case, an equivalence, $F : \mathcal{C} \xrightarrow{\sim} \mathcal{C}$, from a category $\mathcal{C}$ to itself is referred to as an **autoequivalence** of $\mathcal{C}$.

We can also construct an **autoequivalence category**, $\mathrm{Aut}(\mathcal{C})$, defined with objects being autoequivalences $F : \mathcal{C} \xrightarrow{\sim} \mathcal{C}$, and with morphisms given by $\mathrm{Hom}_{\mathrm{Aut}(\mathcal{C})}(F, F') := \mathrm{NatIsom}_{\mathcal{C},\mathcal{C}}(F, F')$.

§2.4.3. Examples of equivalent categories

One example of an autoequivalence includes the following.

- $(-)^{**} : \mathrm{FdVec} \xrightarrow{\sim} \mathrm{FdVec}$ [Exercise 2.23, Example 2.7]

Also, note that for categories $\mathcal{C}$ and $\mathcal{D}$, the following statements hold.

- $\mathcal{C}$ is equivalent to its skeleton $\mathrm{Skel}(\mathcal{C})$ [Exercise 2.33(b)]

- $\mathcal{C} \simeq \mathcal{D}$ if and only if $\mathrm{Skel}(\mathcal{C}) \cong \mathrm{Skel}(\mathcal{D})$ [Exercise 2.33(c)]

More concretely, for any algebra A, and any group G with group algebra $\Bbbk G$, we have the category isomorphisms below, which yield category equivalences.

- $A\text{-Mod} \cong \mathrm{Rep}(A)$ [Exercise 2.29]

- $G\text{-Mod} \cong \mathrm{Rep}(G) \cong \mathrm{Rep}(\Bbbk G) \cong \Bbbk G\text{-Mod}$ [Exercise 2.30, Example 2.12]

On a related note, take a group G, and take $\mathrm{G}_{\mathrm{cat}}$ to be the category with one object with morphisms identified as elements of G. We then get that

- $\mathrm{Rep}(G) \simeq \mathrm{Fun}(\mathrm{G}_{\mathrm{cat}}, \mathrm{Vec})$ [Exercise 2.31]

Since $\mathrm{G}_{\mathrm{cat}}$ is a special case of a groupoid [Exercise 2.4], this is why, by definition, a **representation of a groupoid** $\mathcal{G}$ is a functor of the form $\mathcal{G} \to \mathrm{Vec}$.

More interestingly, there are equivalences between categories that lie in different areas of mathematics. A prevalent model for this lies in Algebraic Geometry as follows; see Section 2.5 of Smith et al. [2000] for more details.

- We have an equivalence of categories,

$$\mathsf{Aff} \simeq (\mathsf{FgRedComAlg})^{\mathrm{op}},$$

between the category of affine varieties, and the opposite category of finitely generated, *reduced*, commutative algebras. This is given by the contravariant *coordinate algebra* functor $\mathcal{O} : \mathsf{Aff} \to \mathsf{ComAlg}$ [§2.3.2] (here, the essential image is in $\mathsf{FgRedComAlg}$); its quasi-inverse is the *spectrum functor* denoted by $\mathsf{Spec} : \mathsf{FgRedComAlg} \to \mathsf{Aff}$.

- Likewise, we have a geometric category that corresponds to the category ComAlg. Namely, we need to replace affine varieties with *affine schemes* to yield:

$$\mathsf{Scheme} \simeq (\mathsf{ComAlg})^{\mathrm{op}}.$$

Many other equivalences that stem from relating a geometric/topological object X (e.g., a certain space, a type of manifold, a surface) with its algebra of functions on X. See Chapter 1 on Khalkhali [2013] for further reading about this and about the category of schemes discussed above.

Lastly, abelian categories (from §2.2.2) are understood in terms of categories of modules over rings, up to equivalence. This is due to the work of Mitchell [1964].

Theorem 2.16 (Mitchell's Embedding Theorem). *Every small abelian category is equivalent to a full subcategory of left modules over some ring R.* □

See Section 1.6 of Weibel [1994] for a discussion of this result. Here, the left R-modules have the underlying structure of an abelian group, not of a vector space (see Exercise 1.14), and we denote this by $R\text{-Mod}_{\mathrm{ab}}$ for emphasis.

Remark 2.17. For a $\Bbbk$-algebra A, the category $A\text{-Mod}$ is abelian since, after forgetting scalar multiplication, the category $A\text{-Mod}_{\mathrm{ab}}$ is abelian. With this result, we can then derive various examples of abelian categories.

- $\mathsf{Ab} \cong \mathbb{Z}\text{-Mod}_{\mathrm{ab}}$.
- $\mathsf{Vec} \cong \Bbbk\text{-Mod}_{\mathrm{ab}}$.
- $\mathsf{Mod}\text{-}A \cong A^{\mathrm{op}}\text{-Mod}$.
- $(A, B)\text{-Bimod} \cong (A \otimes B^{\mathrm{op}})\text{-Mod}$.

§2.4.4. Morita equivalence

Now we will address the following question:

For algebras A and B, when are $A\text{-Mod}$ and $B\text{-Mod}$ equivalent as categories?

We say that A and B are **Morita equivalent** when this is true, and write $A \simeq_{\mathrm{Mor}} B$.

This provides us with another notion of sameness for algebras, which is weaker than isomorphism (which, in turn, is weaker than equality). It is indeed an equivalence relation for algebras [Exercise 2.28]. The terminology is due to the characterizations of this condition by the work of Morita [1958]. One version is in terms of bimodules; see Section 18.D of Lam [1999] for further details.

Theorem 2.18 (Morita's Theorem). *Let A and B be algebras. Then, we have that*

$$A\text{-Mod} \simeq B\text{-Mod}$$

if and only if there exist bimodules $_A P_B$ and $_B Q_A$ such that $P \otimes_B Q \cong A_{\mathrm{reg}}$ as A-bimodules and $Q \otimes_A P \cong B_{\mathrm{reg}}$ as B-bimodules.

In this case, we say that the bimodules P and Q are **invertible**.

Proof. We will discuss the steps below, and will leave the details to the reader; see Exercise 2.35. For the backward direction, consider the functors:

$$F := (_B Q_A) \otimes_A - \,:\ A\text{-Mod} \to B\text{-Mod},$$

$$G := (_A P_B) \otimes_B - \,:\ B\text{-Mod} \to A\text{-Mod}.$$

These functors yield an equivalence of categories, A-Mod $\simeq$ B-Mod (more readily obtained via the condition in Theorem 2.14(a)).

The forward direction takes more steps. Suppose that $F : A\text{-Mod} \xrightarrow{\sim} B\text{-Mod}$ is an equivalence of categories. Let $F(_A(A_{\mathrm{reg}})) =: {}_B Q \in B\text{-Mod}$. We have that

$$A^{\mathrm{op}} \cong \mathrm{End}_{A\text{-Mod}}(A_{\mathrm{reg}}) \cong \mathrm{End}_{B\text{-Mod}}(F(A_{\mathrm{reg}})) = \mathrm{End}_{B\text{-Mod}}(_B Q)$$

as algebras, by Exercise 1.26(a) and by F being fully faithful. Label this isomorphism by f, and we obtain that $_B Q$ is a right A-module via the action $q \triangleleft a := f(a)(q)$ for $a \in A$, $q \in Q$. With this, we get that $Q \in (B, A)$-Bimod.

Claim. $F \cong (_B Q_A) \otimes_A -$ as functors.

Proof of Claim. Take $X \in A$-Mod. Consider the morphism:

$$\sigma_X : X \cong \mathrm{Hom}_{A\text{-Mod}}(A, X) \xrightarrow{F} \mathrm{Hom}_{B\text{-Mod}}(F(A), F(X)) = \mathrm{Hom}_{B\text{-Mod}}(Q, F(X)),$$

which is an iso in A-Mod; see also Proposition 1.20(a). Next, by Tensor-Hom adjunction [Exercise 1.25(b)], we have the bijection below:

$$\mathrm{Hom}_{A\text{-Mod}}(X, \mathrm{Hom}_{B\text{-Mod}}(Q, F(X))) \cong \mathrm{Hom}_{B\text{-Mod}}(Q \otimes_A X, F(X)).$$

Let $\sigma'_X : Q \otimes_A X \to F(X)$ be the morphism in B-Mod corresponding to σ_X. We then have that σ'_X is an iso in B-Mod. Lastly, these isos form the components of a natural isomorphism $\sigma' : (Q \otimes_A -) \xRightarrow{\sim} F$. *Claim Q.E.D.*

Next, there exists a quasi-inverse $G : B\text{-Mod} \to A\text{-Mod}$ to F with natural isomorphisms $\phi : \text{Id}_{A\text{-Mod}} \overset{\cong}{\Rightarrow} GF$ and $\psi : FG \overset{\cong}{\Rightarrow} \text{Id}_{B\text{-Mod}}$ by assumption. Denote $G(_B(B_{\text{reg}}))$ by P. By repeating the arguments above, we obtain that P is in (A, B)-Bimod, and that $G \cong (P \otimes_B -)$ as functors. Finally, one can check that $P \otimes_B Q \cong A_{\text{reg}}$ as A-bimodules, and $Q \otimes_A P \cong B_{\text{reg}}$ as B-bimodules, by using the isomorphisms $\phi_A : A \overset{\sim}{\to} GF(A)$ and $\psi_B : FG(B) \overset{\sim}{\to} B$. $\qquad\square$

Remark 2.19. The characterization of Morita equivalence above is conveniently framed in the category Bim mentioned in §2.1.2, where objects are algebras and morphisms are isoclasses of bimodules over algebras. In particular, isos in Bim are precisely invertible bimodules, and two algebras A and B are Morita equivalent if and only if $A \to B$ is an iso in Bim.

Remark 2.20. There is another characterization of Morita equivalence given in terms of endomorphism algebras. Take algebras A and B, and consider the following terminology.

- $M \in A\text{-Mod}$ is **projective** if the functor $\text{Hom}_{A\text{-Mod}}(M, -) : A\text{-Mod} \to \text{Set}$ sends an epi to an epi; see §2.8.3 later for details.

- $M \in A\text{-Mod}$ is a **generator** if $\text{Hom}_{A\text{-Mod}}(M, -) : A\text{-Mod} \to \text{Set}$ is faithful.

Then, $A \simeq_{\text{Mor}} B$ if and only if $B^{\text{op}} \cong \text{End}_{A\text{-Mod}}(M)$ as algebras for some finitely generated, projective generator M of A-Mod. We defer the proof to Exercise 2.36.

Now if A and B are isomorphic as algebras, then A and B are Morita equivalent. But the converse does not hold as we will see in the next example.

Example 2.21. An algebra A is Morita equivalent to $\text{Mat}_n(A)$, for any n in $\mathbb{N}_{\geq 1}$; see §1.2.1. To obtain this result, take the bimodules below:

$$P := \{(a_1, \ldots, a_n) \mid a_i \in A\} \cong A^{\oplus n} \quad \text{and} \quad Q := \{(a_1, \ldots, a_n)^T \mid a_i \in A\} \cong A^{\oplus n},$$

where T is transpose. Here, the A-action is given by scalar multiplication and $\text{Mat}_n(A)$-action given by matrix multiplication. Verifying this is Exercise 2.37.

Notice that the center of $\text{Mat}_n(A)$ is isomorphic to the center of A as algebras. In fact, we have the following results; the proofs of which comprise Exercise 2.38.

Proposition 2.22. *Take algebras A and B. If $A \simeq_{\text{Mor}} B$, then $Z(A) \cong Z(B)$ as algebras.* $\quad\square$

Corollary 2.23. *Take commutative algebras C and C'. If $C \simeq_{\text{Mor}} C'$, then $C \cong C'$ as algebras.* $\qquad\square$

So commutativity is not preserved under Morita equivalence, but many properties that are preserved; we call these **Morita invariant properties**. Examples of Morita invariant properties include simplicity, semisimplicity, separability, left/right Artinianity, and left/right Noetherianity. See Exercise 2.39.

§2.5. Adjunction

In this section, we weaken the requirement for two functors to yield an equivalence of categories to gain a rich theory of functors themselves. Arguably, the most interesting aspect of category theory is not the way that it frames structures, but rather the way that it frames how to move from one structure to another structure.

Recall from §2.4.2 that two categories C and D are equivalent if there exist functors $F : C \to D$ and $G : D \to C$ such that $\mathrm{Id}_C \cong GF$ and $FG \cong \mathrm{Id}_D$. We will loosen these natural isomorphisms to natural transformations, and require a useful compatibility condition between such natural transformations next.

§2.5.1. Characterization of adjunction

We say that the functors $F : C \to D$ and $G : D \to C$ form an **adjunction**, or are an **adjoint pair**, if there exist natural transformations

$$\eta : \mathrm{Id}_C \Rightarrow GF \qquad \text{and} \qquad \varepsilon : FG \Rightarrow \mathrm{Id}_D,$$

such that the **triangle identities** hold:

$$(\varepsilon * F) \circ^{\mathrm{ver}} (F * \eta) = \mathrm{ID}_F \qquad \text{and} \qquad (G * \varepsilon) \circ^{\mathrm{ver}} (\eta * G) = \mathrm{ID}_G.$$

These identities are also written as $\varepsilon F \circ F\eta = \mathrm{ID}$ and $G\varepsilon \circ \eta G = \mathrm{ID}$, respectively, for short. In particular, the triangle identities are the requirement that, for each $X \in C$ and $Y \in D$, the following two diagrams commute in D and C, respectively.

$$
\begin{array}{ccc}
F(X) \xrightarrow{\;F(\eta_X)\;} FGF(X) & \qquad & G(Y) \xrightarrow{\;\eta_{G(Y)}\;} GFG(Y) \\
\quad\searrow_{\,\mathrm{id}_{F(X)}} \quad \downarrow{\varepsilon_{F(X)}} & & \quad\searrow_{\,\mathrm{id}_{G(Y)}} \quad \downarrow{G(\varepsilon_Y)} \\
\qquad\qquad F(X) & & \qquad\qquad G(Y)
\end{array}
$$

Here, we are using the fact that the composition of functors is associative.

In the above, we refer to F as the **left adjoint** of G, and G as the **right adjoint** of F, and write

$$(F : C \to D) \dashv (G : D \to C).$$

Moreover, η is called the **unit** of the adjunction $F \dashv G$, and ε is the **counit** of $F \dashv G$. Pictorially, we illustrate adjunctions as follows.

$$
C \underset{G}{\overset{F}{\rightleftarrows}} {\perp}\; D
$$

Adjoints need not exist, but if they do, then they are unique up to natural isomorphism. See Sections 4.4 and 4.6 of Riehl [2016] for details.

Next, we consider a useful characterization for the existence of adjunctions.

Proposition 2.24. *Let $F : C \to D$ and $G : D \to C$ be functors between categories C and D. Then, we have an adjunction $F \dashv G$ if and only if for each pair of objects, $X \in C$ and $Y \in D$, there is a bijection of sets,*

$$\zeta := \zeta_{X,Y} : \operatorname{Hom}_D(F(X), Y) \xrightarrow{\sim} \operatorname{Hom}_C(X, G(Y)),$$

that is natural in X and Y.

Here, naturality means that we have the natural isomorphism below.

$$C^{\mathrm{op}} \times D \underset{\operatorname{Hom}_C(-, G(-))}{\overset{\operatorname{Hom}_D(F(-), -)}{\rightrightarrows}} \operatorname{Set} \qquad \sim \Downarrow \zeta$$

In other words, for a fixed object $X \in C$ with arbitrary morphism $g : Y' \to Y''$ in D, and for a fixed object $Y \in D$ with arbitrary morphism $f : X' \to X''$ in C, the diagrams below commute. See also §1.1.4v.

$$
\begin{array}{ccc}
\operatorname{Hom}_D(F(X), Y') & \xrightarrow{\operatorname{Hom}_D(F(X), g)} & \operatorname{Hom}_D(F(X), Y'') \\
{\scriptstyle \zeta_{X,Y'}} \downarrow & & \downarrow {\scriptstyle \zeta_{X,Y''}} \\
\operatorname{Hom}_C(X, G(Y')) & \xrightarrow{\operatorname{Hom}_C(X, G(g))} & \operatorname{Hom}_C(X, G(Y''))
\end{array}
$$

$$
\begin{array}{ccc}
\operatorname{Hom}_D(F(X''), Y) & \xrightarrow{\operatorname{Hom}_D(F(f), Y)} & \operatorname{Hom}_D(F(X'), Y) \\
{\scriptstyle \zeta_{X'',Y}} \downarrow & & \downarrow {\scriptstyle \zeta_{X',Y}} \\
\operatorname{Hom}_C(X'', G(Y)) & \xrightarrow{\operatorname{Hom}_C(f, G(Y))} & \operatorname{Hom}_C(X', G(Y))
\end{array}
$$

Proof of Proposition 2.24. We will sketch the argument below, leaving the details to the reader; see Exercise 2.40. Suppose that $F \dashv G$ via the natural transformations $\eta : \operatorname{Id}_C \Rightarrow GF$ and $\varepsilon : FG \Rightarrow \operatorname{Id}_D$ satisfying the triangle axioms. For a morphism $h : F(X) \to Y$ in D, define the morphism in C:

$$\zeta_{X,Y}(h) := G(h) \circ \eta_X : X \to G(Y).$$

Check that $\zeta_{X,Y}$ natural in X and Y. Moreover, for $\ell : X \to G(Y)$ in C, the inverse of $\zeta_{X,Y}$ is given by

$$\zeta_{X,Y}^{-1}(\ell) := \varepsilon_Y \circ F(\ell) : F(X) \to Y.$$

Now suppose that there exists a bijection $\zeta_{X,Y}$ as given. Define morphisms:

$$\eta_X := \zeta_{X,F(X)}(\operatorname{id}_{F(X)}) : X \to GF(X) \quad \text{and} \quad \varepsilon_Y := \zeta_{G(Y),Y}^{-1}(\operatorname{id}_{G(Y)}) : FG(Y) \to Y.$$

These morphisms yield natural transformations $\eta : \operatorname{Id}_C \Rightarrow GF$ and $\varepsilon : FG \Rightarrow \operatorname{Id}_D$ satisfying the triangle axioms. $\qquad \square$

Faithfulness and fullness of adjoint functors can be detected by studying the components of the adjunction unit and counit; see Exercise 2.41. Moreover, an adjunction, with unit and counit being natural isomorphisms, is precisely an equivalence of categories; we see this as follows.

Proposition 2.25. *For categories C and $\mathcal{D}$, the following statements hold.*

(a) *If $F : C \to \mathcal{D}$ has a left (or right) adjoint with the unit and counit being natural isomorphisms, then F is an equivalence of categories.*

(b) *If $F : C \to \mathcal{D}$ is an equivalence of categories, then there exists a left and right adjoint $G : \mathcal{D} \to C$ of F with unit and counit being natural isomorphisms.*

Proof. (a) This follows from the definition of an equivalence of categories.

(b) Suppose $F : C \to \mathcal{D}$ has quasi-inverse $G : \mathcal{D} \to C$, with given natural isomorphisms

$$\phi : \mathrm{Id}_C \overset{\cong}{\Rightarrow} GF \quad \text{and} \quad \psi : FG \overset{\cong}{\Rightarrow} \mathrm{Id}_{\mathcal{D}}.$$

By Proposition 2.24, we will display a bijection of sets,

$$\zeta_{X,Y} : \mathrm{Hom}_{\mathcal{D}}(F(X), Y) \overset{\sim}{\to} \mathrm{Hom}_C(X, G(Y)),$$

for each $X \in C$ and $Y \in \mathcal{D}$. To do so, consider the assignment

$$\mathrm{Hom}_{\mathcal{D}}(F(X), Y) \overset{\sim}{\to} \mathrm{Hom}_C(GF(X), G(Y)), \quad f \mapsto G(f),$$

which is a bijection because G is fully faithful. Then, precompose $G(f)$ with the iso $\phi_X : X \overset{\sim}{\to} GF(X)$ to get the bijection $\zeta_{X,Y}$. This yields the desired adjunction $F \dashv G$, and we get the other desired adjunction $G \dashv F$ by swapping the roles of F and G. Completing the details of this proof is part of Exercise 2.41. $\qquad\square$

§2.5.2. Properties preserved under adjunction

Next, we discuss how adjoint functors preserve useful constructions.

Proposition 2.26. *Take adjoint functors, $F \dashv G$, with adjunction unit $\eta : \mathrm{Id} \Rightarrow GF$ and adjunction counit $\varepsilon : FG \Rightarrow \mathrm{Id}$. Then, the following statements hold.*

(a) *F preserves pushouts, and G preserves pullbacks.*

(b) *F preserves coequalizers, and G preserves equalizers.*

(c) *F preserves cokernels, and G preserves kernels.*

Proof. We will establish the result that F preserves cokernels, and leave the remaining parts to the reader as an exercise; see Exercise 2.42.

Take a morphism $f : X \to Y$ with cokernel $\alpha_f : Y \to \mathrm{coker}(f)$.

We want to show that the cokernel of $F(f) : F(X) \to F(Y)$ is equal to the morphism $F(\alpha_f) : F(Y) \to F(\mathrm{coker}(f))$. Namely, we aim to establish that F sends the cokernel $\alpha_f : Y \to \mathrm{coker}(f)$ to the cokernel $\alpha_{F(f)} : F(Y) \to \mathrm{coker}(F(f))$.

Take a morphism $\beta_{F(f)} : F(Y) \to C$ with $\beta_{F(f)} F(f) = \vec{0}$. It suffices to construct a unique morphism $\gamma_{F(f)} : F(\mathrm{coker}(f)) \to C$ to make the diagram below commute.

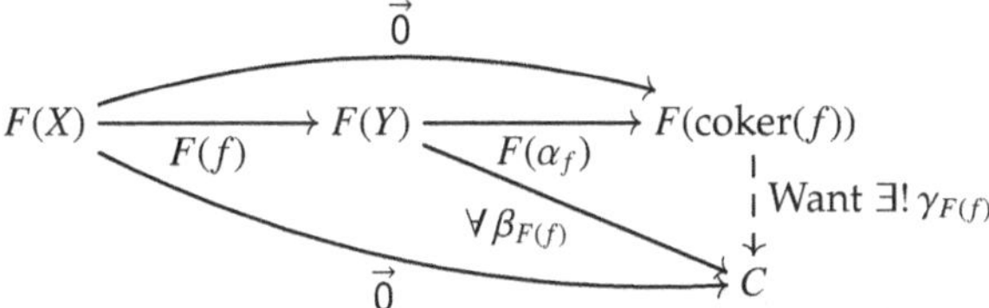

Apply G to the equation $\beta_{F(f)} F(f) = \vec{0}$ to yield commutative diagram below. Here, the square commutes by the naturality of η.

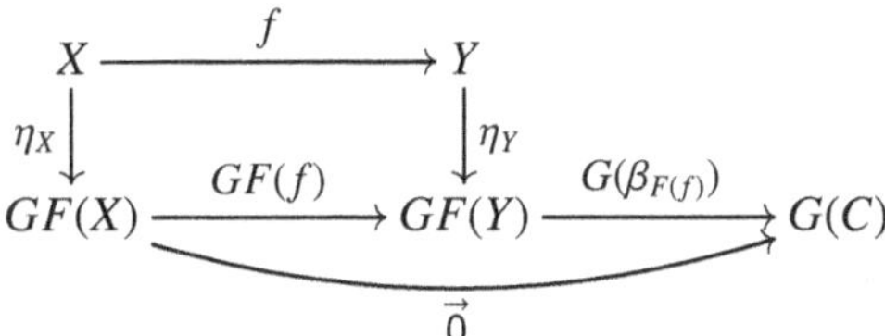

Next, we obtain a unique morphism $\widetilde{\gamma}_f : \mathrm{coker}(f) \to G(C)$ to make the diagram below commute, due to the universal property of $\mathrm{coker}(f)$.

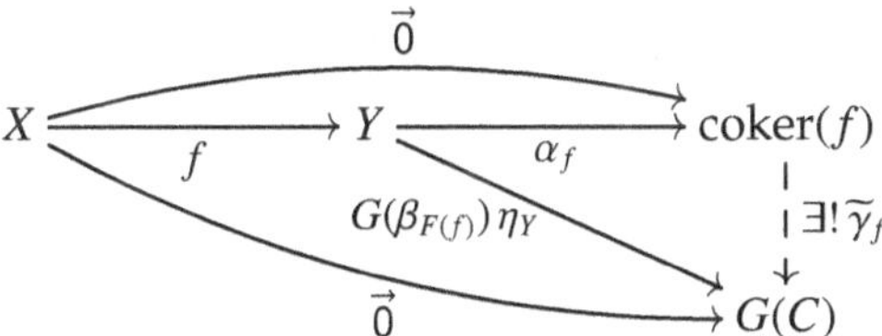

Now, using the bijection in Proposition 2.24, define:

$$\gamma_{F(f)} := \zeta^{-1}_{\mathrm{coker}(f),C}(\widetilde{\gamma}_f) : F(\mathrm{coker}(f)) \to C.$$

This morphism satisfies the desired property, $\beta_{F(f)} = \gamma_{F(f)} F(\alpha_f)$, due to the commutative diagram below.

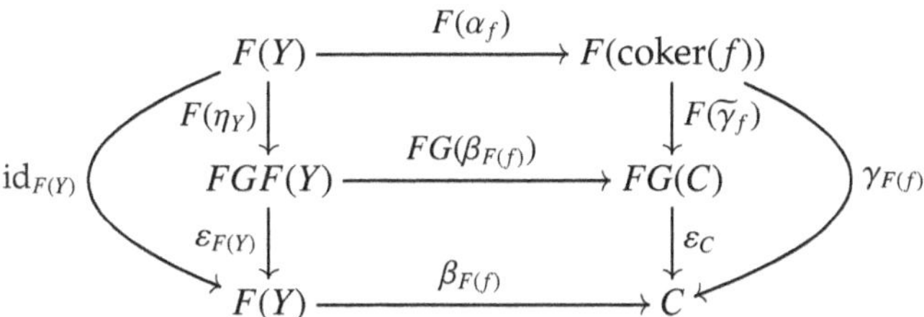

Here, the left region commutes by a triangle identity; the top region commutes by F applied to the previous diagram; the right region is the definition of $\gamma_{F(f)}$; and the bottom region commutes by the naturality of ε. $\qquad\square$

These are special cases of a powerful theorem given below on when *(co)limits* [§2.3.6] are preserved under functors. See Section 4.5 of Riehl [2016] for details.

Theorem 2.27. *Left adjoints preserve colimits, and right adjoints preserve limits.* $\qquad\square$

§2.5.3. Key examples of adjunction

We now discuss vital examples of adjunction.

Example 2.28 (Tensor-Hom adjunction). A crucial example of adjoint functors is given by the Tensor-Hom adjunction for vector spaces. Recall §1.1.4iv to get the adjunction diagrams below, for a fixed vector space V.

$$
\mathrm{Vec} \underset{\mathrm{Hom}_{\Bbbk}(V,-)}{\overset{-\otimes_{\Bbbk} V}{\rightleftarrows}} \mathrm{Vec} \quad\rightsquigarrow\quad \mathrm{Vec}\times\mathrm{Vec} \underset{\mathrm{Hom}_{\mathrm{Vec}}(-,\,\mathrm{Hom}_{\Bbbk}(V,-))}{\overset{\mathrm{Hom}_{\mathrm{Vec}}(-\otimes_{\Bbbk} V,\,-)}{\rightrightarrows}} \mathrm{Set}
$$

$$
\mathrm{Vec} \underset{\mathrm{Hom}_{\Bbbk}(V,-)}{\overset{V\otimes_{\Bbbk} -}{\rightleftarrows}} \mathrm{Vec} \quad\rightsquigarrow\quad \mathrm{Vec}\times\mathrm{Vec} \underset{\mathrm{Hom}_{\mathrm{Vec}}(-,\,\mathrm{Hom}_{\Bbbk}(V,-))}{\overset{\mathrm{Hom}_{\mathrm{Vec}}(V\otimes_{\Bbbk} -,\,-)}{\rightrightarrows}} \mathrm{Set}
$$

This can be upgraded to a Tensor-Hom adjunction for bimodules by way of Exercise 1.25; see Exercise 2.43.

Example 2.29 (Free-Forget adjunction). Now we construct a left adjoint to a forgetful functor $\mathrm{Forg} : \mathcal{D} \to \mathcal{C}$ (see §2.3.2).

A **free object** on $X \in \mathcal{C}$ consists of the following data:

(a) An object, Free_X, in $\mathcal{D}$,

(b) A mono, $\alpha_X : X \to \mathrm{Forg}(\mathrm{Free}_X)$, in $\mathcal{C}$,

such that, for every morphism $\beta_X : X \to \mathrm{Forg}(Z)$ in $\mathcal{C}$ with $Z \in \mathcal{D}$, there exists a unique morphism $\gamma_X : \mathrm{Free}_X \to Z$ in $\mathcal{D}$ with $\beta_X = \mathrm{Forg}(\gamma_X) \circ \alpha_X$ in $\mathcal{C}$. In other words, the diagram below commutes.

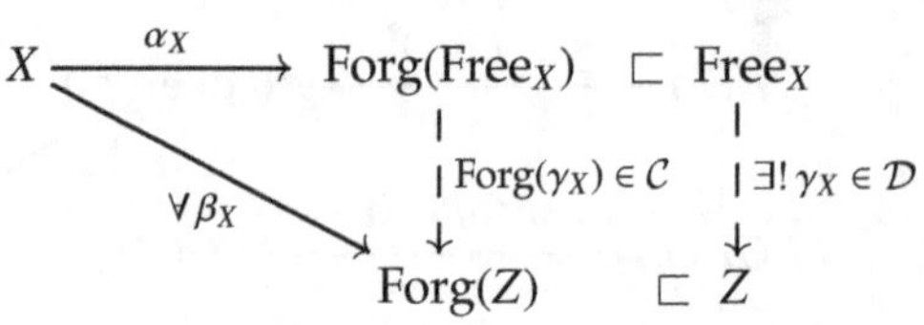

$$
\begin{array}{ccc}
X \xrightarrow{\ \alpha_X\ } & \mathrm{Forg}(\mathrm{Free}_X) & \sqsubset\ \mathrm{Free}_X \\
& \Big\downarrow{\scriptstyle \mathrm{Forg}(\gamma_X)\in\mathcal{C}} & \Big\downarrow{\scriptstyle \exists!\,\gamma_X\in\mathcal{D}} \\
{\scriptstyle \forall\beta_X}\searrow & \mathrm{Forg}(Z) & \sqsubset\ Z
\end{array}
$$

Here, $\sqsubset$ means 'forget structure', or apply the functor Forg.

We refer to $\mathrm{Free} : \mathcal{C} \to \mathcal{D}$, $X \mapsto \mathrm{Free}(X) := \mathrm{Free}_X$ as a **free functor** if this assignment does indeed yield a functor.

The bijection between the morphisms $\beta_X \in \mathcal{C}$ and $\gamma_X \in \mathcal{D}$ yields an adjunction:

$$(\mathrm{Free} : \mathcal{C} \to \mathcal{D}) \dashv (\mathrm{Forg} : \mathcal{D} \to \mathcal{C}).$$

Moreover, the morphisms α_X can be assembled into the unit $\mathrm{Id}_{\mathcal{C}} \Rightarrow \mathrm{Forg} \circ \mathrm{Free}$ of this adjunction; these components are the process of inserting generators.

Specific examples of Free-Forget adjunction include the following, details of which are left as Exercise 2.44.

- $\mathrm{Forg} : \mathrm{Group} \to \mathrm{Set}$ (forgets operation),
  $\mathrm{Free} : \mathrm{Set} \to \mathrm{Group}$ (free group).

- $\mathrm{Forg} : A\text{-}\mathrm{Mod} \to \mathrm{Vec}$ (forgets A-action),
  $\mathrm{Free} := A \otimes_{\Bbbk} - : \mathrm{Vec} \to A\text{-}\mathrm{Mod}$ (free A-module).
  For instance, $\mathrm{Free}(V) \cong A^{\oplus \dim_{\Bbbk} V}$.

- $\mathrm{Forg} : \mathrm{Vec} \to \mathrm{Set}$ (forgets addition and scalar multiplication),
  $\mathrm{Free} : \mathrm{Set} \to \mathrm{Vec}$, $X \mapsto \{f : X \to \Bbbk \mid f^{-1}(\Bbbk^{\times}) \text{ is finite}\}$.
  For instance, $\mathrm{Free}(\{1, \ldots, n\}) \cong \Bbbk^n$.

- $\mathrm{Forg} : \mathrm{Alg} \to \mathrm{Vec}$ (forgets multiplication),
  $\mathrm{Free} := T(-) : \mathrm{Vec} \to \mathrm{Alg}$ (tensor algebra).

Other interesting examples of adjunction appear in Exercises 2.45 and 2.46.

§2.6. Representable functors

In this section, we formalize the discussion of universal property from §1.1.4iii within the categorical setting introduced in this chapter. This leads to the notion of a *representable functor*, and a key result of Yoneda that provides a framework for studying such functors.

§2.6.1. Universal property revisited

Recall from §1.1.4iii that for a gadget X, we say that $\mathrm{Univ}(X)$ is a universal structure attached to X if it is equipped with a connection α_X to X, such that for all connections β_X between X and an arbitrary structure $\mathrm{Arb}(X)$, there is a unique structure-preserving map γ_X between $\mathrm{Univ}(X)$ and $\mathrm{Arb}(X)$ compatible with α_X and β_X.

This takes one of two forms.

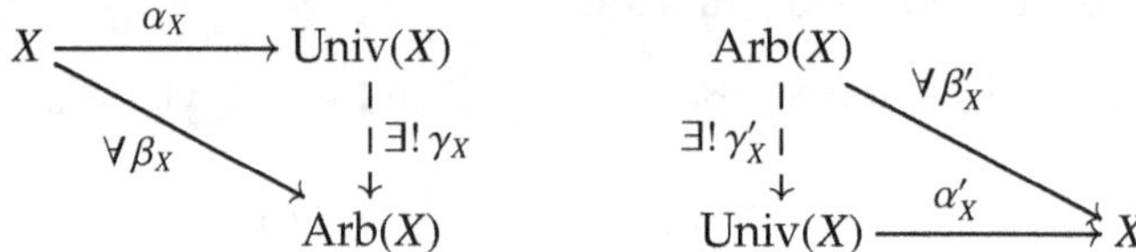

We will skip the discussion of the second type. But for the first type, universality yields the following bijection of sets:

$$\mathrm{Hom}_{\mathrm{structure}}(\mathrm{Univ}(X),\ \mathrm{Arb}(X)) \cong \mathrm{Hom}_{\mathrm{gadget}}(X,\ \mathrm{Arb}(X)_{\mathrm{gadget}}),\quad \gamma_X \leftrightarrow \beta_X. \quad (2.30)$$

In particular, X connects to an 'underlying gadget' $\mathrm{Arb}(X)_{\mathrm{gadget}}$ of $\mathrm{Arb}(X)$.

We have seen numerous examples of (2.30), such as the following.

- $\mathrm{Hom}_{\mathrm{Vec}}(V \otimes_{\Bbbk} W,\ Z) \cong \mathrm{Bilin}(V \times W,\ Z_{\mathrm{vs}})$ (see §1.1.4iii)

- $\mathrm{Hom}_{(B_1,B_2)\text{-Bimod}}(V \otimes_A W,\ Z) \cong A\text{-Balan}(V \times W,\ Z_{(B_1,B_2)\text{-bimod}})$ (see §1.4.2iii)

- $\mathrm{Hom}_{\mathrm{Alg}}(T(V),\ Z) \cong \mathrm{Hom}_{\mathrm{Vec}}(V,\ Z_{\mathrm{vs}})$ (see §1.2.2)

- $\mathrm{Hom}_{A\text{-BimodAlg}}(T_A(V),\ Z) \cong \mathrm{Hom}_{A\text{-Bimod}}(V,\ Z_{A\text{-bimod}})$ (see §1.4.2iv)

- $\mathrm{Hom}_{\mathrm{Alg}}(\Bbbk G,\ Z) \cong \mathrm{Hom}_{\mathrm{Group}}(G,\ Z^{\times})$ (see §1.2.6)

- $\mathrm{Hom}_{\mathrm{Group}}(\mathrm{Free}(X),\ Z) \cong \mathrm{Hom}_{\mathrm{Set}}(X,\ Z_{\mathrm{set}})$ (see Example 2.29)

- $\mathrm{Hom}_{A\text{-Mod}}(A \otimes_{\Bbbk} V,\ Z) \cong \mathrm{Hom}_{\mathrm{Vec}}(V,\ Z_{\mathrm{vs}})$ (see Example 2.29)

Let us formalize these examples given the tools in this chapter. We say that a covariant (resp., contravariant) functor $H : \mathcal{D} \to \mathrm{Set}$ is **representable** if

$$H(-) \cong \mathrm{Hom}_{\mathcal{D}}(U, -) \quad (\text{resp.,}\ H(-) \cong \mathrm{Hom}_{\mathcal{D}}(-, U))$$

for some object $U \in \mathcal{D}$. In this case, U is called the **universal object that represents the functor H**, or is referred to as a **universal representing object**. Universal (representing) objects are unique up to iso, as discussed in §2.6.2 below; cf. Exercise 1.2. We focus on the case when H is covariant below.

Consider the following examples.

- For vector spaces V and W, let $H := H_{V \times W} : \mathrm{Vec} \to \mathrm{Set}$ send a vector space Z to the set of bilinear maps $V \times W \to Z$. Then, $H_{V \times W}(-) = \mathrm{Bilin}(V \times W, -)$ is represented by the tensor product $V \otimes_{\Bbbk} W$ as the universal object. One needs to verify the naturality of the isomorphism between $H(-)$ and $\mathrm{Hom}_{\mathrm{Vec}}(V \otimes_{\Bbbk} W, -)$.

- For a vector space V, take $H := H_V : \mathrm{Alg} \to \mathrm{Set}$ to be the functor that sends an algebra Z to the set $\mathrm{Hom}_{\mathrm{Vec}}(V, Z_{\mathrm{vs}})$. Now $H_V(-) := \mathrm{Hom}_{\mathrm{Vec}}(V, (-)_{\mathrm{vs}})$ is represented by the tensor algebra $T(V)$ as the universal object.

- For a set X, take $H := H_X : \text{Group} \to \text{Set}$ to be the functor that sends a group Z to the set $\text{Hom}_{\text{Set}}(X, Z_{\text{set}})$. Here, $H_X(-) := \text{Hom}_{\text{Set}}(X, (-)_{\text{set}})$ is represented by the free group $\text{Free}(X)$ as the universal object.

In general, if $F : \mathcal{C} \to \mathcal{D}$ has a right adjoint $G : \mathcal{D} \to \mathcal{C}$, then for any object $X \in \mathcal{C}$, we get that

$$H_X(-) := \text{Hom}_{\mathcal{C}}(X, G(-)) : \mathcal{D} \to \text{Set}$$

is represented by the universal object $F(X)$ in $\mathcal{D}$. Indeed, $F \dashv G$ yields

$$\text{Hom}_{\mathcal{D}}(F(X), -) \cong \text{Hom}_{\mathcal{C}}(X, G(-))$$

by Proposition 2.24. The converse statement holds as well; see Exercise 2.47.

§2.6.2. Yoneda's Lemma

Now we introduce a result attributed to Yoneda [1954] (see also Section III.2 of MacLane [1971]) that provides a framework for studying representable functors.

Proposition 2.31 (Yoneda's Lemma). *Let* $F : \mathcal{D} \to \text{Set}$ *be a functor from a category* $\mathcal{D}$ *to* Set*. Then, for each object* $U \in \mathcal{D}$*, there is a bijection:*

$$\Phi_{F,U} : \text{Nat}_{\mathcal{D},\text{Set}}(\text{Hom}_{\mathcal{D}}(U, -), F) \xrightarrow{\sim} F(U).$$

Proof. We will sketch the proof of this result, and leave the details to the reader. Take a natural transformation ϕ in $\text{Nat}_{\mathcal{D},\text{Set}}(\text{Hom}_{\mathcal{D}}(U, -), F)$. For $Z \in \mathcal{D}$, the component of ϕ at Z is a set morphism: $\phi_Z : \text{Hom}_{\mathcal{D}}(U, Z) \to F(Z)$. Now define

$$\Phi_{F,U}(\phi) := \phi_U(\text{id}_U) \in F(U).$$

Next, let us define a set morphism $\Psi_{F,U} : F(U) \to \text{Nat}_{\mathcal{D},\text{Set}}(\text{Hom}_{\mathcal{D}}(U, -), F)$. Take an element $x \in F(U)$ and object $Z \in \mathcal{D}$, and define the set morphism:

$$(\Psi_{F,U}(x))_Z : \text{Hom}_{\mathcal{D}}(U, Z) \to F(Z), \quad f \mapsto F(f)(x).$$

These form the components of a natural transformation $\Psi_{F,U}(x) : \text{Hom}_{\mathcal{D}}(U, -) \Rightarrow F$. Therefore, $\Psi_{F,U}$ is a set morphism from $F(U)$ to $\text{Nat}_{\mathcal{D},\text{Set}}(\text{Hom}_{\mathcal{D}}(U, -), F)$. Lastly, the set morphisms $\Phi_{F,U}$ and $\Psi_{F,U}$ are mutually inverse. $\square$

We refer the reader to Section 2.2 of Richter [2020] and Section 2.2 of Riehl [2016] for more discussion. For instance, the result above yields a natural transformation

$$\Phi : \text{Nat}_{\mathcal{D},\text{Set}}(\text{Hom}_{\mathcal{D}}(*, -), F) \xRightarrow{\sim} F(*)$$

with the components $\Phi_{*=U} := \Phi_{F,U}$.

Here are some interesting consequences of Yoneda's Lemma; the proof of which we leave as an exercise; see Exercise 2.48.

Corollary 2.32. *Take a category $\mathcal{D}$, with $U, U' \in \mathcal{D}$. Then, the following statements hold.*

(a) *There is a bijection:*

$$\Phi_{U',U} : \mathrm{Nat}_{\mathcal{D},\mathsf{Set}}(\mathrm{Hom}_{\mathcal{D}}(U, -), \mathrm{Hom}_{\mathcal{D}}(U', -)) \xrightarrow{\sim} \mathrm{Hom}_{\mathcal{D}}(U', U).$$

(b) *Suppose that a functor $F : \mathcal{D} \to \mathsf{Set}$ is represented by universal objects U and U' in $\mathcal{D}$, that is, $\mathrm{Hom}_{\mathcal{D}}(U, -) \cong F \cong \mathrm{Hom}_{\mathcal{D}}(U', -)$. Then, $U \cong U'$ in $\mathcal{D}$.*

(c) **(Yoneda embedding)** *The functor below is fully faithful:*

$$\begin{aligned}
\Psi : \mathcal{D}^{\mathrm{op}} &\to \mathsf{Fun}(\mathcal{D}, \mathsf{Set}) \\
U &\mapsto \mathrm{Hom}_{\mathcal{D}}(U, -) \\
(g : U \to U') &\mapsto (\, \mathrm{Hom}_{\mathcal{D}}(U', -) \to \mathrm{Hom}_{\mathcal{D}}(U, -),\ \textit{precompose with } g\,).
\end{aligned}$$

(d) *We have that $\mathrm{Hom}_{\mathcal{D}}(U, -) \cong \mathrm{Hom}_{\mathcal{D}}(U', -)$ if and only if $U \cong U'$ in $\mathcal{D}$.* $\qquad\square$

The last consequence above is the most used in practice, as it says that representable functors can be understood precisely in terms of their universal representing objects. A contravariant version of this result is given as follows.

Lemma 2.33. *Take a category $\mathcal{D}$, with $U, U' \in \mathcal{D}$. Then, $\mathrm{Hom}_{\mathcal{D}}(-, U) \cong \mathrm{Hom}_{\mathcal{D}}(-, U')$ if and only if $U \cong U'$ in $\mathcal{D}$.* $\qquad\square$

One needs to dualize Proposition 2.31 and Corollary 2.32 to verify this result.

§2.7. Simplicity and semisimplicity

Now we focus on the 'building blocks' for objects in categories $\mathcal{C}$, namely we discuss indecomposability and simplicity in the categorical setting. (Compare this to §§1.4.1, 1.5 on these notions for algebras over a field and their modules.) This will also lead to the definition of a semisimple category.

Standing hypothesis. We assume here that $\mathcal{C}$ is abelian; see §2.2.2.

---

§2.7.1. Indecomposable and simple objects

We call a nonzero object $X \in \mathcal{C}$ **decomposable** if $X \cong X_1 \sqcup X_2$ for some nonzero subobjects X_1 and X_2 of X. Else, we say that X is **indecomposable**.

Similar to Proposition 1.22, we have the following characterization of indecomposable objects; see Exercise 2.49(a).

Proposition 2.34. *A nonzero object $X \in \mathcal{C}$ is indecomposable if and only if the only idempotent morphisms in $\mathrm{Hom}_{\mathcal{C}}(X, X)$ are $\vec{0}_{X,X}$ and id_X.* $\qquad\square$

A nonzero object $X \in \mathcal{C}$ is **simple** if the only nonzero subobject of X is X itself.

Using Remark 1.27, a simple object must be indecomposable, but the converse need not hold.

We also have a version of Schur's Lemma in this setting presented as follows; see Exercise 2.49(b). Compare to Proposition 1.28.

Proposition 2.35 (Schur's Lemma). *Let X and Y be simple objects in $\mathcal{C}$, and take $f \in \mathrm{Hom}_{\mathcal{C}}(X, Y)$. Then, either $f = \vec{0}_{X,Y}$, or f is an iso.* $\qquad\square$

Now the consequence below is straightforward; see Exercise 2.49(c).

Corollary 2.36. *If $\mathcal{C}$ is $\Bbbk$-linear and X is a simple object in $\mathcal{C}$, then we have that $\mathrm{Hom}_{\mathcal{C}}(X, X)$ is a division algebra over $\Bbbk$.* $\qquad\square$

§2.7.2. Finite length objects

Now we discuss how to measure how far an object $X \in \mathcal{C}$ is from being simple.

A **composition series** for an object X in $\mathcal{C}$ is a sequence of monos,

$$0 = X_0 \xrightarrow{f_0} X_1 \xrightarrow{f_1} X_2 \longrightarrow \cdots \xrightarrow{f_{n-1}} X_n \xrightarrow{f_n} \cdots \longrightarrow X,$$

such that each **composition factor** $X_{i+1}/X_i := \mathrm{coker}(f_i)$ is a simple object in $\mathcal{C}$.

If $X = X_n$ above for some n, then we say that X has **finite length**, and we refer to the minimum such n as the **length** of X.

Length is well-defined by a version of the Jordan-Hölder theorem for abelian categories, similar to the result for modules [Theorem 1.29]; see the examples after Proposition IV.5.3, along with §III.2-3, in Stenström [1975] for more details. In any case, an object is simple precisely when its length is 1.

Theorem 2.37 (Jordan-Hölder Theorem). *Let X be an object of $\mathcal{C}$ of finite length. If X has two composition series*

$$0 = X_0 \to X_1 \to \cdots \to X_n = X \quad \text{and} \quad 0 = Y_0 \to Y_1 \to \cdots \to Y_m = X,$$

then $n = m$ and there exists a permutation σ of $\{1, \ldots, n\}$ such that $X_{\sigma(i)+1}/X_{\sigma(i)} \cong Y_{i+1}/Y_i$ as objects, for all i. $\qquad\square$

Given an object X of finite length, the number of times a simple object Y is isomorphic to a composition factor of X is called the **multiplicity** of Y in X, denoted by $[X : Y]$. If Y is a simple object in $\mathcal{C}$ and $\mathcal{C}$ is $\Bbbk$-linear, then

$$[X : Y] = \dim_{\Bbbk} \mathrm{Hom}_{\mathcal{C}}(X, Y). \tag{2.38}$$

We also have a decomposition result for objects of finite length, using indecomposability. The proof holds similarly to that for Theorem 1.31 for modules over a $\Bbbk$-algebra.

Theorem 2.39 (Krull-Schmidt Theorem). *Take $X \in C$ of finite length. Then, up to iso, $X \cong X_1 \sqcup \cdots \sqcup X_n$, for a unique choice of indecomposable subobjects X_i of X.* □

§2.7.3. Semisimple objects and semisimple categories

Now we consider the decomposability of objects in terms of simple objects.

We say that a nonzero object $X \in C$ is **semisimple** if $X \cong \coprod_{i \in I} X_i$, for some simple objects X_i in C. Else, we say that X is **nonsemisimple**.

A category C is said to be **semisimple** if each of its objects is semisimple.

Note that we are only guaranteed to have that a coproduct of finitely many objects exists in an abelian category; infinite coproducts may not exist. So, for the notion above, one may need to impose that a semisimple object has a finiteness condition (e.g., finite length) to use decompositions in terms of a coproduct of simple objects.

Recall that simple objects are indecomposable; we will next discuss the converse in semisimple categories (cf. Proposition 1.48).

Given a subobject Y of an object X in C, we say that a subobject Y' of X is a **complement** to Y if $Y \sqcup Y' \cong X$.

The result below follows in a similar manner to the proof for Lemma 1.30; see Section V.6 of Stenström [1975] for a discussion in a more general categorical setting.

Proposition 2.40. *A subobject of a semisimple object of finite length in C has a complement in C. Thus, indecomposable objects of finite length in a semisimple category are simple objects.* □

Moreover, we have the useful result below.

Lemma 2.41. *If $X \cong \coprod_{i \in I} X_i$ and $Y \cong \coprod_{j \in J} Y_j$ are semisimple objects in C of finite length, then there exists a bijection: $\mathrm{Hom}_C(X, Y) \cong \mathrm{Hom}_C(Y, X)$.* □

Proof. By Schur's Lemma [Proposition 2.35], a morphism $\phi_{i,j}$ between simple objects X_i and Y_j is an iso or a zero morphism. So,

$$\mathrm{Hom}_C(X_i, Y_j) \cong \mathrm{Hom}_C(Y_j, X_i), \tag{2.42}$$

where $\phi_{i,j} : X_i \to Y_j$ gets sent to its inverse if $\phi_{i,j}$ is an iso, or to $\vec{0}_{Y_j, X_i}$ if $\phi_{i,j} = \vec{0}_{X_i, Y_j}$.

Next, since X and Y have finite length, we can take the index sets I and J to be finite. Therefore, $X \cong \coprod_{i \in I} X_i$ and $Y \cong \prod_{j \in J} Y_j$. So, we get that

$$
\begin{aligned}
\mathrm{Hom}_C(X, Y) \quad &\cong \mathrm{Hom}_C(\textstyle\coprod_{i \in I} X_i, \prod_{j \in J} Y_j) \quad \cong \prod_{i \in I, j \in J} \mathrm{Hom}_C(X_i, Y_j) \\
&\cong \prod_{i \in I, j \in J} \mathrm{Hom}_C(Y_j, X_i) \quad \cong \mathrm{Hom}_C(Y, X).
\end{aligned}
$$

Here, the second and last isomorphisms hold by Exercise 2.8, and the third isomorphism holds by (2.42). $\qquad\square$

Finally, we discuss examples of semisimple abelian categories; see also Exercise 2.50.

- A-Mod is semisimple if and only if the algebra A is semisimple (see §1.6). This is due to Proposition 1.48.

- Vec $\cong$ k-Mod is semisimple because k is a semisimple algebra.

- (A, B)-Bimod $\cong (A \otimes B^{\mathrm{op}})$-Mod is semisimple when A and B are separable algebras, due to Exercise 1.34(b) and Proposition 1.57.

- G-Mod $\cong$ kG-Mod is semisimple if and only if G is finite; see Theorem 1.47 and Exercise 2.30.

- Ab is an abelian category, but it is not semisimple. The best way to see this is to employ the homological tools in the next section (see Example 2.47 later).

§2.8. Snippet of Homological Algebra

Next, we discuss how close a category C is to being semisimple. This, in turn, leads us to the field, Homological Algebra, which examines how sequences of morphisms behave under functors. We will only cover a snippet of this field here to illustrate that semisimplicity is a strong hypothesis to put on a category, and to display some of the tools that are used when semisimplicity is not available.

Standing hypothesis. We assume here that C is abelian; see §2.2.2.

---

§2.8.1. Exactness and splitting

We refer the reader to Section 2.1 in Cohn [2003], and to Sections I.15 and I.19 in Mitchell [1965], for details about the material below. The reader may also wish to view Section 2.1 of Rotman [2009] when C is a category of modules. Understanding the proofs of some of the results below is Exercise 2.51.

A sequence of morphisms in $\mathcal{C}$,

$$\cdots \longrightarrow X_{i-1} \xrightarrow{f_{i-1}} X_i \xrightarrow{f_i} X_{i+1} \longrightarrow \cdots,$$

is **exact at** X_i, for some i, if $\ker(f_i) = \mathrm{im}(f_{i-1})$. If the sequence is exact at all X_i, then the sequence is said to be **exact**.

Lemma 2.43. *We have the following facts about exact sequences.*

(a) *The sequence* $0 \xrightarrow{\vec{0}_{X'}} X' \xrightarrow{f} X$ *is exact if and only if f is monic.*

(b) *The sequence* $X \xrightarrow{g} X'' \xrightarrow{\vec{X''0}} 0$ *is exact if and only if g is epic.*

(c) *The sequence* $0 \xrightarrow{\vec{0}_X} X \xrightarrow{h} Y \xrightarrow{\vec{Y0}} 0$ *is exact if and only if h is an iso.* $\qquad\square$

Often the zero morphisms are omitted from the notation above.

A sequence of morphisms in $\mathcal{C}$ of particular importance is of the form,

$$0 \longrightarrow X' \xrightarrow{f} X \xrightarrow{g} X'' \longrightarrow 0, \tag{2.44}$$

called a **short exact sequence**. Here, f is monic, g is epic, and $\ker(g) = \mathrm{im}(f)$. Also, X' is a subobject of X, and we have that $X'' \cong X/X'$ as a quotient object of X.

In fact, we can study any exact sequence in terms of short exact sequences via a method called **splicing**, as illustrated below. Here, $K_i := \ker(f_i) = \mathrm{im}(f_{i-1})$.

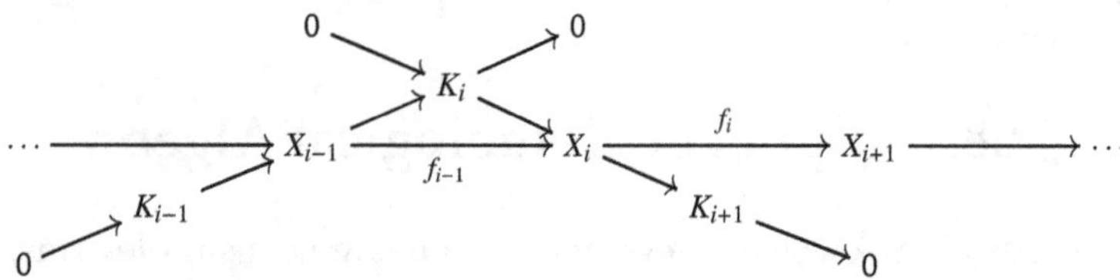

Next, we consider some convenient conditions for short exact sequences.

Proposition 2.45. *The following statements are equivalent for the sequence (2.44):*

(a) *There exists a morphism $s : X'' \to X$ in $\mathcal{C}$ such that $gs = \mathrm{id}_{X''}$;*

(b) *There exists a morphism $r : X \to X'$ in $\mathcal{C}$ such that $rf = \mathrm{id}_{X'}$.*

In either case, we obtain that $X \cong X' \,\square\, X''$. $\qquad\square$

Above, the morphism s is called a **section** of g, and r is called a **retraction** of f. When either part (a) or (b) hold, we say that the sequence (2.44) is **split**.

Now we have that semisimplicity implies the split condition by the following result. (This result was previously considered "folklore". But all results need a careful proof, folklore or not, and it is good that we have one available here.)

Proposition 2.46 (Lemma 2.1 of Positselski and Šťovíček [2022])**.** *If C is a semisimple category, then every short exact sequence in C splits.* □

See also Remarks 2.2 and 2.4 of Positselski and Šťovíček [2022] for a discussion about the converse of this result failing, if you are interested. In any case, the result above helps rule out the semisimple condition for many categories.

Example 2.47. Recall Remark 2.17. In $\mathrm{Ab} \simeq \mathbb{Z}\text{-}\mathrm{Mod}_{ab}$, take the short exact sequence

$$0 \to \mathbb{Z} \xrightarrow{f} \mathbb{Z} \to \mathbb{Z}/2\mathbb{Z} \to 0.$$

Here, the morphism f is given by multiplication by 2. Suppose that $r : \mathbb{Z} \to \mathbb{Z}$ is a group homomorphism with $rf = \mathrm{id}_\mathbb{Z}$. Then, $rf(1) = r(2) = 2n = 1$ for some $n \in \mathbb{Z}$, which yields a contradiction. Thus, our given short exact sequence is not split, and the category Ab cannot be semisimple.

We are also interested in how short exact sequences behave under functors. Given a covariant (resp., contravariant) functor $F : C \to D$, we say that:

- F is **left exact** if for every short exact sequence (2.44) in C, we have that the following sequence is exact in D,

$$0 \to F(X') \xrightarrow{F(f)} F(X) \xrightarrow{F(g)} F(X'') \qquad (\text{resp.}, 0 \to F(X'') \xrightarrow{F(g)} F(X) \xrightarrow{F(f)} F(X'));$$

- F is **right exact** if for every short exact sequence (2.44) in C, we have that the following sequence is exact in D,

$$F(X') \xrightarrow{F(f)} F(X) \xrightarrow{F(g)} F(X'') \to 0 \qquad (\text{resp.}, F(X'') \xrightarrow{F(g)} F(X) \xrightarrow{F(f)} F(X') \to 0);$$

- F is **exact** if it is left and right exact.

Next, we recall some useful facts about commutative diagrams in abelian categories with exact rows. See Section VIII.4 of MacLane [1971], which explicitly covers part (a) of the result below; we leave parts (b) and (c) to the reader.

Lemma 2.48. *Consider the statements below.*

(a) (**Short-Five Lemma**) *Take the commutative diagram below in C with exact rows:*

$$
\begin{array}{ccccccccc}
0 & \longrightarrow & X' & \longrightarrow & X & \longrightarrow & X'' & \longrightarrow & 0 \\
& & \downarrow{\scriptstyle h'} & & \downarrow{\scriptstyle h} & & \downarrow{\scriptstyle h''} & & \\
0 & \longrightarrow & Y' & \longrightarrow & Y & \longrightarrow & Y'' & \longrightarrow & 0.
\end{array}
$$

If h' and h'' are monic (resp., epic, isos), then so is h.

(b) (**Four Lemma I**) *Take the commutative diagram below in C with exact rows:*

$$
\begin{array}{ccccccc}
X' & \longrightarrow & X & \longrightarrow & X'' & \longrightarrow & W \\
\downarrow{\scriptstyle h'} & & \downarrow{\scriptstyle h} & & \downarrow{\scriptstyle h''} & & \downarrow{\scriptstyle \ell} \\
Y' & \longrightarrow & Y & \longrightarrow & Y'' & \longrightarrow & Z.
\end{array}
$$

If h' and h'' are epic and ℓ is monic, then h is epic.

(c) (**Four Lemma II**) *Take the commutative diagram below in C with exact rows:*

$$
\begin{array}{ccccccc}
W & \longrightarrow & X' & \longrightarrow & X & \longrightarrow & X'' \\
\downarrow{\scriptstyle \ell} & & \downarrow{\scriptstyle h'} & & \downarrow{\scriptstyle h} & & \downarrow{\scriptstyle h''} \\
Z & \longrightarrow & Y' & \longrightarrow & Y & \longrightarrow & Y''.
\end{array}
$$

If h' and h'' are monic and ℓ is epic, then h is monic. $\qquad\square$

We will study exactness for tensor and Hom functors in §2.8.2 below. Before this, here are some useful facts.

Proposition 2.49. *The following statements hold for an additive functor $F : C \to D$.*

(a) *F is left (resp., right) exact if and only if it preserves kernels (resp., cokernels).*

(b) *F is left (resp., right) exact when F has a left (resp., right) adjoint $G : D \to C$.*

(c) *F preserves split, short exact sequences.*

Proof. For part (a), suppose that F is left exact. Take a morphism $f : X \to Y$, and factor it as

$$
f : X \xrightarrow{\tilde{\alpha}} \operatorname{im}(f) \xrightarrow{\alpha} Y,
$$

where α is monic and $\tilde{\alpha}$ is epic [Exercise 2.14]. This yields short exact sequences:

$$
0 \to \ker(\tilde{\alpha}) \longrightarrow X \xrightarrow{\tilde{\alpha}} \operatorname{im}(f) \longrightarrow 0, \qquad 0 \to \operatorname{im}(f) \xrightarrow{\alpha} Y \longrightarrow \operatorname{coker}(\alpha) \longrightarrow 0.
$$

Since F is left exact, we obtain the exact sequences below:

$$
0 \to F(\ker(\tilde{\alpha})) \longrightarrow F(X) \xrightarrow{F(\tilde{\alpha})} F(\operatorname{im}(f)),
$$

$$
0 \to F(\operatorname{im}(f)) \xrightarrow{F(\alpha)} F(Y) \longrightarrow F(\operatorname{coker}(f)).
$$

This implies that $F(\alpha)$ is monic, and that $\ker(F(\tilde{\alpha})) \cong F(\ker(\tilde{\alpha}))$ [Lemma 2.43(a)]. Now using the fact that $\ker(gg') = \ker(g')$ if g is monic, we obtain:

$$
F(\ker(f)) = F(\ker(\alpha\tilde{\alpha})) = F(\ker(\tilde{\alpha})) \cong \ker(F(\tilde{\alpha})) = \ker(F(\alpha)F(\tilde{\alpha})) = \ker(F(f)).
$$

Therefore, F preserves kernels.

Now suppose that F preserves kernels, and take, $0 \to X' \xrightarrow{f} X \xrightarrow{g} X'' \to 0$, a short exact sequence. Then, $\ker(F(f)) \cong F(\ker(f)) \cong F(0) \cong 0$; so $F(f)$ is monic by Lemma 2.43(a). Moreover,

$$\ker(F(g)) \cong F(\ker(g)) \cong F(\operatorname{im}(f)) \cong F(X') \cong \operatorname{im}(F(f)).$$

Thus, the sequence of morphisms, $0 \to F(X') \xrightarrow{F(f)} F(X) \xrightarrow{F(g)} F(X'')$, is exact. Thus, we obtain that F is left exact. Moreover, one can work in the opposite category to obtain the dual statement for the rest of part (a).

Part (b) follows from part (a) and Proposition 2.26(c). Moreover, part (c) follows from Lemma 2.6. $\qquad\square$

Example 2.50. Let A and B be $\Bbbk$-algebras, and take a bimodule $Q \in (B, A)$-Bimod. See also §§1.4.2iii, 1.4.3i. Then, by Exercise 2.43, we have that

$$(Q \otimes_A - : A\text{-Mod} \to B\text{-Mod}) \dashv (\operatorname{Hom}_{B\text{-Mod}}(Q, -) : B\text{-Mod} \to A\text{-Mod}).$$

So, by Proposition 2.49(b), $(Q \otimes_A -)$ is right exact, and $\operatorname{Hom}_{B\text{-Mod}}(Q, -)$ is left exact.

§2.8.2. Eilenberg-Watts Theorem

In the more down-to-earth setting for categories of finite-dimensional modules over finite-dimensional $\Bbbk$-algebras, we have the result on exact functors below due to Eilenberg [1960] and Watts [1960].

Theorem 2.51 (Eilenberg-Watts Theorem). *Let A, B be finite-dimensional $\Bbbk$-algebras, and let $F : A\text{-FdMod} \to B\text{-FdMod}$ be a $\Bbbk$-linear functor. Then, for the statements below:*

(a)	*F is left exact;*	(a')	*F is right exact;*
(b)	*F has a left adjoint;*	(b')	*F has a right adjoint;*
(c)	*$F \cong \operatorname{Hom}_{A\text{-FdMod}}(P, -)$,*	(c')	*$F \cong Q \otimes_A -$,*
	for some $P \in (A, B)$-FdBimod;		*for some $Q \in (B, A)$-FdBimod;*

we have that (a) $\Leftrightarrow$ (b) $\Leftrightarrow$ (c), and that (a') $\Leftrightarrow$ (b') $\Leftrightarrow$ (c').

Proof. Proposition 2.49(b) implies that (b) $\Rightarrow$ (a) and (b') $\Rightarrow$ (a'). Also, (c) $\Rightarrow$ (b) and (c') $\Rightarrow$ (b') hold by Exercise 2.43.

For (a') $\Rightarrow$ (c'), take $V \in A\text{-FdMod}$, and denote $Q := F({}_A(A_{\mathrm{reg}}))$. Then, we obtain that $Q \in (B, A)$-FdBimod (see the proof of Theorem 2.18). Then,

$$\phi_V := F_{A_{\mathrm{reg}}, V} : V \cong \operatorname{Hom}_{A\text{-FdMod}}({}_A(A_{\mathrm{reg}}), V) \longrightarrow \operatorname{Hom}_{B\text{-FdMod}}(Q, F(V)).$$

Now $\phi_V \in \operatorname{Hom}_{A\text{-FdMod}}(V, \operatorname{Hom}_{B\text{-FdMod}}(Q, F(V))) \cong \operatorname{Hom}_{B\text{-FdMod}}(Q \otimes_A V, F(V))$ by Exercise 2.43, and this is natural in V. So, we get a natural transformation:

$$A\text{-FdMod} \underset{F}{\overset{Q \otimes_A -}{\rightrightarrows}} {\Downarrow \phi}\; B\text{-FdMod}.$$

Since $V \in A\text{-FdMod}$, there exists an epimorphism $g : A^{\oplus n} \to V$ for a positive integer n, such that the following sequence is exact in $A\text{-FdMod}$:

$$0 \longrightarrow \ker(g) \xrightarrow{f} A^{\oplus n} \xrightarrow{g} V \xrightarrow{h} 0.$$

Now by the right exactness of F and $(Q \otimes_A -)$ (see Example 2.50), and by the fact that F and $(Q \otimes_A -)$ both commute with finite direct sums [Lemma 2.6], we get the commutative diagram below in $B\text{-FdMod}$.

$$
\begin{array}{ccccccc}
Q \otimes_A \ker(g) & \xrightarrow{\ \mathrm{id}_Q \otimes_A f\ } & Q \otimes_A (A^{\oplus n}) & \xrightarrow{\ \mathrm{id}_Q \otimes_A g\ } & Q \otimes_A V & \xrightarrow{\ \mathrm{id}_Q \otimes_A h\ } & 0 \\
\phi_{\ker(g)} \downarrow & & \phi_{A^{\oplus n}} \downarrow \cong & & \phi_V \downarrow & & \downarrow \phi_0 \\
F(\ker(g)) & \xrightarrow{\ F(f)\ } & F(A^{\oplus n}) & \xrightarrow{\ F(g)\ } & F(V) & \xrightarrow{\ F(h)\ } & 0
\end{array}
$$

Since g is epic and F is right exact, $F(g)$ is epic; namely, note that epis are cokernels and use Proposition 2.49(a). So, $\phi_V (\mathrm{id}_Q \otimes_A g) = F(g)\phi_{A^{\oplus n}}$ is epic. Thus, ϕ_V is epic. Recall that V is arbitrary, so by taking $V = \ker(g)$, we get that $\phi_{\ker(g)}$ is epic. Also, $\phi_0 = \vec{0}_{0,0} = \mathrm{id}_0$, an iso. We then get that ϕ_V is monic by a Four Lemma [Lemma 2.48(c)]. Now ϕ_V is an iso by Proposition 2.4. Therefore, (a') $\Rightarrow$ (c').

A similar argument establishes that (a) $\Rightarrow$ (c). $\qquad\square$

§2.8.3. Projective and injective objects

Now we summarize how Hom functors preserve short exact sequences. Details are in Section 2.2 of Cohn [2003], in Sections I.14, V.7, VII.6 of Mitchell [1965], in Section 3.2 of Popescu [1973], and in Sections 3.1, 3.2 of Rotman [2009] when C is a category of modules. We encourage the reader to explore these references. The details of two of the next results are left to Exercise 2.52.

Proposition 2.52. *Given objects P, Q in C, the following statements hold.*

(a) *The functor $\mathrm{Hom}_C(P, -) : C \to \mathrm{Ab}$ is covariant and left exact.*

(b) *The functor $\mathrm{Hom}_C(-, Q) : C \to \mathrm{Ab}$ is contravariant and left exact.* $\qquad\square$

Next, we have precise conditions to obtain when $\mathrm{Hom}_C(P, -)$ is exact.

Proposition 2.53. *The following statements are equivalent for an object $P \in C$:*

(a) *The functor $\mathrm{Hom}_C(P, -) : C \to \mathrm{Ab}$ is right exact;*

(b) *Every short exact sequence of the form, $0 \to X' \to X \to P \to 0$, in C is split;*

(c) *For every epi $p : Y \to Z$ in C, and every morphism $f : P \to Z$ in C, there exists a morphism $\tilde{f} : P \to Y$ in C such that $f = p\tilde{f}$.*

Proof. For (a) $\Rightarrow$ (b), take a short exact sequence $0 \to X' \to X \xrightarrow{g} P \to 0$. By part (a) and Proposition 2.52(a), we obtain a short exact sequence

$$0 \longrightarrow \operatorname{Hom}_{\mathcal{C}}(P, X') \longrightarrow \operatorname{Hom}_{\mathcal{C}}(P, X) \xrightarrow{\operatorname{Hom}_{\mathcal{C}}(P,g)} \operatorname{Hom}_{\mathcal{C}}(P, P) \longrightarrow 0.$$

Since $\operatorname{Hom}_{\mathcal{C}}(P, g)$ is an epic morphism [Lemma 2.43(b)], it is a surjection in Ab [Exercise 2.1(b)]. Therefore, for $\mathrm{id}_P \in \operatorname{Hom}_{\mathcal{C}}(P, P)$, there exists $h \in \operatorname{Hom}_{\mathcal{C}}(P, X)$ such that $\operatorname{Hom}_{\mathcal{C}}(P, g)(h) = gh = \mathrm{id}_P$. So, $0 \to X' \to X \xrightarrow{g} P \to 0$ splits and part (b) holds.

For (b) $\Rightarrow$ (c), consider the set-up of part (c) and take the short exact sequence:

$$0 \longrightarrow \ker(p) \longrightarrow Y \xrightarrow{p} Z \longrightarrow 0.$$

This sequence splits by part (b), so there is a morphism $s : Z \to Y$ such that $ps = \mathrm{id}_Z$. Now take $\tilde{f} := sf : P \to Y$. Then, we get that $p\tilde{f} = psf = f$, as desired.

For (c) $\Rightarrow$ (a), it suffices to show that $\operatorname{Hom}_{\mathcal{C}}(P, -)$ preserves cokernels (or epis by normality) by Proposition 2.49(a). Take an epi $p : Y \to Z$, and consider the morphism

$$\operatorname{Hom}_{\mathcal{C}}(P, p) : \operatorname{Hom}_{\mathcal{C}}(P, Y) \to \operatorname{Hom}_{\mathcal{C}}(P, Z) \in \mathsf{Ab}.$$

Then, for $f \in \operatorname{Hom}_{\mathcal{C}}(P, Z)$, there exists a morphism $\tilde{f} \in \operatorname{Hom}_{\mathcal{C}}(P, Y)$ such that $\operatorname{Hom}_{\mathcal{C}}(P, p)(\tilde{f}) = f$ by part (c). Therefore, $\operatorname{Hom}_{\mathcal{C}}(P, p)$ is surjective, and is epic [Exercise 2.1(b)], as required. $\square$

We call the object P in Proposition 2.53 a **projective** object of $\mathcal{C}$.

We also have precise conditions to obtain when the functor $\operatorname{Hom}_{\mathcal{C}}(-, Q)$ is exact.

Proposition 2.54. *The following statements are equivalent for an object $Q \in \mathcal{C}$:*

(a) *The functor $\operatorname{Hom}_{\mathcal{C}}(-, Q) : \mathcal{C} \to \mathsf{Ab}$ is right exact;*

(b) *Every short exact sequence of the form, $0 \to Q \to X \to X'' \to 0$, in $\mathcal{C}$ is split;*

(c) *For every mono $q : Z \to Y$ in $\mathcal{C}$ and every morphism $g : Z \to Q$ in $\mathcal{C}$ there exists a morphism $\tilde{g} : Y \to Q$ in $\mathcal{C}$ such that $g = \tilde{g}q$.* $\square$

We call the object Q in Proposition 2.54 an **injective** object of $\mathcal{C}$.

Projective and injective objects in $\mathcal{C}$ are visualized as follows. Recall from Lemma 2.43 that the maps to (resp., from) zero objects indicate that p is an epi (resp., q is a mono) in $\mathcal{C}$. That is, we have exactness at Z for the horizontal maps below.

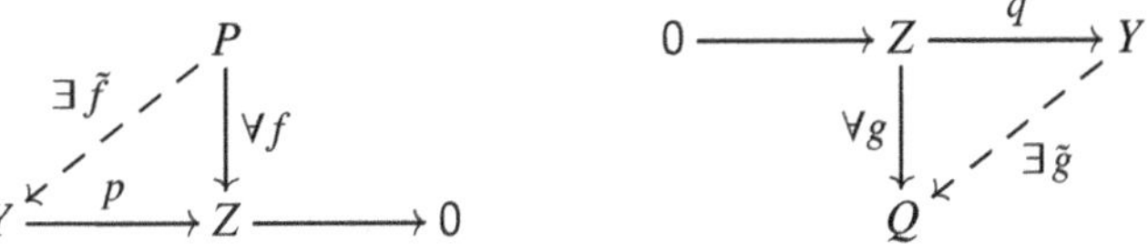

We say that C has **enough projectives** if, for every object $Z \in C$, there exists a projective object $P(Z) \in C$ along with an epi, $\pi_Z : P(Z) \to Z$, in C.

Dually, C has **enough injectives** if, for every object $Z \in C$, there exists an injective object $I(Z) \in C$ along with a mono, $\lambda_Z : Z \to I(Z)$, in C.

Sometimes one wants for a given object Z, a projective object that maps to Z, or an injective object that maps from Z, in a minimal fashion.

A **projective cover** of $Z \in C$ is a projective object $P(Z) \in C$ equipped with an epi $\pi_Z : P(Z) \to Z$ in C, such that if $f : P \to Z$ is an epi from a projective object P in C, there is an epi $\tilde{f} : P \to P(Z)$ with $f = \pi_Z \circ \tilde{f}$ in C.

Dually, an **injective hull** of $Z \in C$ is an injective object $I(Z) \in C$ equipped with a mono $\lambda_Z : Z \to I(Z)$ in C, such that if $g : Z \to Q$ is a mono to an injective object Q in C, there is a mono $\tilde{g} : I(Z) \to Q$ with $g = \tilde{g} \circ \lambda_Z$ in C.

Now we discuss some interesting consequences of the propositions above.

Corollary 2.55. *Let P_1 and P_2 be objects in C, and take $P := P_1 \sqcup P_2$. Then, P_1 and P_2 are projective when P is a projective object in C.*

Proof. Let $\alpha_i : P_i \to P$, for $i = 1, 2$, be the morphisms from the universal property of the coproduct. Also, let $\gamma_i : P \to P_i$, for $i = 1, 2$, be the morphisms derived by the universal property of the coproduct by using the morphisms $\beta_i := \mathrm{id}_{P_i}$ and $\beta_j := \vec{0}_{P_j, P_i}$, for $j \neq i$.

Suppose that P is projective. Take an epi $p : Y \to Z$ in C, along with a morphism $f : P_1 \to Z$. We want a morphism $\tilde{f}$ in C such that $f = p\tilde{f}$. Consider the composition $f_1 := f\gamma_1 : P \to Z$. Since P is projective, there exists a morphism $\tilde{f}_1 : P \to Y$ such that $f_1 = p\tilde{f}_1$. Now take $\tilde{f} := \tilde{f}_1\alpha_1 : P_1 \to Y$. We then get that:

$$p\tilde{f} \;=\; p\tilde{f}_1\alpha_1 \;=\; f_1\alpha_1 \;=\; f\gamma_1\alpha_1 \;=\; f.$$

Thus, P_1 is projective by Proposition 2.53. Likewise, P_2 is projective. $\qquad\square$

Moreover, the next result is a consequence of Propositions 2.46, 2.53, and 2.54.

Corollary 2.56. *Each object in a semisimple abelian category is both projective and injective.* $\qquad\square$

We also have the following useful result when $C = A$-Mod.

Proposition 2.57. *The following statements are equivalent for a $\Bbbk$-algebra A:*

(a) *A is semisimple;*

(b) *Every object M in A-Mod (or in Mod-A) is projective;*

(c) *Every object M in A-Mod (or in Mod-A) is injective.* $\qquad\square$

Proof. We have that part (a) implies both parts (b) and (c) by Corollary 2.56.

Next, assume that part (c) holds; we will establish part (b). For $M \in A$-Mod, take an arbitrary short exact sequence $0 \to X' \to X \to M \to 0$ in A-Mod. By assumption, X' is injective. Thus, the sequence splits by Proposition 2.54. Since the sequence is arbitrary, we obtain that M is projective by Proposition 2.53. There is a similar argument for right A-modules.

Now to get that part (b) implies part (a), it suffices to show that any left (or right) ideal I of A is a direct summand of A; this follows similarly to the proof of Proposition 1.48. We have a short exact sequence

$$0 \to I \to A \to A/I \to 0$$

in A-Mod (or in Mod-A). By part (b), A/I is projective. So, the sequence splits by Proposition 2.53. Therefore, $A \cong I \oplus (A/I)$ by Proposition 2.45, as desired. $\quad\square$

Remark 2.58. The distance from an algebra being semisimple is measured by the failure of the Hom functors above being right exact. This is captured by *Ext groups*, and numerically by *global dimension*, in Homological Algebra. Indeed, semisimple algebras are precisely the algebras that have global dimension 0.

Remark 2.59. Let us consider the adjoint counterpart of Hom functors for categories of (bi)modules, namely tensor functors. For a general (bi)module W over an algebra A, the functors $(W \otimes_A -)$ and $(- \otimes_A W)$ are covariant and right exact (see Exercise 1.20 and Proposition 2.49(a)). But these functors are not necessarily left exact (see Exercise 1.19 and Proposition 2.49(a)). If the functors $(W \otimes_A -)$ and $(- \otimes_A W)$ are left exact, then we call W a **flat** module. The failure of flatness is captured by *Tor groups* in Homological Algebra.

§2.9. Finiteness for linear categories

In this part, we briefly discuss finiteness conditions that are used often in conjunction with, or in place of, semisimplicity for $\Bbbk$-linear categories. See Section 1.8 of Etingof et al. [2015] for the details of the discussion below.

Standing hypothesis. Assume that $\mathcal{C}$ is $\Bbbk$-linear and abelian; see §2.2.2.

We say that $\mathcal{C}$ is **locally finite** if the following conditions hold:

- The $\Bbbk$-vector space $\mathrm{Hom}_{\mathcal{C}}(X, Y)$ is finite-dimensional, for each $X, Y \in \mathcal{C}$;
- Each object of $\mathcal{C}$ has finite length.

Proposition 2.60. *Assume that C is locally finite, and take an object $X \in C$. Then, the following statements hold.*

(a) *If X is a simple object, then $\mathrm{End}_C(X) \cong \Bbbk$ as algebras.*

(b) *If $\mathrm{End}_C(X) \cong \Bbbk$ as vector spaces, and if C is semisimple, then X is a simple object.*

Proof. (a) This follows from Corollary 2.36 and Lemma 1.36.

(b) If X is not simple, then there exists nonzero simple subobject Y of X, where $X \cong Y \sqcup Z \cong Y \sqcap Z$, for some nonzero object Z in C [Proposition 2.40]. By Exercise 2.8, $\mathrm{End}_C(X)$ has a subspace $\mathrm{End}_C(Y) \oplus \mathrm{End}_C(Z)$ of vector space dimension greater than 1, as required. $\qquad\square$

An object $X \in C$ that satisfies the condition that $\mathrm{End}_C(X) \cong \Bbbk$ as algebras (or, equivalently, as vector spaces) is said to be **absolutely simple**.

We also say that C is **finite** if the following conditions hold:

- C is locally finite;

- C has enough projectives;

- There are finitely many isoclasses of simple objects in C.

In fact, we have the following characterization of finite categories.

Proposition 2.61. *We have that C is finite if and only if it is equivalent to the category, A-FdMod, for some finite-dimensional algebra A.*

Proof. We sketch the proof here. First, the category A-FdMod, for A a finite-dimensional algebra, is finite. On the other hand, take a finite category C, with a complete set of isoclasses representatives $\{X_1, \ldots, X_n\}$ of simple objects in C. Then, there exist projective covers $\{P(X_i)\}_{i=1}^n$ in C. Now take $A := \bigoplus_{i=1}^n \mathrm{End}_C(P(X_i))$, which is a finite-dimensional $\Bbbk$-algebra. We then get that A^{op}-FdMod is equivalent to C via the functor:

$$C \longrightarrow A^{\mathrm{op}}\text{-FdMod}, \quad Y \mapsto \mathrm{Hom}_C(\textstyle\coprod_{i=1}^n P(X_i), Y). \qquad\square$$

For instance, FdVec is a key example of a finite category.

Observe from the result above that we can treat finite linear categories like categories of modules, and understanding such categories boils down to examining finitely many simple objects and their projective covers.

By Proposition 2.61 and the Eilenberg-Watts Theorem [Theorem 2.51], we also obtain in the finite setting a converse of Proposition 2.49(b) on the connection between exactness and adjunction for functors.

Corollary 2.62. *A left (resp., right) exact functor between finite categories has a left (resp., right) adjoint.* $\qquad\square$

§2.10. Summary

We introduced the concept of a category, which provides a framework for studying objects and structure-preserving maps between objects. Many examples of categories were presented, especially derived from structures in Chapter 1. To do computations in categories, unlike in Chapter 1, we do not necessarily have elements within objects to manipulate. Instead, we need to perform operations on the objects themselves, especially by using universal constructions. Thus, it is convenient to work with categories in which many universal constructions exist, such as in abelian categories; see Figure 2.1.

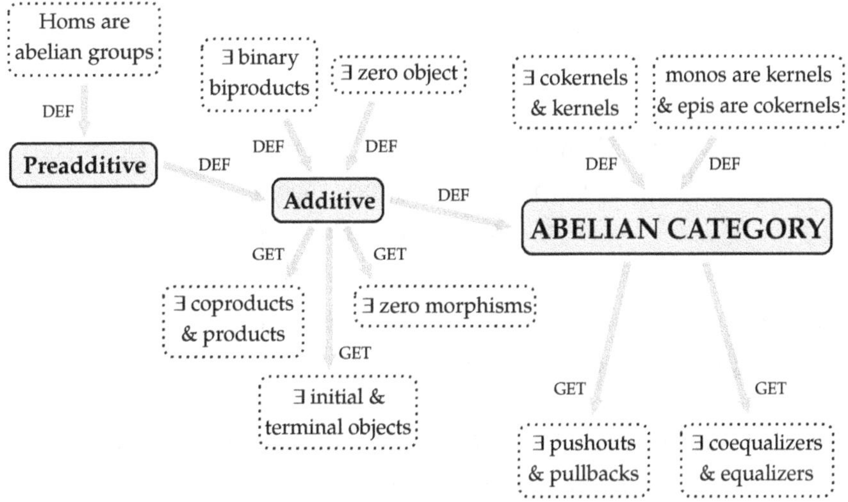

Figure 2.1: Universal constructions in abelian categories.

We also discussed the various notions of "sameness" in category theory. Just like we need to generalize equalities to isomorphisms in abstract algebra, we need to generalize isomorphisms to equivalences in category theory. Moreover, one moves from a category to another category via functors, and it is interesting to study sameness for functors as well– for this, one moves from natural isomorphisms to adjunctions to have a rich theory. See Figure 2.2 below.

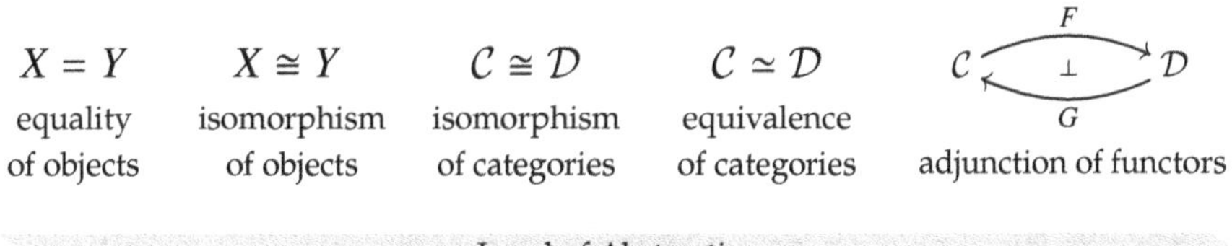

Figure 2.2: "Sameness" in abstract algebra and in category theory.

§2.11. Modern applications

We now illustrate how various notions that were introduced in this chapter on categories are used in modern mathematics. A full understanding of the resources here is not expected. Instead, we aim to put the chapter's material into context by offering videos and content to casually explore.

A welcoming invitation to **higher category theory** is presented in the videos below. The notion of an ∞-category is of particular interest.

Emily Riehl's 2020 Johns Hopkins President's Frontier Award lecture on
"What is Category Theory in mathematics?"
https://youtu.be/WLkMBMUk48E

Emily Riehl's 2021 Mathematical Picture Language Seminar lecture on
"Elements of ∞-Category Theory"
https://youtu.be/ZVreRhrtUyM

An engaging lecture on category theory and its role in **information science** is given below; the slides are also available below.

Peter Hines' 2020 ForML Lab Colloquium at Augusta University on
"Category Theory in Communication, Cryptography, and Security"
https://youtu.be/Njw5Aad-gBQ

A great course on **applied category theory** is available below. This may be of interest to those in computer science and to applied mathematicians.

David Spivak and Brendan Fong's 2019 MIT Independent Activities Period
Course 18.S097 on "Applied Category Theory"
https://youtu.be/UusLtx9fIjs

An intriguing lecture on category theory in **software engineering** is presented below. The speaker's background in engineering is also highlighted.

Angeline Aguinaldo's 2021 Topos Institute Berkeley Seminar lecture on
"Diary of a software engineer using categories"
https://youtu.be/gbP5ww3U10g

§2.12. References for further exploration

- The textbooks by Richter [2020] and Riehl [2016] are excellent introductions to category theory, particularly with a view towards Homotopy Theory in Algebraic Topology.

- The classic textbook by MacLane [1971] is absolutely a must-have resource for those interested in category theory.

- The textbooks by Mitchell [1965], Popescu [1973], and Stenström [1975] are vital resources for learning category theory with an algebraic point of view.

- On the other hand, for applications of category theory in Noncommutative Geometry, check out the user-friendly textbook by Khalkhali [2013].

- Fong and Spivak [2019] provides an intriguing introduction to Applied Category Theory, with a view toward computer science and applied mathematics.

- Eugenia Cheng once said, "Category theory is the mathematics of mathematics." Her recent book, Cheng [2022], on category theory for a mainstream audience is a great one.

- Tai-Danae Bradley's blog, Math3ma, has excellent explanations of categorical concepts in layman's terms. It is very highly recommended.

    ```
    https://www.math3ma.com/categories/category-theory
    ```

§2.13. Exercises

2.1   Recall the notions of monos, epis, and isos in a category C from §2.1.1.

    (a) Show that the composition of two monos (resp., two epis, two isos) in C is a mono (resp., an epi, an iso) in C.

    (b) Verify that an iso in C is both monic and epic in C.

    (c) We say that a morphism $g : X \to Y$ in C is **split-monic** if there exists a morphism $h : Y \to X$ in C with $hg = \mathrm{id}_X$, and is **split-epic** if there exists a morphism $f : Y \to X$ in C with $gf = \mathrm{id}_Y$.

        (i) Show that a split-epic mono in C is an iso in C.

        (ii) Show that a split-monic epi in C is an iso in C.

2.2   Here, we compare the categorical notions of monos and epis from §2.1.1 with injective and surjective maps in specific settings.

   (a) Show that in the category Set from §2.1.2i, we have that monos are precisely injective functions, and epis are precisely surjective functions.

   (b) Show that in the category Ab from §2.1.2i, we have that monos are precisely injective group homomorphisms, and epis are precisely surjective group homomorphisms.

   (c) Take the category Ring from §2.1.2i. Show that the inclusion morphism $\mathbb{Z} \to \mathbb{Q}$ in Ring is both monic and epic, but is not an iso.

2.3   For a category $\mathcal{C}$, recall the notion of its opposite category $\mathcal{C}^{\mathrm{op}}$ from §2.1.1. Show that the definition of $\mathrm{Ob}(\mathcal{C}^{\mathrm{op}})$ and $\mathrm{Hom}_{\mathcal{C}^{\mathrm{op}}}(X, Y)$ in §2.1.1 does indeed give $\mathcal{C}^{\mathrm{op}}$ the structure of a category.

2.4   Recall the definition of a groupoid $\mathcal{G}$ in §2.1.2i. Explain why if $\mathrm{Ob}(\mathcal{G})$ consists of a single object X, then $\mathcal{G}$ can be identified as a group.

2.5   [Open-ended] Recall the examples of categories from §2.1.2i-v.

   (a) Derive more examples of categories in addition to what is included in §2.1.2i-v; try to construct at least one for each mathematical type.

   (b) Do the examples in part (a) arise as subcategories of a category in §2.1.2i-v? If so, are they full subcategories?

2.6   [Open-ended] Determine if each of the amusing collections of objects and morphisms from §2.1.2vi forms a category. Derive more (non-) examples of 'amusing' categories.

2.7   [Open-ended] Justify the examples of initial, terminal, and zero objects in §2.2.1i. Then examine the initial, terminal, and zero objects for some other categories in §2.1.2i-v; in particular, discuss if such objects exist.

2.8   Recall the coproduct and product constructions from §2.2.1ii. Take objects $X, Y, Z, X_1, \ldots, X_m, Y_1, \ldots, Y_n$ in a locally small category $\mathcal{C}$.

   (a) Verify that $\mathrm{Hom}_{\mathcal{C}}(X \sqcup Y, Z) \cong \mathrm{Hom}_{\mathcal{C}}(X, Z) \times \mathrm{Hom}_{\mathcal{C}}(Y, Z)$.

   (b) Verify that $\mathrm{Hom}_{\mathcal{C}}(Z, X \sqcap Y) \cong \mathrm{Hom}_{\mathcal{C}}(Z, X) \times \mathrm{Hom}_{\mathcal{C}}(Z, Y)$.

   (c) Likewise, show that:

$$\mathrm{Hom}_{\mathcal{C}}(\textstyle\coprod_{i=1}^{m} X_i, Z) \cong \textstyle\prod_{i=1}^{m} \mathrm{Hom}_{\mathcal{C}}(X_i, Z),$$

$$\mathrm{Hom}_{\mathcal{C}}(Z, \textstyle\prod_{j=1}^{n} X_j) \cong \textstyle\prod_{j=1}^{n} \mathrm{Hom}_{\mathcal{C}}(Z, X_j).$$

2.9 Recall the pushouts and pullbacks constructions from §2.2.1iii.

    (a) Verify that for X, Y, Z in Set, we have the following statements.

        (i) $X \sqcup_Z Y$ is the quotient set, $(X \uplus Y)/\!\sim$, with $f(z) \sim g(z)$ for each $z \in Z$, and $\uplus$ is disjoint union.

        (ii) $X \sqcap_Z Y = \{(x, y) \in X \times Y \mid f(x) = g(y)\}$.

    (b) Show that $X \sqcup_Z Y$ and $X \sqcap_Z Y$ also exist in Top by concretely describing these topological spaces.

2.10 Recall the universal constructions in a category $\mathcal{C}$ from §2.2.1i-iii.

    (a) Assume that $\mathcal{C}$ has an initial object I and a terminal object T. Show $X \sqcup_I Y \cong X \sqcup Y$ and $X \sqcap_T Y \cong X \sqcap Y$ as objects in $\mathcal{C}$.

    (b) Given a pushout $(X \sqcup_Z Y, f : Z \to X, g : Z \to Y, C, \alpha_X, \alpha_Y)$, show that if f is epic, then α_Y is epic.

    (c) For a pullback $(X \sqcap_Z Y, f' : X \to Z, g' : Y \to Z, P, \alpha'_X, \alpha'_Y)$, show that if f' is monic, then α'_Y is monic.

    (d) Verify that $(V \sqcup_Y W) \sqcup_Z X \cong V \sqcup_Y (W \sqcup_Z X)$ as objects in $\mathcal{C}$.

2.11 Recall the coequalizers and equalizers defined in §2.2.1iv.

    (a) For parallel morphisms $f, g : X \to Y$ in Set, justify why $\mathrm{coeq}(f, g)$ and $\mathrm{eq}(f, g)$ are as claimed in §2.2.1iv.

    (b) For parallel morphisms $f, g : X \to Y$ in each of the categories below, describe $\mathrm{coeq}(f, g)$ and $\mathrm{eq}(f, g)$:

        (i) Ab   (ii) Vec   (iii) Top.

    (c) For a $\Bbbk$-algebra A, with $V \in$ Mod-A and $W \in A$-Mod, verify that $V \otimes_A W$ from §1.4.2iii arises as a coequalizer of two morphisms in Vec.

2.12 Recall the universal constructions from §2.2.1ii,iii,iv.

    (a) Show that if a category has pushouts and coproducts, then it has coequalizers. Namely, for morphisms $f, g : X \to Y$ in $\mathcal{C}$, take the unique morphisms $\gamma_{f,\mathrm{id}_Y} : X \sqcup Y \to Y$ and $\gamma_{g,\mathrm{id}_Y} : X \sqcup Y \to Y$ defined by the coproduct, where β_X in §2.2.1ii is f and g, respectively. Then, show that $\mathrm{coeq}(f, g)$ is the pushout of γ_{f,id_Y} and γ_{g,id_Y}.

    (b) State and prove a connection between pullbacks, products, and equalizers in a category $\mathcal{C}$ similar to the statement in part (a).

2.13 Recall the coequalizers and equalizers defined in §2.2.1iv, and the cokernels and kernels of morphisms from §2.2.1vi.

    (a) Show that coequalizers are epic, and that equalizers are monic.

    (b) Assume that C has a zero object. Show that cokernels in C are epic, and that kernels in C are monic.

    (c) Assume that C has a zero object. Show that a cokernel in C is a certain coequalizer, and a kernel in C is a certain equalizer.

    (d) Assume that C is preadditive as in §2.2.2i. Show that coequalizers (resp., equalizers) are cokernels (resp., kernels).

2.14 Assume that a category C has equalizers as in §2.2.1iv. Take $f : X \to Y$ in C with factorization $X \xrightarrow{\tilde{\alpha}} \mathrm{im}(f) \xrightarrow{\alpha \ \mathrm{mono}} Y$. Show that $\tilde{\alpha}$ is epic.

2.15 Recall the material in §2.2.1v,vi. Consider the following categories which have a zero object, and thus, has zero morphisms:

        (i) Group;   (ii) Ab;   (iii) Vec.

Describe the cokernel and kernel of a morphism in each of the above.

2.16 Recall §§2.2.2i,ii. Let C be a preadditive category, and take $X, Y \in C$. Complete the details of the proof of Proposition 2.2 on the simultaneous existence of the coproduct $X \sqcup Y$, the product $X \sqcap Y$, and the biproduct $X \square Y$ in C.

2.17 Take an additive category C with biproduct $\square$ and zero object 0 as in §2.2.2ii. Show that for any object $X \in C$, we get that $X \square 0 \cong X \cong 0 \square X$ as objects in C.

2.18 Complete the details of the proof of Proposition 2.5 in §2.2.2iii about pushouts existing in abelian categories. Complete the proof for pullbacks as well, if you are curious.

2.19 Recall the notion of a functor from §2.3.1. We say a functor F **preserves a property P** for morphisms if a morphism f having property P implies that so does $F(f)$. On the other hand, F **reflects a property P** for morphisms if a morphism $F(f)$ having property P implies that so does f. A similar notion holds for functors preserving or reflecting properties of objects.

    (a) Show that all functors preserve isos.

    (b) Provide an example of a functor that does not preserve monos, and of a functor that does not preserve epis.

    (c) Provide an example of a functor that does not reflect isos.

(d) Show that fully faithful functors reflect isos.

(e) Provide an example of a functor that does not reflect monos, and of a functor that does not reflect epis.

2.20 Recall the discussion about functors $F : C \to D$ in §2.3.1, and equivalence of categories from §2.4.2.

(a) Show that the image of F is not necessarily a subcategory of D.

(b) Explain why the essential image $\mathrm{Im}^{\mathrm{ess}}(F)$ is a subcategory of D.

(c) Show that when F is fully faithful, then $\mathrm{Im}^{\mathrm{ess}}(F)$ is equivalent to C.

2.21 Prove Lemma 2.6 in §2.3.1 establishing that a functor between additive categories is additive if and only if it preserves biproducts.

2.22 [Open-ended]   Recall the discussion of functors in §§2.3.1, 2.3.2.

(a) Determine if the functors in §2.3.2 are covariant or contravariant.

(b) Derive examples of functors besides those in §2.3.2, including some contravariant functors. Feel free to use outside resources.

(c) Explain in detail why Forg : Ring $\to$ Ab is not essentially surjective.

(d) Explain in detail why Forg : Ring $\to$ Ab is not full.

(e) For (some of) the other examples in §2.3.2, and along with those in part (b), discuss whether these functors are faithful, full, or essentially surjective.

2.23 Recall Exercise 1.4 on duals and doubles of vector spaces, and recall natural transformations from §2.3.4. Let us continue Example 2.7.

(a) For a vector space V, consider the morphism in Vec:

$$j_V : V \to V^{**}, \; v \mapsto [f \mapsto f(v)].$$

Verify that these morphisms form the components of a natural transformation j from the identity functor Id : Vec $\to$ Vec to the double dual functor $(-)^{**}$: Vec $\to$ Vec.

(b) Show that if we replace Vec with FdVec, then the natural transformation j in part (a) is a natural isomorphism.

(c) Explain why there does not exist a natural transformation from $\mathrm{Id}_{\mathsf{Vec}}$ to the dual functor $(-)^*$: Vec $\to$ Vec.

2.24 Consider the natural transformations from §2.3.4, and let us continue Example 2.8. Consider the functors $\otimes \circ (\otimes \times \mathrm{Id})$ and $\otimes \circ (\mathrm{Id} \times \otimes)$ from $\mathsf{Vec} \times \mathsf{Vec} \times \mathsf{Vec}$ to Vec. Verify that there exists a natural isomorphism between these functors with components,

$$a_{U,V,W} : (U \otimes V) \otimes W \xrightarrow{\sim} U \otimes (V \otimes W),$$

for each $U, V, W \in \mathsf{Vec}$.

2.25 Verify that $\mathsf{Fun}(\mathcal{C}, \mathcal{D})$ discussed in §2.3.4 is indeed a category.

2.26 Establish the interchange law in (2.9) of §2.3.5 between the vertical and horizontal compositions of natural transformations.

2.27 Verify the details of Examples 2.10 and 2.11 in §2.3.6 showing that many of the universal constructions in §2.2.1 arise as colimits and limits.

2.28 Show that the notion of two categories being isomorphic [§2.4.1] is an equivalence relation (i.e., it is reflexive, symmetric, and transitive). Verify the same statement for the notion of two categories being equivalent [§2.4.2].

2.29 Recall the notion of isomorphic categories in §2.4.1, and also see Exercise 1.12. For any algebra A, establish that

$$A\text{-}\mathsf{Mod} \cong \mathsf{Rep}(A).$$

2.30 Recall the notion of isomorphic categories from §2.4.1, and also see Exercise 1.13. Let us continue Example 2.12. Take a group G, with its group algebra $\Bbbk G$, and verify that

$$G\text{-}\mathsf{Mod} \cong \mathsf{Rep}(G) \cong \mathsf{Rep}(\Bbbk G) \cong \Bbbk G\text{-}\mathsf{Mod}.$$

2.31 Recall the material about functors in §2.3.1, functor categories in §2.3.4, and representations of groups in §1.3.4. Also recall the notions of isomorphic and equivalent categories in §§2.4.1, 2.4.2.

For a group G, take $\mathsf{G}_{\mathrm{cat}}$ to be the category with one object X, and with morphisms identified as elements of G (cf. Exercise 2.4). Show that there is an equivalence of categories:

$$\mathsf{Rep}(G) \simeq \mathsf{Fun}(\mathsf{G}_{\mathrm{cat}}, \mathsf{Vec}).$$

Is this an isomorphism of categories?

2.32 Recall the material about category isomorphism, category equivalence, and skeletons of categories from §§2.4.1, 2.4.2.

    (a) Take the natural numbers $n \in \mathbb{N}$ as objects, with morphisms $n \to m$ given by matrices in $\mathrm{Mat}_{m \times n}(\Bbbk)$, and show that this forms a category: Mat.

    (b) Take the category FdVec, and verify that $\mathrm{Skel}(\mathrm{FdVec}) \cong \mathrm{Mat}$.

    (c) Prove that $\mathrm{Mat} \not\cong \mathrm{FdVec}$.

    (d) Prove that $\mathrm{Mat} \simeq \mathrm{FdVec}$.

2.33 Recall the material about category isomorphism, category equivalence, and skeletons of categories from §§2.4.1, 2.4.2. Take categories $\mathcal{C}, \mathcal{D}$, and show:

    (a) $\mathrm{Skel}(\mathcal{C}) \cong \mathcal{C}$ if and only if $\mathrm{Skel}(\mathcal{C}) = \mathcal{C}$;

    (b) $\mathrm{Skel}(\mathcal{C}) \simeq \mathcal{C}$ always;

    (c) $\mathcal{C} \simeq \mathcal{D}$ if and only if $\mathrm{Skel}(\mathcal{C}) \cong \mathrm{Skel}(\mathcal{D})$.

2.34 Complete the details of the proof for Theorem 2.14 in §2.4.2 on reconciling the two notions of an equivalence of categories.

2.35 Complete the details of the proof for Theorem 2.18 in §2.4.4 on conditions when two algebras A and B are Morita equivalent.

2.36 Complete the details of Remark 2.20 in §2.4.4 on verifying that two algebras A and B are Morita equivalent if and only if there exists a finitely generated, projective generator M of A-Mod such that $B^{\mathrm{op}} \cong \mathrm{End}_{A\text{-Mod}}(M)$ as algebras.

Hint. Use (the proof of) Morita's Theorem [Theorem 2.18].

2.37 Complete the details of Example 2.21 in §2.4.4 on verifying that an algebra A is always Morita equivalent to $\mathrm{Mat}_n(A)$, for any $n \in \mathbb{N}_{\geq 1}$.

2.38 Proposition 2.22 in §2.4.4 states that if two algebras A and B are Morita equivalent, then $Z(A) \cong Z(B)$ as algebras. Complete the steps below to derive a proof of this result, and of its consequence, Corollary 2.23.

    (a) Recall natural transformations from §2.3.4. Define the **center** of $\mathcal{C}$ to be the collection of natural transformations of $\mathrm{Id}_{\mathcal{C}} : \mathcal{C} \to \mathcal{C}$ to itself:

$$\underline{\mathcal{Z}}(\mathcal{C}) := \mathrm{Nat}_{\mathcal{C},\mathcal{C}}(\mathrm{Id}_{\mathcal{C}}, \mathrm{Id}_{\mathcal{C}}).$$

Show that $\underline{\mathcal{Z}}(\mathcal{C})$ is closed under composition of natural transformations, and there is an identity element with respect to this composition. (That is, if $\underline{\mathcal{Z}}(\mathcal{C})$ is a set, then it is a monoid under composition.)

(b) Show that if C is an abelian category, then $\underline{Z}(C)$ is a ring.

(c) Assume that C and D are abelian categories. Show that if C and D are equivalent categories, then $\underline{Z}(C) \cong \underline{Z}(D)$ as rings.

(d) Establish that $\underline{Z}(A\text{-Mod})$ forms an algebra.

(e) Identify the center of $Z(A)$ with the algebra $\underline{Z}(A\text{-Mod})$ by having $a \in Z(A)$ correspond to a natural transformation $\phi^a \in \underline{Z}(A\text{-Mod})$ that has components, $\phi^a_M : M \to M$, $m \mapsto a \triangleright m$, for $M \in A\text{-Mod}$.

(f) Draw the conclusions of Proposition 2.22 and Corollary 2.23.

2.39 Recall the discussion of Morita equivalence in §2.4.4, especially Theorem 2.18. Show that the algebraic properties below are Morita invariant:

(a) Simplicity (from §1.5.2);

(b) Semisimplicity (from §1.6).

2.40 Complete the details of the proof for Proposition 2.24 in §2.5.1 on characterizing adjoint functors.

2.41 Recall the various types of morphisms in Exercise 2.1, and the notion of adjunction from §2.5.1. Take adjoint functors $F \dashv G$ with unit η and counit ε.

(a) Show that F is faithful (resp., is full, is fully faithful) if and only if each component of η is monic (resp., is split-epic, is an iso).

(b) Show that G is faithful (resp., is full, is fully faithful) if and only if each component of ε is epic (resp., is split-monic, is an iso).

(c) Complete the details of the proof of Proposition 2.25.

2.42 Complete the proof of Proposition 2.26 in §2.5.2 on how functors that arise as a left (resp., right) adjoint preserve:

(a) pushouts (resp., pullbacks);

(b) coequalizers (resp., equalizers);

(c) cokernels (resp., kernels).

2.43 Recall Exercise 1.25. Complete the details of Example 2.28 in §2.5.3 on expressing the Tensor-Hom adjunction for bimodules diagrammatically.

2.44 Verify that each of the specific pairs of functors, Free and Forg listed in Example 2.29 in §2.5.3, forms an adjunction Free $\dashv$ Forg.

2.45 Recall the discussion of adjunction in §2.5.1. By using one of the characterizations of adjunction in Proposition 2.24, explain why each of the pairs of functors F and G below yields an adjunction $F \dashv G$.

(a) $F := \mathrm{Ind}_A^B$ and $G := \mathrm{Res}_A^B$, for $\phi : A \to B \in \mathsf{Alg}$ (see §1.4.4).

(b) $F := \mathrm{Res}_A^B$ and $G := \mathrm{Coind}_A^B$, for $\phi : A \to B \in \mathsf{Alg}$ (see §1.4.4).

(c) $F := \Bbbk(-) : \mathsf{Group} \to \mathsf{Alg}$ (form the group algebra), and
$G := (-)^\times : \mathsf{Alg} \to \mathsf{Group}$ (take the group of units).

(d) For those with background in topology, take
$F : \mathsf{Set} \to \mathsf{Top}$ (impose discrete topology), and
$G : \mathsf{Top} \to \mathsf{Set}$ (forget topology).

(e) For those with background in topology, take
$F : \mathsf{Top} \to \mathsf{Set}$ (forget topology), and
$G : \mathsf{Set} \to \mathsf{Top}$ (impose indiscrete/ trivial topology).

An aside. Note from parts (a,b) that $F := \mathrm{Ind}_A^B \dashv \mathrm{Res}_A^B \dashv \mathrm{Coind}_A^B$. The adjunction $\mathrm{Ind}_A^B \dashv \mathrm{Res}_A^B$ is referred to as **Frobenius reciprocity**. Moreover, functors $L, R : \mathcal{C} \to \mathcal{D}$ and $G : \mathcal{D} \to \mathcal{C}$, satisfying the condition that $L \dashv G \dashv R$, are called an **adjoint triple**.

2.46 [Open-ended] If $L \dashv G \dashv R$ is an adjoint triple [Exercise 2.45] with $L \cong R$, then G is called a **Frobenius functor**. (In this case, $G \dashv R$ is also called a **strongly adjoint pair**.) That is, Frobenius functors are those that have a left adjoint that coincides with its right adjoint, up to natural isomorphism. Explore and discuss instances of Frobenius functors in this chapter and in the broader literature.

2.47 Recall the discussion of adjoint functors and representability from §§2.5 and 2.6.1. Prove that a functor $G : \mathcal{D} \to \mathcal{C}$ has a left adjoint $F : \mathcal{C} \to \mathcal{D}$ if and only if for any object $X \in \mathcal{C}$ the functor H_X below is representable:

$$H_X(-) := \mathrm{Hom}_{\mathcal{C}}(X, G(-)) : \mathcal{D} \to \mathsf{Set}.$$

2.48 Recall the discussion of Yoneda's Lemma from §2.6.2, and establish the proof of Corollary 2.32.

Hint. You can consult Section 2.2 of Richter [2020] for further details, but it is best to do this only as needed.

2.49 Recall the discussion of indecomposability and simplicity in §2.7 in the setting of abelian categories.

    (a) Verify the characterization of an object being indecomposable in this setting as stated in Proposition 2.34.

    (b) Verify the statement of Schur's Lemma in this setting, namely Proposition 2.35, on morphisms between simple objects.

    (c) Verify the consequence of Schur's Lemma given in Corollary 2.36.

2.50 [Open-ended] Explore whether the various categories listed in §2.1.2 are semisimple (see §§2.7.3 and 2.8). Feel free to use outside resources!

2.51 Provide a proof of the results in §2.8.1 below on features of exact sequences.

    (a) Lemma 2.43.

    (b) Proposition 2.45.

Feel free to use the references provided in §2.8.1 if you get stuck.

2.52 Provide a (detailed) proof of the results in §2.8.3 below on projective and injective objects.

    (a) Proposition 2.52.

    (b) Proposition 2.54.

Feel free to use the references provided in §2.8.3 if you get stuck.

$\cdot$ CHAPTER 3 $\cdot$

MONOIDAL CATEGORIES

History

A *monoidal category* is a category C that is equipped with an operation $\otimes$ and
an object $\mathbb{1}$, such that the triple $(C, \otimes, \mathbb{1})$ mimics the structure of a monoid. They
were introduced by Bénabou [1963] as *catégories avec multiplication*, and were
formalized by Mac Lane [1963] and Kelly [1964]. Their current name was coined
by Eilenberg. Duality (or *rigidity*) was introduced by Saavedra Rivano [1972].
Substantial advances were also made by Joyal and Street [1993] on *braided* monoidal
categories (covered in a future volume), and by Etingof et al. [2005] on *fusion*
categories– such monoidal categories are prevalent in representation theory, in
mathematical physics, in quantum information theory, and in many other fields.

Overview

An introduction to monoidal categories is covered in §3.1; many examples are
provided. We compare monoidal categories via *monoidal functors* in §3.2. *Module
categories* over monoidal categories are presented in §3.3. A way of simplifying
axioms for these structures is presented in §3.4, and graphical diagrams are
then defined in §3.5 for computations in these structures. Then, various types of
monoidal categories, and module categories over them, are examined, including
those that are *rigid* in §3.6, *pivotal* in §3.7, *spherical* in §3.8, *fusion* in §3.9, and *tensor*
in §3.10. We also discuss *enriched* categories in §3.11. The chapter ends with a
summary discussion in §3.12, modern applications in §3.13, references in §3.14,
and several exercises.

Standing hypotheses. Linear structures are over an algebraically closed
field $\Bbbk$ of characteristic 0, and algebras over $\Bbbk$ are associative and unital.

---

§3.1. Monoidal categories

Building on our discussion of categories C in §2.1, we present here a convenient operation to combine objects and morphisms within C. This endows C with a rich structure that mimics the definition of a monoid, with which one can generalize various algebraic structures in the categorical context (as done in Chapter 4).

§3.1.1. Monoidal categories

A **monoidal category** consists of the following data.

(a) A category C.

(b) (**monoidal product**) A bifunctor $\otimes : C \times C \to C$.

(c) (**monoidal unit object**) A distinguished (i.e., a particular) object $\mathbb{1}$ in C.

(d) (**associativity constraint**) A natural isomorphism

$$a : \otimes \circ (\otimes \times \mathrm{Id}_C) \;\overset{\sim}{\Rightarrow}\; \otimes \circ (\mathrm{Id}_C \times \otimes)$$

of functors from $C \times C \times C$ to C. That is, there is a collection of isomorphisms

$$\{a_{X,Y,Z} : (X \otimes Y) \otimes Z \overset{\sim}{\to} X \otimes (Y \otimes Z)\}_{X,Y,Z \in C}$$

in C, natural (i.e., respects morphisms) in each variable X, Y, Z.

(e) (**left unitality constraint**)   A natural isomorphism

$$\ell : \mathbb{1} \otimes \mathrm{Id}_C \;\overset{\sim}{\Rightarrow}\; \mathrm{Id}_C$$

of functors from C to C. That is, there is a collection of isos $\{\ell_X : \mathbb{1} \otimes X \overset{\sim}{\to} X\}_{X \in C}$ in C, natural in X.

(f) (**right unitality constraint**) A natural isomorphism

$$r : \mathrm{Id}_C \otimes \mathbb{1} \;\overset{\sim}{\Rightarrow}\; \mathrm{Id}_C$$

of functors from C to C. That is, there is a collection of isos $\{r_X : X \otimes \mathbb{1} \overset{\sim}{\to} X\}_{X \in C}$ in C, natural in X.

This data must satisfy the commutative diagrams below, for all $W, X, Y, Z \in C$.

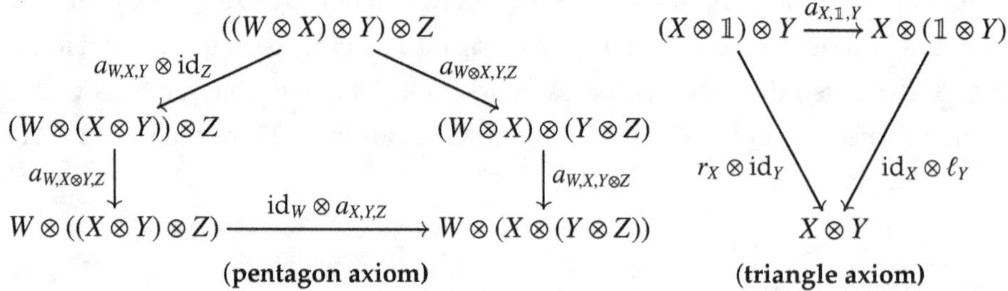

(pentagon axiom)   (triangle axiom)

See Exercise 3.1 for identities for the associativity and unitality constraints.

On the notation $\otimes$, and $\mathcal{C}$ versus $\mathcal{A}$. From now on, $\otimes$ denotes the monoidal product, and the tensor product over $\Bbbk$ is denoted by $\otimes_{\Bbbk}$.

Moreover, we write $\mathcal{A}$ (instead of $\mathcal{C}$) when working with an ordinary category that is not necessarily monoidal.

Remark 3.1. A category becoming monoidal involves equipping it with extra *structure*. See the chat about structure versus property from §1.1.1. There may be more than one way of imposing a monoidal structure on a category.

We say that a monoidal category is **strict** if its associativity and unitality constraints are identity maps; we will study this condition in §3.4.

Given a monoidal category $\mathcal{C} := (\mathcal{C}, \otimes, \mathbb{1}, a, \ell, r)$, a **monoidal subcategory** of $\mathcal{C}$ is a subcategory $\mathcal{D}$ such that we have:

- closure under $\otimes$, that is, $X \otimes Y \in \mathcal{D}$, for all $X, Y \in \mathcal{D}$;

- $\mathbb{1} \in \mathcal{D}$; and

- the constraints a, ℓ, r restrict to $\mathcal{D}$ making $\mathcal{D}$ itself a monoidal category.

A monoidal subcategory is **full** if the underlying subcategory is full.

Moreover, for a monoidal category $(\mathcal{C}, \otimes, \mathbb{1}, a, \ell, r)$, there are numerous notions of an **opposite monoidal category**. Towards this, recall the opposite category $\mathcal{A}^{\mathrm{op}}$ from §2.1.1, where we have $f^{\mathrm{op}} : Y \to X$ in $\mathcal{A}^{\mathrm{op}}$ for every morphism $f : X \to Y$ in $\mathcal{A}$. Also, consider the **opposite monoidal product**,

$$\otimes^{\mathrm{op}} : \mathcal{C} \times \mathcal{C} \to \mathcal{C}, \quad (X, Y) \mapsto X \otimes^{\mathrm{op}} Y := Y \otimes X.$$

Now the following are monoidal categories; see Exercise 3.2.

- $\mathcal{C}^{\mathrm{op}} := \left(\mathcal{C}^{\mathrm{op}}, \otimes, \mathbb{1}, a^{\mathrm{op}} := \{a_{X,Y,Z}^{-1}\}, \ell^{\mathrm{op}} := \{\ell_X^{-1}\}, r^{\mathrm{op}} := \{r_X^{-1}\} \right)$.

- $\mathcal{C}^{\otimes\mathrm{op}} := \left(\mathcal{C}, \otimes^{\mathrm{op}}, \mathbb{1}, a^{\otimes\mathrm{op}} := \{a_{Z,Y,X}^{-1}\}, \ell^{\otimes\mathrm{op}} := \{r_X\}, r^{\otimes\mathrm{op}} := \{\ell_X\} \right)$.

- $\mathcal{C}^{\mathrm{rev}} := \left(\mathcal{C}^{\mathrm{op}}, \otimes^{\mathrm{op}}, \mathbb{1}, a^{\mathrm{rev}} := \{a_{Z,Y,X}\}, \ell^{\mathrm{rev}} := \{r_X^{-1}\}, r^{\mathrm{rev}} := \{\ell_X^{-1}\} \right)$.

Note that this notation varies across the literature; we are using the version from Turaev and Virelizier [2017] here.

§3.1.2. Examples of monoidal categories

We present examples of monoidal categories, many building on the examples of categories from §2.1.2. Verifying the details and adding more examples is Exercise 3.3. Recall that all vector spaces and algebras are over $\Bbbk$ below.

i. Algebraic monoidal categories

- Vec: The category of $\Bbbk$-vector spaces is monoidal, with $\otimes := \otimes_\Bbbk$ and $\mathbb{1} := \Bbbk$, and with associativity and unitality constraints as in Exercise 1.3(a,b). See also Example 2.8 and Exercise 2.24. This monoidal category is not strict.

- $\text{Vec}_\oplus$: The category of $\Bbbk$-vector spaces admits another monoidal structure, where $\otimes := \oplus$ and $\mathbb{1} := 0_{\text{Vec}}$ (the zero vector space). This is also not strict.

- FdVec, $\text{FdVec}_\oplus$: Likewise, the category of finite-dimensional $\Bbbk$-vector spaces admits monoidal structures, which, again, are not strict.

- $\text{Alg}_\circledast$: The category of $\Bbbk$-algebras is monoidal with $\otimes := \circledast$ and $\mathbb{1} := \Bbbk$. See §1.4.2vi. Indeed, $\Bbbk \circledast A \cong A \cong A \circledast \Bbbk$, for any $A \in \text{Alg}$.

- $\text{Alg}_{\otimes_\Bbbk}$: Alg is also monoidal with $\otimes := \otimes_\Bbbk$ and $\mathbb{1} := \Bbbk$. See §1.4.2v.

- Ab: The category of abelian groups (or, of $\mathbb{Z}$-modules with underlying abelian group structure, as in Exercise 1.14) is monoidal. Here, $\otimes := \otimes_\mathbb{Z}$ and $\mathbb{1} := \mathbb{Z}$.

- Ring: Further, the category of unital rings is monoidal with $\otimes := \otimes_\mathbb{Z}$ and $\mathbb{1} := \mathbb{Z}$.

- G-Mod: The category of (left) modules over a group G is monoidal. For G-modules $(V, \triangleright)$ and $(V', \triangleright')$, define $(V, \triangleright) \otimes (V', \triangleright') := (V \otimes_\Bbbk V', \blacktriangleright)$, where

$$g \blacktriangleright (v \otimes_\Bbbk v') := (g \triangleright v) \otimes_\Bbbk (g \triangleright' v'),$$

for $g \in G$, $v \in V$, $v' \in V'$. Moreover, $\mathbb{1}$ is the trivial G-module $\Bbbk$, where $g \triangleright \lambda = \lambda$, for $g \in G$ and $\lambda \in \Bbbk$.

- A-Bimod: The category of bimodules over an arbitrary $\Bbbk$-algebra A admits a monoidal structure, where $\otimes := \otimes_A$ and $\mathbb{1} := A_{\text{reg}}$, with the associativity and unitality constraints given in Exercise 1.18.

- Vec_N: For an additive monoid N with identity element 0, the category of N-graded $\Bbbk$-vector spaces is monoidal. For $V := \bigoplus_{n \in N} V_n$ and $W := \bigoplus_{n' \in N} W_{n'}$, define:

$$V \otimes W := \bigoplus_{m \in N} (V \otimes W)_m, \quad \text{where } (V \otimes W)_m := \bigoplus_{n+n'=m} (V_n \otimes_\Bbbk W_{n'}).$$

Also, $\mathbb{1} := \bigoplus_{m \in N} \mathbb{1}_m$, with $\mathbb{1}_0 = \Bbbk$, and with $\mathbb{1}_{m \neq 0}$ being the zero vector space.

- Vec_G: Likewise, for a group G, the category of G-graded $\Bbbk$-vector spaces is monoidal, where for $V := \bigoplus_{h \in G} V_h$ and $W := \bigoplus_{h' \in G} W_{h'}$ we define:

$$V \otimes W := \bigoplus_{g \in G} (V \otimes W)_g, \quad \text{where } (V \otimes W)_g := \bigoplus_{hh'=g} (V_h \otimes_\Bbbk W_{h'}).$$

Also, $\mathbb{1} := \bigoplus_{g \in G} \mathbb{1}_g$, with $\mathbb{1}_e = \Bbbk$, and with $\mathbb{1}_{g \neq e}$ being the zero vector space.

- Vec_G': One can also equip the category Vec_G with a monoidal structure via

$$(\bigoplus_{g\in G} V_g) \otimes (\bigoplus_{g\in G} W_g) := \bigoplus_{g\in G}(V_g \otimes_{\Bbbk} W_g)$$

  and $\mathbb{1} := \bigoplus_{g\in G} \Bbbk$. Again, this shows that a category can admit different monoidal structures. (By 'different', we mean *not monoidally equivalent*; see Example 3.10 later).

- $\underline{N}$: Take an additive monoid N with identity element 0. Define a monoidal category with elements of N as objects, with identity morphisms id_m as morphisms, with $n \otimes n' := n + n'$, for $m, n, n' \in N$, and with $\mathbb{1} := 0$. This monoidal category is strict.

- $N_{\leq}$: Take a monoid N with a partial ordering $\leq$. Define a monoidal category with objects being elements of N, with a morphism $n \to n'$ existing if and only if $n \leq n'$, with $n \otimes n' := n + n'$, and with $\mathbb{1} := 0$.

- $\underline{G}, G_{\leq}$: One can define similar monoidal categories for a group G.

Another source of algebraic monoidal categories are the categories of modules over *bialgebras* and *Hopf algebras* discussed in a future volume. We also have monoidal categories of *comodules* over such algebras (also discussed later).

There is also an extended exercise, Exercise 3.35, that introduces a monoidal category that has interesting associativity and unitality constraints:

- Vec_G^{ω}: Take $\Bbbk^{\times}$ to be the multiplicative group of nonzero elements of $\Bbbk$. With a $\Bbbk^{\times}$-valued *3-cocycle* ω on a group G, we can modify the associativity and unitality constraints of the monoidal category Vec_G to form another monoidal category Vec_G^{ω}. See Exercise 3.35(a-d).

ii. Logical and categorical monoidal categories

- Set: The category of sets admits a monoidal structure, with $\otimes$ being cartesian product $\times$, and with $\mathbb{1}$ being the singleton set $\{\cdot\}$.

- Cat: The category of small categories is monoidal, with $\otimes$ being the product of categories introduced in §2.1.3. Also, $\mathbb{1}$ is the category 1 consisting of a single object X with $\mathrm{Hom}_1(X, X) = \{\mathrm{id}_X\}$.

- $\mathcal{A}_{\sqcup}$: A category $\mathcal{A}$ that admits binary coproducts (so, finitely many coproducts) and an initial object I is monoidal. Here, $\otimes = \sqcup$ and $\mathbb{1} = \mathsf{I}$. This is called a **cocartesian monoidal category**.

- $\mathcal{A}_{\sqcap}$: A category $\mathcal{A}$ that admits binary products (so, finitely many products) and a terminal object T is monoidal. Here, $\otimes = \sqcap$ and $\mathbb{1} = \mathsf{T}$. This is called a **cartesian monoidal category**.

- $\mathrm{End}(\mathcal{A}) := \mathrm{Fun}(\mathcal{A}, \mathcal{A})$: The category of endofunctors of a category $\mathcal{A}$ is strict monoidal, with $\otimes$ being composition $\circ$, and with $\mathbb{1} = \mathrm{Id}_{\mathcal{A}}$.

- $\mathrm{Aut}(\mathcal{A})$: The category of autoequivalences of a category $\mathcal{A}$ is strict monoidal, with $\otimes$ being composition $\circ$, and with $\mathbb{1} = \mathrm{Id}_{\mathcal{A}}$.

iii. Topological monoidal categories

- nCob: The category of $(n-1)$-manifolds with cobordisms of dimension n is monoidal, with $\otimes$ being disjoint union, and $\mathbb{1}$ being the empty manifold.

- Top: The category of topological spaces is a monoidal category, where $\otimes$ is cartesian product, and with $\mathbb{1}$ being a one-point space.

- Braid: The objects are natural numbers $\mathbb{N}$. Moreover, for an object $n \in \mathbb{N}$, consider the n-th **braid group** B_n, which has generators $\sigma_1, \ldots, \sigma_{n-1}$ and relations:

$$\sigma_i \sigma_{i+1} \sigma_i = \sigma_{i+1} \sigma_i \sigma_{i+1}, \quad \text{for } i = 1, \ldots, n-2;$$
$$\sigma_i \sigma_j = \sigma_j \sigma_i, \quad \text{for } |i - j| > 1.$$

Then, $\mathrm{Hom}_{\mathrm{Braid}}(n, m) = B_n$ if $n = m$, and is empty otherwise. Morphisms are drawn as braids; see Figure 3.1. Here, B_0 is the empty braid. The monoidal product on objects is given by $n \otimes m := n + m$, with $\mathbb{1} := 0$, and the monoidal product on morphisms is given by putting braids side-by-side.

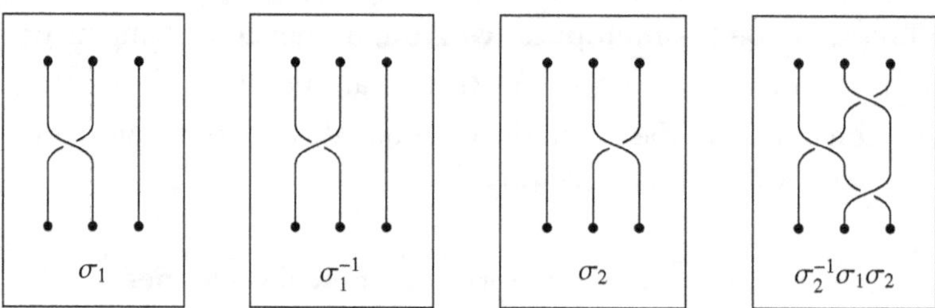

Figure 3.1: Morphisms from [3] to [3] in Braid, read from the top down.

iv. Analytic monoidal categories

- Hilb: The category of Hilbert spaces over field $\Bbbk$ is monoidal; see Definitions 0.52 and 1.3 of Heunen and Vicary [2019] for details.

- FdHilb: The category of finite-dimensional Hilbert spaces is also monoidal.

v. A combinatorial monoidal category

- V-Quiv: The category of quivers $Q := (V, Q_1, s, t)$ (see §1.2.5) for a fixed vertex set V is monoidal. A morphism from (V, Q_1, s, t) to (V, Q_1', s', t') is a function

$f : Q_1 \to Q_1'$ with $s = s'f$ and $t = t'f$. Here, $(Q \otimes Q')_1$ is the set of composable arrows, $\{(a, a') \in Q_1 \times Q_1' \mid s(a) = t'(a')\}$. Also, $(\mathbb{1})_1 = \varnothing$.

§3.1.3. Additive monoidal categories

Now we discuss the interactions between monoidal and additive (or abelian, linear) features of a category.

A monoidal category $(\mathcal{C}, \otimes, \mathbb{1}, a, \ell, r)$ is **additive** if the underlying category $\mathcal{C}$ is additive, and the endofunctors $(X \otimes -)$ and $(- \otimes X)$ on $\mathcal{C}$ are additive, for each $X \in \mathcal{C}$. The latter means that for any object X in $\mathcal{C}$, and any morphisms f and f' in $\mathcal{C}$, we get that as morphisms in $\mathcal{C}$:

$$\mathrm{id}_X \otimes (f + f') = (\mathrm{id}_X \otimes f) + (\mathrm{id}_X \otimes f'), \quad (f + f') \otimes \mathrm{id}_X = (f \otimes \mathrm{id}_X) + (f' \otimes \mathrm{id}_X). \quad (3.2)$$

A monoidal category $(\mathcal{C}, \otimes, \mathbb{1}, a, \ell, r)$ is **abelian** if the underlying category $\mathcal{C}$ is abelian, and the endofunctors $(X \otimes -)$ and $(- \otimes X)$ on $\mathcal{C}$ are additive, for each $X \in \mathcal{C}$.

Further, a monoidal category $(\mathcal{C}, \otimes, \mathbb{1}, a, \ell, r)$ is $(\mathbb{k}\text{-})$**linear** if the underlying category $\mathcal{C}$ is linear, and the endofunctors $(X \otimes -)$ and $(- \otimes X)$ on $\mathcal{C}$ are linear, for each $X \in \mathcal{C}$. The latter means that for any object X in $\mathcal{C}$, any morphisms f and f' in $\mathcal{C}$, and any scalar $\lambda \in \mathbb{k}$, we get that (3.2) holds, along with:

$$\mathrm{id}_X \otimes (\lambda f) = \lambda(\mathrm{id}_X \otimes f), \quad (\lambda f) \otimes \mathrm{id}_X = \lambda(f \otimes \mathrm{id}_X).$$

It is straightforward to see that if $\mathcal{C}$ is additive (resp., abelian, or linear) monoidal, then so is the product category $\mathcal{C} \times \mathcal{C}$.

Additive monoidal categories have the following convenient features.

Lemma 3.3. *Suppose that $\mathcal{C}$ is an additive monoidal category. Then, for any object $X \in \mathcal{C}$:*

$$\mathrm{id}_X \otimes \vec{0} = \vec{0}, \quad \vec{0} \otimes \mathrm{id}_X = \vec{0}, \quad X \otimes 0 \cong 0, \quad 0 \otimes X \cong 0.$$

Moreover, $g \otimes \vec{0} = \vec{0} \otimes g = \vec{0}$, for any morphism g.

Proof. By (3.2), $\mathrm{id}_X \otimes \vec{0} = \mathrm{id}_X \otimes (f - f) = (\mathrm{id}_X \otimes f) - (\mathrm{id}_X \otimes f) = \vec{0}$, for any morphism $f \in \mathcal{C}$. Now for any object $Y \in \mathcal{C}$, a morphism $g : X \otimes 0 \to Y$ in $\mathcal{C}$ must be equal to $g\,(\mathrm{id}_X \otimes \mathrm{id}_0) = g\,(\mathrm{id}_X \otimes \vec{0}) = \vec{0}$. So by uniqueness, $X \otimes 0$ is an initial object. Similarly, $X \otimes 0$ is terminal. Thus, $X \otimes 0$ is a zero object. Likewise, $\vec{0} \otimes \mathrm{id}_X = \vec{0}$ and $0 \otimes X \cong 0$. The last statement follows from the previous statements. $\square$

Moreover, we have the following application of Lemma 2.6.

Lemma 3.4. *When $\mathcal{C}$ is additive monoidal, $\otimes$ distributes over $\square$. That is, for all $X, Y, Z \in \mathcal{C}$:*

$$X \otimes (Y \square Z) \cong (X \otimes Y) \square (X \otimes Z), \quad (Y \square Z) \otimes X \cong (Y \otimes X) \square (Z \otimes X). \qquad \square$$

§3.2. Monoidal functors and equivalence

We introduce monoidal functors, and then discuss isomorphisms and equivalence between monoidal categories. Many examples are also presented. Take $\mathcal{C} := (C, \otimes^{\mathcal{C}}, \mathbb{1}^{\mathcal{C}}, a^{\mathcal{C}}, \ell^{\mathcal{C}}, r^{\mathcal{C}})$ and $\mathcal{D} := (D, \otimes^{\mathcal{D}}, \mathbb{1}^{\mathcal{D}}, a^{\mathcal{D}}, \ell^{\mathcal{D}}, r^{\mathcal{D}})$ to be monoidal categories throughout; superscripts may be omitted later.

§3.2.1. Monoidal functors

A **monoidal functor** from $\mathcal{C}$ to $\mathcal{D}$ consists of the following data.

(a) A functor between the underlying categories $F : \mathcal{C} \to \mathcal{D}$.

(b) (**monoidal product constraint**) A natural transformation

$$F^{(2)} : F(-) \otimes^{\mathcal{D}} F(-) \implies F(- \otimes^{\mathcal{C}} -)$$

of functors from $\mathcal{C} \times \mathcal{C} \to \mathcal{D}$. That is, we have a collection of morphisms $\{F^{(2)}_{X,Y} : F(X) \otimes^{\mathcal{D}} F(Y) \to F(X \otimes^{\mathcal{C}} Y)\}_{X,Y \in \mathcal{C}}$, natural in X and Y.

(c) (**monoidal unit constraint**) A morphism $F^{(0)} : \mathbb{1}^{\mathcal{D}} \to F(\mathbb{1}^{\mathcal{C}})$ in $\mathcal{D}$.

This data must satisfy the commutative diagrams below, for all $X, Y, Z \in \mathcal{C}$.

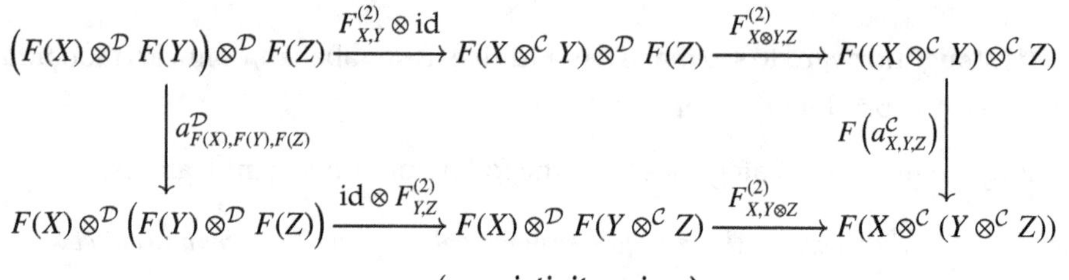

$$
\begin{array}{ccc}
\left(F(X) \otimes^{\mathcal{D}} F(Y)\right) \otimes^{\mathcal{D}} F(Z) \xrightarrow{F^{(2)}_{X,Y} \otimes \mathrm{id}} F(X \otimes^{\mathcal{C}} Y) \otimes^{\mathcal{D}} F(Z) \xrightarrow{F^{(2)}_{X \otimes Y, Z}} F((X \otimes^{\mathcal{C}} Y) \otimes^{\mathcal{C}} Z) \\
\Big\downarrow a^{\mathcal{D}}_{F(X),F(Y),F(Z)} \qquad\qquad\qquad\qquad\qquad\qquad F\left(a^{\mathcal{C}}_{X,Y,Z}\right) \Big\downarrow \\
F(X) \otimes^{\mathcal{D}} \left(F(Y) \otimes^{\mathcal{D}} F(Z)\right) \xrightarrow{\mathrm{id} \otimes F^{(2)}_{Y,Z}} F(X) \otimes^{\mathcal{D}} F(Y \otimes^{\mathcal{C}} Z) \xrightarrow{F^{(2)}_{X, Y \otimes Z}} F(X \otimes^{\mathcal{C}} (Y \otimes^{\mathcal{C}} Z))
\end{array}
$$

(associativity axiom)

$$
\begin{array}{cc}
\mathbb{1}^{\mathcal{D}} \otimes^{\mathcal{D}} F(X) \xrightarrow{\ell^{\mathcal{D}}_{F(X)}} F(X) & \qquad F(X) \otimes^{\mathcal{D}} \mathbb{1}^{\mathcal{D}} \xrightarrow{r^{\mathcal{D}}_{F(X)}} F(X) \\
F^{(0)} \otimes \mathrm{id} \Big\downarrow \qquad \Big\uparrow F(\ell^{\mathcal{C}}_X) & \mathrm{id} \otimes F^{(0)} \Big\downarrow \qquad \Big\uparrow F(r^{\mathcal{C}}_X) \\
F(\mathbb{1}^{\mathcal{C}}) \otimes^{\mathcal{D}} F(X) \xrightarrow{F^{(2)}_{1,X}} F(\mathbb{1}^{\mathcal{C}} \otimes^{\mathcal{C}} X) & F(X) \otimes^{\mathcal{D}} F(\mathbb{1}^{\mathcal{C}}) \xrightarrow{F^{(2)}_{X,1}} F(X \otimes^{\mathcal{C}} \mathbb{1}^{\mathcal{C}})
\end{array}
$$

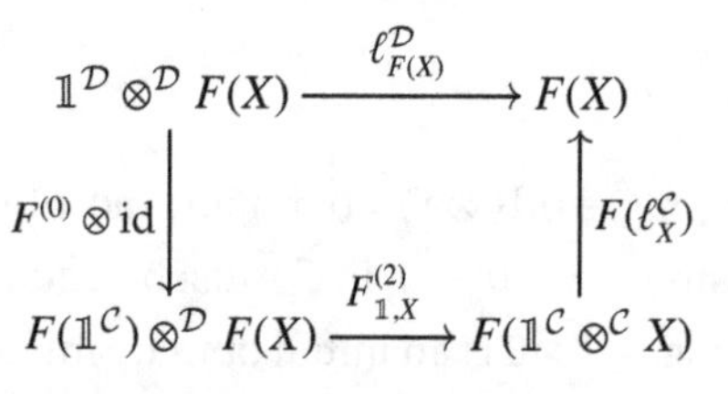

(left unitality axiom) 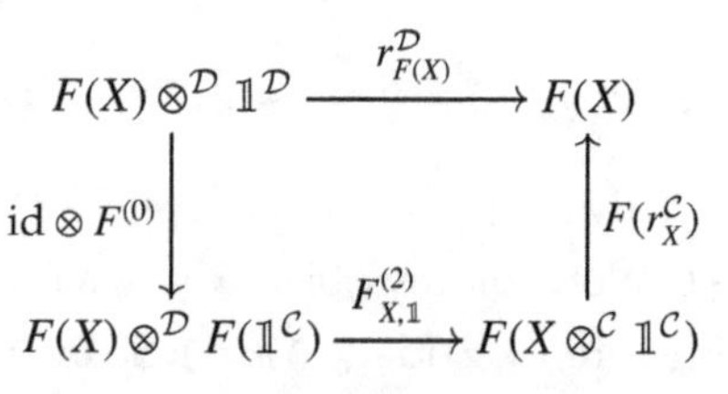   (right unitality axiom)

Remark 3.5. The notation for the monoidal product and unit constraints varies across the literature; sometimes they are given by F^2, F^0 or F_2, F_0. But we use the notation above to not confuse them with self-compositions (e.g., as in F^2), and to avoid using iterated subscripts (e.g., as in $(F_2)_{X,Y}$).

140

A monoidal functor $(F, F^{(2)}, F^{(0)})$ is called **strict** if $\{F^{(2)}_{X,Y}\}_{X,Y \in \mathcal{C}}$ and $F^{(0)}$ are all identity morphisms in $\mathcal{D}$.

A monoidal functor $(F, F^{(2)}, F^{(0)})$ is called **strong** if $\{F^{(2)}_{X,Y}\}_{X,Y \in \mathcal{C}}$ and $F^{(0)}$ are all isos in $\mathcal{D}$. Here, we denote $(F^{(2)}_{X,Y})^{-1}$ by $F^{(-2)}_{X,Y}$, and $(F^{(0)})^{-1}$ by $F^{(-0)}$.

Remark 3.6. Sometimes in the literature, a monoidal functor as defined above is called **lax (monoidal)**, and monoidal functors may be assumed to be strong.

One example of a monoidal functor is the **identity monoidal functor**:

$$\mathrm{Id}_{(\mathcal{C}, \otimes)} := (\mathrm{Id}_{\mathcal{C}}, \ \mathrm{Id}^{(2)}, \ \mathrm{Id}^{(0)}) : \mathcal{C} \to \mathcal{C},$$

where $\mathrm{Id}^{(2)}_{X,Y} := \mathrm{id}_{X \otimes_{\mathcal{C}} Y}$ for $X, Y \in \mathcal{C}$, and $\mathrm{Id}^{(0)} := \mathrm{id}_{\mathbb{1}_{\mathcal{C}}}$; this is strict and strong.

The composition of monoidal functors is also monoidal; see Exercise 3.4.

§3.2.2. Isomorphism and equivalence of monoidal categories

We have the following notions of 'sameness' for monoidal categories.

We say that monoidal categories $\mathcal{C}$ and $\mathcal{D}$ are **isomorphic as monoidal categories** (or are **monoidally isomorphic**) if there exists a strong monoidal functor $(F, F^{(2)}, F^{(0)}) : \mathcal{C} \to \mathcal{D}$, such that the functor F is an isomorphism of categories. Here, we write $\mathcal{C} \overset{\otimes}{\cong} \mathcal{D}$.

Likewise, $\mathcal{C}$ and $\mathcal{D}$ are **equivalent as monoidal categories** (or are **monoidally equivalent**) if there is a strong monoidal functor $(F, F^{(2)}, F^{(0)}) : \mathcal{C} \to \mathcal{D}$, such that F is an equivalence of categories. In this case, we write $\mathcal{C} \overset{\otimes}{\simeq} \mathcal{D}$.

One can characterize these notions by preserving the structure between monoidal functors as follows.

Take monoidal functors $(F, F^{(2)}, F^{(0)})$ and $(F', F'^{(2)}, F'^{(0)})$ between monoidal categories $\mathcal{C}$ and $\mathcal{D}$. Then, a **monoidal natural transformation** (resp., **monoidal natural isomorphism**) from $(F, F^{(2)}, F^{(0)})$ to $(F', F'^{(2)}, F'^{(0)})$ is a natural transformation $\phi : F \Rightarrow F'$ (resp., natural isomorphism $\phi : F \overset{\sim}{\Rightarrow} F'$), such that the following diagrams commute for all $X, Y \in \mathcal{C}$.

$$
\begin{array}{ccc}
F(X) \otimes^{\mathcal{D}} F(Y) & \xrightarrow{\ F^{(2)}_{X,Y}\ } & F(X \otimes^{\mathcal{C}} Y) \\
{\scriptstyle \phi_X \otimes \phi_Y} \downarrow & & \downarrow {\scriptstyle \phi_{X \otimes Y}} \\
F'(X) \otimes^{\mathcal{D}} F'(Y) & \xrightarrow{\ F'^{(2)}_{X,Y}\ } & F'(X \otimes^{\mathcal{C}} Y)
\end{array}
\qquad
\begin{array}{ccc}
 & \xrightarrow{F^{(0)}} & F(\mathbb{1}^{\mathcal{C}}) \\
\mathbb{1}^{\mathcal{D}} & & \downarrow {\scriptstyle \phi_{\mathbb{1}^{\mathcal{C}}}} \\
 & \xrightarrow[F'^{(0)}]{} & F'(\mathbb{1}^{\mathcal{C}})
\end{array}
$$

In this case, we write $\phi : F \overset{\otimes}{\Rightarrow} F'$ (resp., $\phi : F \overset{\otimes}{\underset{\sim}{\Rightarrow}} F'$, or just $F \overset{\otimes}{\cong} F'$).

Now consider the next result, the proof of which is Exercise 3.5.

Proposition 3.7. *Take monoidal categories C and D. Then, the statements below hold.*

(a) *We have that $C \overset{\otimes}{\cong} D$ if and only if there exists monoidal functors $(F, F^{(2)}, F^{(0)}) : C \to D$ and $(G, G^{(2)}, G^{(0)}) : D \to C$ such that as monoidal functors:*

$$GF \overset{\otimes}{\cong} \mathrm{Id}_{(C, \otimes^C)} \qquad and \qquad FG \overset{\otimes}{\cong} \mathrm{Id}_{(D, \otimes^D)}.$$

(b) *We have that $C \overset{\otimes}{\simeq} D$ if and only if there exists monoidal functors $(F, F^{(2)}, F^{(0)}) : C \to D$ and $(G, G^{(2)}, G^{(0)}) : D \to C$ such that as monoidal functors:*

$$GF \overset{\otimes}{\cong} \mathrm{Id}_{(C, \otimes^C)} \qquad and \qquad FG \overset{\otimes}{\cong} \mathrm{Id}_{(D, \otimes^D)}. \qquad \square$$

See also Exercise 3.16 for practice with monoidal equivalence.

§3.2.3. Examples of monoidal functors and equivalence

Consider the following examples of monoidal functors and monoidal equivalence.

Example 3.8. For a group G, the forgetful functor

$$F := \mathrm{Forg} \colon\ G\text{-Mod} \to \mathsf{Vec}, \quad (V, \triangleright : G \times V \to V) \mapsto V$$

is monoidal. Here, the monoidal product constraint is given by

$$F^{(2)}_{(V,\triangleright),(V',\triangleright')} : F(V,\triangleright) \otimes^{\mathsf{Vec}} F(V',\triangleright') \to F\big((V,\triangleright) \otimes^{G\text{-Mod}} (V',\triangleright')\big),$$

where we have that

- $F(V,\triangleright) \otimes^{\mathsf{Vec}} F(V',\triangleright') = F(V,\triangleright) \otimes_{\Bbbk} F(V',\triangleright') = V \otimes_{\Bbbk} V'$, and

- $F\big((V,\triangleright) \otimes^{G\text{-Mod}} (V',\triangleright')\big) = F(V \otimes_{\Bbbk} V', \blacktriangleright) = V \otimes_{\Bbbk} V'$.

Now we define $F^{(2)}_{(V,\triangleright),(V',\triangleright')} := \mathrm{id}_{V \otimes_{\Bbbk} V'}$. Moreover, for $F^{(0)} : \mathbb{1}^{\mathsf{Vec}} \to F(\mathbb{1}^{G\text{-Mod}})$, where $\mathbb{1}^{\mathsf{Vec}} = \Bbbk$ and $F(\mathbb{1}^{G\text{-Mod}}) = F(\Bbbk, \triangleright) = \Bbbk$, we also define $F^{(0)} := \mathrm{id}_{\Bbbk}$. So, $(F, F^{(2)}, F^{(0)})$ is strict and strong monoidal. See Exercise 3.6 for similar examples.

Not all forgetful monoidal functors are strong; consider the example below.

Example 3.9. The forgetful functor $F : \mathsf{FdVec} \to \mathsf{Set}$ is monoidal, but not strong monoidal. In particular, $F^{(0)} : \{\cdot\} = \mathbb{1}^{\mathsf{Set}} \to F(\mathbb{1}^{\mathsf{FdVec}}) = \Bbbk_{\mathsf{set}}$ cannot be an isomorphism of sets.

Example 3.10. For a group G, the monoidal categories, Vec_G and Vec'_G are isomorphic (in fact, equal) as categories. But the functor $\mathrm{Id} : \mathsf{Vec}_G \to \mathsf{Vec}'_G$ does not admit a monoidal structure. Indeed, let us consider the left unitality axiom,

$$\ell_V^{\mathsf{Vec}'_G} = \ell_V^{\mathsf{Vec}_G} \circ F^{(2)}_{\mathbb{1}^{\mathsf{Vec}_G}, V} \circ (F^{(0)} \otimes^{\mathsf{Vec}'_G} \mathrm{id}_V),$$

which is supposed to hold in Vec'_G, for $V \in \mathsf{Vec}_G$. Now for any $g \in G$, the degree g component of $\ell_V^{\mathsf{Vec}'_G}$ is given by $(\ell_V^{\mathsf{Vec}'_G})_g : \Bbbk \otimes_{\Bbbk} V_g \xrightarrow{\sim} V_g$. On the other hand, for $g \neq e$, we have that $(\mathbb{1}^{\mathsf{Vec}_G})_g$ is the zero vector space. So $(F^{(0)})_g$ is the zero linear map for $g \neq e$, and the left unitality axiom cannot hold in degree $g \neq e$.

For the examples below, recall the tensor algebra and symmetric algebra constructions from §§1.2.2, 1.2.3, and also see Exercise 1.24.

Example 3.11. We denote the monoidal category Vec by $\mathsf{Vec}_{\otimes_\Bbbk}$ to emphasize its monoidal structure, and recall $\mathsf{Vec}_\oplus = \mathsf{Vec}_\sqcup$. Consider the functors below:

$$T : \mathsf{Vec} \to \mathsf{Alg} \ \ \text{(tensor algebra)}, \qquad S : \mathsf{Vec} \to \mathsf{Alg} \ \ \text{(symmetric algebra)}.$$

(a) We have a strong monoidal functor, $(T, T^{(2)}, T^{(0)}) : \mathsf{Vec}_\oplus \to \mathsf{Alg}_\circledast$, where $T^{(2)}_{V,W}$ is given in Exercise 1.24(iv) for all $V, W \in \mathsf{Vec}$, and $T^{(0)} = \mathrm{id}_\Bbbk$.

(b) Moreover, the isomorphism in Exercise 1.24(i) gives the monoidal product constraint for a monoidal functor, $(S, S^{(2)}, S^{(0)}) : \mathsf{Vec}_\oplus \to \mathsf{Alg}_{\otimes_\Bbbk}$. One can also show that $S^{(0)} = \mathrm{id}_\Bbbk$ satisfies the monoidal unit axioms.

The next set of examples involve actions on categories. In Chapter 1, an action of one algebraic structure S on another algebraic structure U is a structure-preserving map ρ from S to a certain collection of endomorphisms of U. We will mimic this here, where ρ will be replaced by a monoidal functor.

Example 3.12. For a monoidal category $(\mathcal{C}, \otimes, \mathbb{1}, a, \ell, r)$, we will write $(\mathcal{C}, \otimes)$ for short, and we will write $\mathcal{C}$ for its underlying category.

(a) The **left regular action** on $(\mathcal{C}, \otimes)$ on itself is a strong monoidal functor,

$$\rho : (\mathcal{C}, \otimes) \to \mathrm{End}(\mathcal{C}), \quad X \mapsto (X \otimes -),$$

where, for all $X, Y, Z \in \mathcal{C}$, we define

$$\rho^{(2)}_{X,Y}(Z) := a^{-1}_{X,Y,Z} : X \otimes (Y \otimes Z) \xrightarrow{\sim} (X \otimes Y) \otimes Z, \qquad \rho^{(0)}(Z) := \ell_Z^{-1} : Z \xrightarrow{\sim} \mathbb{1} \otimes Z.$$

(b) The **right regular action** on $(\mathcal{C}, \otimes)$ on itself is a strong monoidal functor,

$$\rho : \mathcal{C}^{\otimes\mathrm{op}} := (\mathcal{C}, \otimes^{\mathrm{op}}) \to \mathrm{End}(\mathcal{C}), \quad X \mapsto (- \otimes X),$$

where, for all $X, Y, Z \in \mathcal{C}$, we define

$$\rho^{(2)}_{X,Y}(Z) := a_{Z,Y,X} : (Z \otimes Y) \otimes X \xrightarrow{\sim} Z \otimes (Y \otimes X), \qquad \rho^{(0)}(Z) := r_Z^{-1} : Z \xrightarrow{\sim} Z \otimes \mathbb{1}.$$

In general, we define the following notions. See also Exercise 3.7.

- A **left action of a monoidal category** $(\mathcal{C}, \otimes)$ **on a category** $\mathcal{A}$ is, by definition, a strong monoidal functor:

$$\rho : (\mathcal{C}, \otimes) \to \mathrm{End}(\mathcal{A}).$$

- A **right action of a monoidal category** $(C, \otimes)$ **on a category** $\mathcal{A}$ is, by definition, a strong monoidal functor:

$$\rho : C^{\otimes \mathrm{op}} := (C, \otimes^{\mathrm{op}}) \to \mathrm{End}(\mathcal{A}).$$

Next, we will consider actions of monoids and groups on categories.

Example 3.13. Recall the monoidal categories $\underline{N}$ and $\underline{G}$, for an additive monoid N and group G, respectively.

(a) We have a strong monoidal functor, $\rho : \underline{N} \to \mathrm{End}(\mathrm{Vec}_N)$, where for $n \in N$,

$$\rho(n) : \mathrm{Vec}_N \to \mathrm{Vec}_N, \quad \bigoplus_{m \in N} V_m \mapsto \bigoplus_{m \in N} V_{n+m}.$$

(b) We have a strong monoidal functor, $\rho : \underline{G} \to \mathrm{Aut}(\mathrm{Vec}_G)$, where for $g \in G$,

$$\rho(g) : \mathrm{Vec}_G \xrightarrow{\sim} \mathrm{Vec}_G, \quad \bigoplus_{h \in G} V_h \mapsto \bigoplus_{h \in G} V_{ghg^{-1}}.$$

Determining $\rho^{(2)}$ and $\rho^{(0)}$ in each case is Exercise 3.8.

In general, we define the following notions.

- An **action of a monoid** N **on a category** $\mathcal{A}$ is defined to be a strong monoidal functor, $\rho : \underline{N} \to \mathrm{End}(\mathcal{A})$.

- An **action of a group** G **on a category** $\mathcal{A}$ is defined to be a strong monoidal functor, $\rho : \underline{G} \to \mathrm{Aut}(\mathcal{A})$.

- One can also consider the monoidal category $\mathrm{Aut}_\otimes(C)$ of monoidal auto-equivalences of a monoidal category C. Then an **action of a group** G **on a monoidal category** C is a strong monoidal functor, $\rho : \underline{G} \to \mathrm{Aut}_\otimes(C)$.

The monoidal functors below are vital tools in Quantum Topology.

Example 3.14. The functor $Z : n\mathrm{Cob} \to \mathrm{FinHilb}$ (or $\mathrm{Vec}_{\mathbb{C}}$), the *Topological Quantum Field Theory*, mentioned in §2.3.2 can be upgraded to a monoidal functor. See the textbook by Kock [2004] for the case $n = 2$, where the target category is $\mathrm{Vec}_{\mathbb{C}}$.

Now we provide an example of a monoidal equivalence.

Example 3.15. Even though the category A-Mod is not monoidal for an arbitrary algebra A, we have that the endofunctor category $\mathrm{End}(A\text{-Mod})$ is monoidal. In fact, as monoidal categories, we have that

$$\mathrm{End}(A\text{-Mod}) \overset{\otimes}{\cong} A\text{-Bimod}.$$

Here, the underlying functor of the monoidal equivalence is defined by

$$\rho : A\text{-Bimod} \to \mathrm{End}(A\text{-Mod}), \quad {}_A V_A \mapsto ({}_A V_A) \otimes_A -.$$

Namely, by taking $a^A := a^{A\text{-Bimod}}$ and $\ell^A := \ell^{A\text{-Bimod}}$, we then get the following monoidal product and monoidal unit constraints for ρ:

$$\rho^{(2)}_{V,W}(Z) := (a^A_{V,W,Z})^{-1} : V \otimes_A (W \otimes_A Z) \xrightarrow{\sim} (V \otimes_A W) \otimes_A Z,$$

$$\rho^{(0)}(Z) := (\ell^A_Z)^{-1} : Z \xrightarrow{\sim} A \otimes_A Z,$$

for all $V, W \in A\text{-Bimod}$ and $Z \in A\text{-Mod}$.

Example 3.16. Again, A-Mod is not monoidal for an arbitrary algebra A, but we do have that $\Bbbk G$-Mod is monoidal, for a finite group G. Namely, for V, W in $\Bbbk G$-Mod with action $\triangleright$, we define an action of $\Bbbk G$ on $V \otimes W = V \otimes_\Bbbk W$ and on $\mathbb{1} = \Bbbk$:

$$\left(\textstyle\sum_{g\in G} \lambda_g g\right) \blacktriangleright (v \otimes_\Bbbk w) \quad := \textstyle\sum_{g\in G} \lambda_g (g \triangleright v) \otimes_\Bbbk (g \triangleright w),$$

$$\left(\textstyle\sum_{g\in G} \lambda_g g\right) \blacktriangleright 1_\Bbbk \quad := \textstyle\sum_{g\in G} \lambda_g (g \triangleright 1_\Bbbk) = \textstyle\sum_{g\in G} \lambda_g,$$

for $g \in G$, $v \in V$, and $w \in W$. Then, the isomorphism $G\text{-Mod} \cong \Bbbk G\text{-Mod}$ from Exercise 2.30 can be upgraded to a monoidal isomorphism

$$G\text{-Mod} \stackrel{\otimes}{\cong} \Bbbk G\text{-Mod}.$$

See Exercise 3.6(c).

Deriving more examples of monoidal functors is Exercise 3.9.

§3.3. Module categories and bimodule categories

In this part, we discuss module categories over monoidal categories, which are defined analogously to modules over algebras; see §1.3.2. This concept might seem familiar given a certain exercise that was previously assigned in this chapter (find it!). On a related note, module categories can be viewed as *representations of monoidal categories*. Many examples of module categories are provided below, and these concepts are also extended to bimodule categories.

§3.3.1. Module categories and module functors

Fix a monoidal category $\mathcal{C} := (\mathcal{C}, \otimes, \mathbb{1}, a, \ell, r)$.

A **left $\mathcal{C}$-module category** consists of the following data.

(a) A category $\mathcal{M}$.

(b) (**left action bifunctor**) A bifunctor $\triangleright : \mathcal{C} \times \mathcal{M} \to \mathcal{M}$.

(c) (**left module associativity constraint**) A natural isomorphism

$$m : \triangleright \circ (\otimes \times \mathrm{Id}_\mathcal{M}) \xRightarrow{\sim} \triangleright \circ (\mathrm{Id}_\mathcal{C} \times \triangleright)$$

of functors from $\mathcal{C} \times \mathcal{C} \times \mathcal{M}$ to $\mathcal{M}$. That is, we have a collection of isos

$$\{m_{X,Y,M} : (X \otimes Y) \rhd M \xrightarrow{\sim} X \rhd (Y \rhd M)\}_{X,Y \in \mathcal{C}, M \in \mathcal{M}},$$

natural in X, Y, and M.

(d) (**left module unitality constraint**) A natural isomorphism

$$p : \mathbb{1} \rhd \mathrm{Id}_{\mathcal{M}} \ \Rightarrow \ \mathrm{Id}_{\mathcal{M}}$$

of functors from $\mathcal{M}$ to $\mathcal{M}$. That is, we have a collection of isos

$$\{p_M : \mathbb{1} \rhd M \xrightarrow{\sim} M\}_{M \in \mathcal{M}},$$

natural in M.

The following diagrams must also commute, for all $X, Y, Z \in \mathcal{C}$ and $M \in \mathcal{M}$.

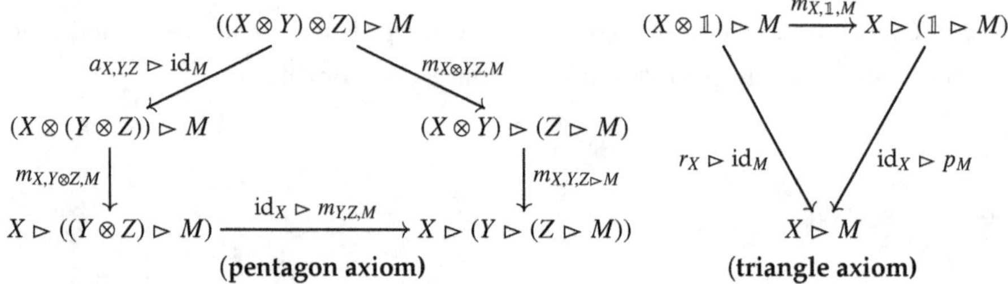

(pentagon axiom)

(triangle axiom)

Likewise, a **right $\mathcal{C}$-module category** consists of the following data.

(a) A category $\mathcal{M}$.

(b) (**right action bifunctor**) A bifunctor $\lhd : \mathcal{M} \times \mathcal{C} \to \mathcal{M}$.

(c) (**right module associativity constraint**) A natural isomorphism

$$n : \lhd \circ (\mathrm{Id}_{\mathcal{C}} \times \otimes) \ \Rightarrow \ \lhd \circ (\lhd \times \mathrm{Id}_{\mathcal{C}}).$$

(d) (**right module unitality constraint**) A natural isom. $q : \mathrm{Id}_{\mathcal{M}} \lhd \mathbb{1} \ \Rightarrow \ \mathrm{Id}_{\mathcal{M}}$.

The following diagrams must also commute, for all $X, Y, Z \in \mathcal{C}$ and $M \in \mathcal{M}$.

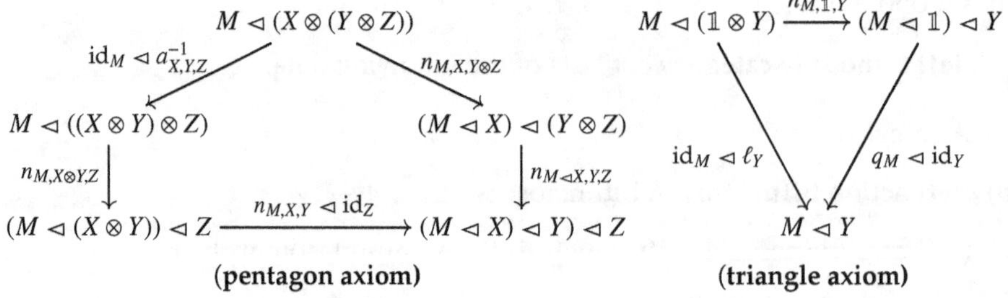

(pentagon axiom)

(triangle axiom)

Structure versus 'structure category'. We have seen previously in §§2.4.1 and 2.4.2 that when moving up a level of abstraction, one replaces equalities with isomorphisms. Similar moves are needed when generalizing algebraic structures to the categorical framework as follows.

- underlying structure $\rightsquigarrow$ category
  (e.g., underlying vector space V of a left A-module $\rightsquigarrow$ category $\mathcal{M}$)

- structure morphisms $\rightsquigarrow$ structure functors
  (e.g., action map $\triangleright$ $\rightsquigarrow$ action bifunctor $\triangleright$)

- compatibility equations amongst structure morphisms $\rightsquigarrow$
  compatibility natural isomorphisms amongst structure functors
  (e.g., module associativity equation $\rightsquigarrow$ module associativity constraint)

In short, this is how one forms a 'structure category'. For instance, complete Exercise 3.10 after finishing §3.3.1 (and before reading §3.3.3).

We also want to transport from one $\mathcal{C}$-module category to another. Fix left $\mathcal{C}$-module categories $(\mathcal{M}, \triangleright, m, p)$ and $(\mathcal{M}', \triangleright', m', p')$.

A **left $\mathcal{C}$-module functor** from $(\mathcal{M}, \triangleright, m, p)$ to $(\mathcal{M}', \triangleright', m', p')$ consists of the following data.

(a) A functor between the underlying categories $F : \mathcal{M} \to \mathcal{M}'$.

(b) (**left module functor constraint**) A natural isomorphism

$$s : F \circ \triangleright \;\overset{\Rightarrow}{\sim}\; \triangleright' \circ (\mathrm{Id}_{\mathcal{C}} \times F)$$

from $\mathcal{C} \times \mathcal{M}$ to $\mathcal{M}'$, i.e., a collection of isos that are natural in each slot:

$$\{s_{X,M} : F(X \triangleright M) \xrightarrow{\sim} X \triangleright' F(M)\}_{X \in \mathcal{C}, M \in \mathcal{M}}.$$

We also require that the diagrams below commute, for all $X, Y \in \mathcal{C}$, $M \in \mathcal{M}$.

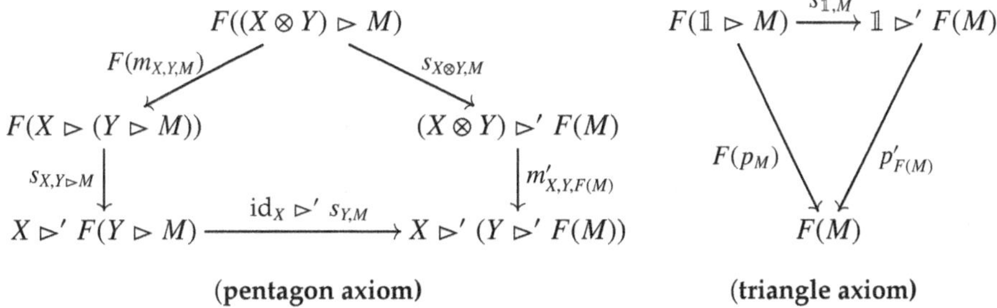

(pentagon axiom)   (triangle axiom)

Likewise, a **right C-module functor** between right C-module categories,

$$(M, \lhd, n, q) \longrightarrow (M', \lhd', n', q'),$$

is a functor $F : M \to M'$ equipped with **right module functor constraint**,

$$t : (F \circ \lhd) \overset{\cong}{\Rightarrow} (\lhd' \circ (F \times \mathrm{Id}_C)),$$

satisfying the commutative diagrams below, for all $X, Y \in C$ and $M \in M$.

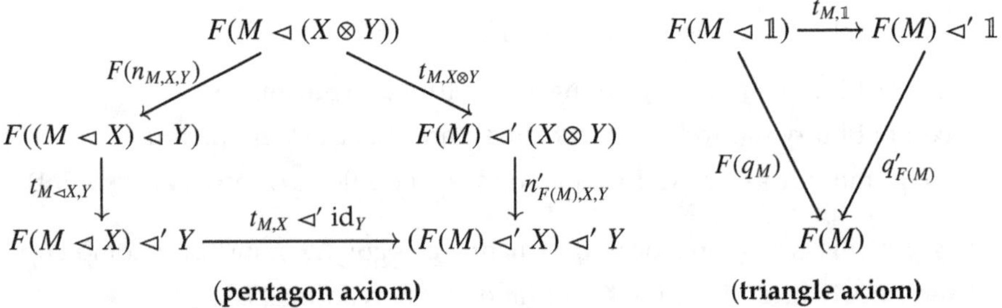

(pentagon axiom)   **(triangle axiom)**

Two left (resp., right) C-module categories M and M' are said to be **equivalent as C-module categories** if there exists a left (resp., right) C-module functor $(F, s) : M \to M'$ (resp., $(F, t) : M \to M'$), where F is an equivalence of the underlying categories.

The collection of C-module categories forms a category itself. Consider the following notation.

- C-Mod: The category of left C-module categories as objects, and left C-module functors as morphisms.

- Mod-C: The category of right C-module categories as objects, and right C-module functors as morphisms.

With module categories, we can also build monoidal categories resembling endomorphism algebras. See Exercise 3.11 after reading Example 3.17.

- $\mathrm{End}_{C\text{-Mod}}(M)$: the category of left C-module endofunctors of $M \in C$-Mod, with $\otimes$ given by composition $\circ$, and with $\mathbb{1} = \mathrm{Id}_M$.

- $\mathrm{End}_{\text{Mod-}C}(M)$: the category of right C-module endofunctors of module category $M \in$ Mod-C, with $\otimes$ given by composition $\circ$, and with $\mathbb{1} = \mathrm{Id}_M$.

§3.3.2. Examples of module categories

Next, we display examples of module categories over a monoidal category. Completing the details of these examples comprise Exercise 3.12.

The first set of examples pertain to an arbitrary monoidal category C.

Example 3.17. We have that $\mathcal{C}$ is a left and a right $\mathcal{C}$-module category with $\rhd = \otimes^{\mathcal{C}}$ and $\lhd = \otimes^{\mathcal{C}}$, resp.; these are **regular $\mathcal{C}$-module categories**, denoted by $\mathcal{C}_{\mathrm{reg}}$.

Example 3.18. Take a strong monoidal functor $F : \mathcal{C} \to \mathcal{D}$, with a left $\mathcal{D}$-module category $(\mathcal{M}, \rhd)$. Then, we can define a left $\mathcal{C}$-module category $(\mathcal{M}, \blacktriangleright)$ with

$$X \blacktriangleright M := F(X) \rhd M,$$

for all $X \in \mathcal{C}$ and $M \in \mathcal{M}$. Here, we say that $(\mathcal{M}, \blacktriangleright) \in \mathcal{C}\text{-Mod}$ is the **restriction of** $(\mathcal{M}, \rhd) \in \mathcal{D}\text{-Mod}$ **to $\mathcal{C}$ along** F. Compare to §1.4.4.

For the next examples, let G be a group, and let A and B be $\Bbbk$-algebras.

Example 3.19. Consider the monoidal category G-Mod from §3.1.2i. Then, for any subgroup H of G, we have that H-Mod is a left (or a right) module category over G-Mod. This holds by applying Exercise 3.6 and Example 3.18 by taking

$$F = \mathrm{Res}^{G}_{H}, \quad \text{with } \mathcal{C} = G\text{-Mod}, \quad \mathcal{D} = \mathcal{M} = H\text{-Mod}.$$

In particular, Vec is a module category over G-Mod.

Example 3.20. Consider the monoidal category Vec_{G} from §3.1.2i, and recall G-actions on categories discussed after Example 3.13. Then, a category $\mathcal{M}$ is a left Vec_{G}-module category if and only if $\mathcal{M}$ admits an action of G, that is, if there exists a strong monoidal functor $\rho : \underline{G} \to \mathrm{Aut}(\mathcal{M})$.

Example 3.21. Consider the monoidal category A-Bimod from §3.1.2i. Then, we obtain that (A, B)-Bimod is a left module category over A-Bimod, and (B, A)-Bimod is a right module category over A-Bimod.

We now discuss module categories over monoidal endofunctor categories.

Example 3.22. Take the monoidal category $\mathrm{End}(\mathcal{A})$ from §3.1.2ii.

(a) We have that $\mathcal{A}$ is a left $\mathrm{End}(\mathcal{A})$-module category with $F \rhd M := F(M)$, for all $F \in \mathrm{End}(\mathcal{A})$ and $M \in \mathcal{A}$.

(b) For another category $\mathcal{B}$, we have that:

- $\mathrm{Fun}(\mathcal{B}, \mathcal{A})$ is a left $\mathrm{End}(\mathcal{A})$-module category with $F \rhd G := F \circ G$, for all $F \in \mathrm{End}(\mathcal{A})$ and $G \in \mathrm{Fun}(\mathcal{B}, \mathcal{A})$;

- $\mathrm{Fun}(\mathcal{A}, \mathcal{B})$ is a right $\mathrm{End}(\mathcal{A})$-module category with $G \lhd F := G \circ F$, for all $F \in \mathrm{End}(\mathcal{A})$ and $G \in \mathrm{Fun}(\mathcal{A}, \mathcal{B})$.

Example 3.23. Take the monoidal category $\mathrm{End}_{\mathrm{Mod}\text{-}\mathcal{C}}(\mathcal{M})$, for $\mathcal{M} \in \mathrm{Mod}\text{-}\mathcal{C}$, from §3.3.1. For any $\mathcal{N} \in \mathrm{Mod}\text{-}\mathcal{C}$, take $\mathrm{Fun}_{\mathrm{Mod}\text{-}\mathcal{C}}(\mathcal{M}, \mathcal{N})$ be the category of right $\mathcal{C}$-module functors from $\mathcal{M}$ to $\mathcal{N}$. Then, we have the following statements.

(a) $\mathrm{Fun}_{\mathrm{Mod}\text{-}\mathcal{C}}(\mathcal{N}, \mathcal{M})$ is a left module category over $\mathrm{End}_{\mathrm{Mod}\text{-}\mathcal{C}}(\mathcal{M})$ with $F \rhd G := F \circ G$, for all $F \in \mathrm{End}_{\mathrm{Mod}\text{-}\mathcal{C}}(\mathcal{M})$ and $G \in \mathrm{Fun}_{\mathrm{Mod}\text{-}\mathcal{C}}(\mathcal{N}, \mathcal{M})$;

(b) $\mathrm{Fun}_{\mathrm{Mod}\text{-}\mathcal{C}}(\mathcal{M}, \mathcal{N})$ is a right module category over $\mathrm{End}_{\mathrm{Mod}\text{-}\mathcal{C}}(\mathcal{M})$ with $G \lhd F := G \circ F$, for all $F \in \mathrm{End}_{\mathrm{Mod}\text{-}\mathcal{C}}(\mathcal{M})$ and $G \in \mathrm{Fun}_{\mathrm{Mod}\text{-}\mathcal{C}}(\mathcal{M}, \mathcal{N})$.

We now discuss a module category over a topological monoidal category.

Example 3.24. Consider the monoidal category Braid from §3.1.2iii, along with a similar monoidal category below.

- Perm: The objects are natural numbers $\mathbb{N}$. Moreover, for an object $n \in \mathbb{N}$, consider the n-th **permutation group** S_n, which has generators $\tau_1, \ldots, \tau_{n-1}$ and relations:

$$\tau_i^2 = e, \qquad \text{for } i = 1, \ldots, n-1;$$
$$(\tau_i \tau_{i+1})^3 = e, \quad \text{for } i = 1, \ldots, n-2;$$
$$\tau_i \tau_j = \tau_j \tau_i, \qquad \text{for } |i - j| > 1.$$

Then, $\mathrm{Hom}_{\mathrm{Perm}}(n, m) = S_n$ if $n = m$ and is empty otherwise. Morphisms are drawn as braids without overlaps, as we depict for $n = 3$ below. The monoidal structure on Perm to similar to that for Braid.

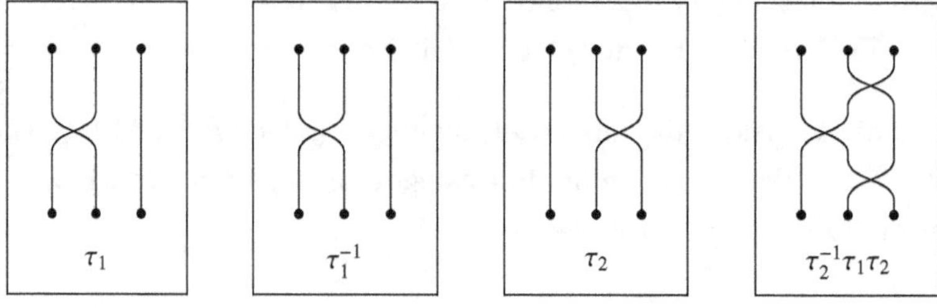

Figure 3.2: Examples of morphisms in the monoidal category Perm, cf. Fig. 3.1.

Since S_n is a quotient group of the braid group B_n, we get a functor from Braid to Perm sending an object n to n, and by sending a morphism in B_n to its homomorphic image in S_n. Moreover, this functor can be upgraded to a strong monoidal functor. Thus, Perm is a left module category over Braid by Example 3.18.

§3.3.3. Bimodule categories

Fix a monoidal categories $\mathcal{C}$ and $\mathcal{D}$. A $(\mathcal{C}, \mathcal{D})$-**bimodule category** is a tuple

$$(\mathcal{M}, \rhd, \lhd, \ m, n, p, q, b),$$

where

(a) $(\mathcal{M}, \rhd, m, p)$ is a left $\mathcal{C}$-module category;

(b) $(\mathcal{M}, \lhd, n, q)$ is a right $\mathcal{D}$-module category; and

(c) (**middle associativity constraint**) There exists a natural isomorphism

$$b : \lhd \circ (\rhd \times \mathrm{Id}_{\mathcal{D}}) \;\overset{\simeq}{\Rightarrow}\; \rhd \circ (\mathrm{Id}_{\mathcal{C}} \times \lhd)$$

from $\mathcal{C} \times \mathcal{M} \times \mathcal{D}$ to $\mathcal{M}$, such that the following diagrams commute, for all $X, Y \in \mathcal{C}$ and $Z, W \in \mathcal{D}$ and $M \in \mathcal{M}$.

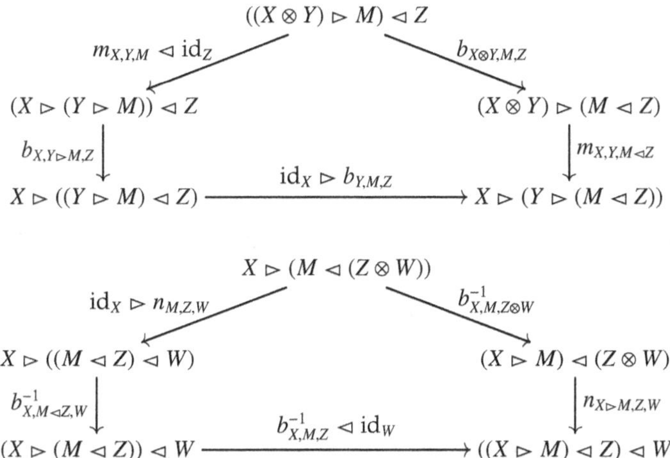

A $(\mathcal{C}, \mathcal{C})$-bimodule category is referred to as a $\mathcal{C}$-**bimodule category**.

A $(\mathcal{C}, \mathcal{D})$-**bimodule functor** between $(\mathcal{C}, \mathcal{D})$-bimodule categories,

$$(\mathcal{M}, \rhd, \lhd, m, n, p, q, b) \to (\mathcal{M}', \rhd', \lhd', m', n', p', q', b'),$$

is a triple (F, s, t), where $(F, s) : (\mathcal{M}, \rhd, m, p) \to (\mathcal{M}', \rhd', m', p')$ is a left $\mathcal{C}$-module functor, and $(F, t) : (\mathcal{M}, \lhd, n, q) \to (\mathcal{M}', \lhd', n', q')$ is a right $\mathcal{D}$-module functor, such that the following diagram commutes.

$$
\begin{array}{ccccc}
F((X \rhd M) \lhd Y) & \overset{t_{X \rhd M, Y}}{\longrightarrow} & F(X \rhd M) \lhd' Y & \overset{s_{X,M} \lhd' \mathrm{id}_Y}{\longrightarrow} & (X \rhd' F(M)) \lhd' Y \\
{\scriptstyle F(b_{X,M,Y})}\big\downarrow & & & & \big\downarrow{\scriptstyle b'_{X,F(M),Y}} \\
F(X \rhd (M \lhd Y)) & \overset{s_{X,M \lhd Y}}{\longrightarrow} & X \rhd' F(M \lhd Y) & \overset{\mathrm{id}_X \lhd' t_{M,Y}}{\longrightarrow} & X \rhd' (F(M) \lhd' Y)
\end{array}
$$

The collection of $(\mathcal{C}, \mathcal{D})$-bimodule categories also form a category:

- $(\mathcal{C}, \mathcal{D})$-Bimod: The category of $(\mathcal{C}, \mathcal{D})$-bimodule categories as objects, and for morphisms we use $(\mathcal{C}, \mathcal{D})$-bimodule functors.

- $\mathcal{C}$-Bimod: This denotes the category $(\mathcal{C}, \mathcal{C})$-Bimod.

Two $(\mathcal{C}, \mathcal{D})$-bimodule categories $\mathcal{M}$ and $\mathcal{M}'$ are **equivalent as $(\mathcal{C}, \mathcal{D})$-bimodule categories** if there exists a $(\mathcal{C}, \mathcal{D})$-bimodule functor $(F, s, t) : \mathcal{M} \to \mathcal{M}'$, where F is an equivalence of the underlying categories.

Consider the example of a bimodule category below.

Example 3.25. We have that $\mathcal{C}$ is a $\mathcal{C}$-bimodule category with $\rhd = \lhd = \otimes^{\mathcal{C}}$ and $b_{X,Y,Z} = a^{\mathcal{C}}_{X,Y,Z}$; this is the **regular $\mathcal{C}$-bimodule category**, also denoted by $\mathcal{C}_{\mathrm{reg}}$.

Deriving more examples of bimodule categories comprises Exercise 3.13.

§3.3.4. Module categories over additive monoidal categories

To have a rich theory of module categories $\mathcal{M}$ over monoidal categories $\mathcal{C}$, it is common to impose conditions on $\mathcal{M}$ that reflect the behavior of $\mathcal{C}$.

For instance, we define a **left module category over an additive monoidal category** $\mathcal{C}$ to be a left $\mathcal{C}$-module category $(\mathcal{M}, \rhd)$ such that $\mathcal{M}$ is additive, and the functors $(X \rhd -) : \mathcal{M} \to \mathcal{M}$ and $(- \rhd M) : \mathcal{C} \to \mathcal{M}$ are additive, for each $X \in \mathcal{C}$ and $M \in \mathcal{M}$. Compare to (3.2).

Likewise, a left module category over an abelian (resp., linear) monoidal category $\mathcal{C}$ is a left $\mathcal{C}$-module category $(\mathcal{M}, \rhd)$, where $\mathcal{M}$ is abelian (resp., linear) and $(X \rhd -)$ and $(- \rhd M)$ are additive (resp., linear), for each $X \in \mathcal{C}$ and $M \in \mathcal{M}$.

Analogues for right module and bimodule categories are defined likewise.

In the additive case, we say that a left $\mathcal{C}$-module category $\mathcal{M}$ is **decomposable** if it is equivalent to a product of nonzero left $\mathcal{C}$-module categories (see §2.2.2ii). Else, we say that $\mathcal{M}$ is **indecomposable**.

For instance, if $\mathcal{C} = \mathsf{FdVec}$, then A-FdMod is a right $\mathcal{C}$-module category for any $\Bbbk$-algebra A. Moreover, A-FdMod is a decomposable module category when A is a decomposable algebra. See Exercise 3.14.

§3.4. Strictness and coherence

It is often convenient to ignore the associativity and unitality constraints when performing calculations in monoidal categories and in module categories. To do so, we use the notions of strictness and coherence discussed below.

§3.4.1. Strictness for monoidal categories

We say a monoidal category $(\mathcal{C}, \otimes, \mathbb{1}, a, \ell, r)$ is **strict** if the components of the associativity constraint a and unitality constraints ℓ, r are identity maps.

In this case, we write:

$$X \otimes Y \otimes Z \quad \text{for} \quad (X \otimes Y) \otimes Z = X \otimes (Y \otimes Z);$$

$$X \quad \text{for} \quad \mathbb{1} \otimes X \text{ and } X \otimes \mathbb{1}.$$

One example of a strict monoidal category is $(\mathrm{End}(\mathcal{A}), \circ, \mathrm{Id}_{\mathcal{A}})$, for a category $\mathcal{A}$.

In fact, we will make this assumption often due to the result below, which is attributed to Mac Lane; see Section XI.3 of Mac Lane [1998].

Theorem 3.26 (Strictification Theorem). *Any monoidal category $\mathcal{C}$ is monoidally equivalent to a strict monoidal category.*

To prove this, we consider the regular right $\mathcal{C}$-module category $\mathcal{C}_{\mathrm{reg}}$ from Example 3.17. We will then construct a strict monoidal category, denoted by $\mathcal{C}^{\mathrm{str}}$, which is modeled on $\mathrm{End}_{\mathrm{Mod}\text{-}\mathcal{C}}(\mathcal{C}_{\mathrm{reg}})$. Moreover, we will show that $\mathcal{C}$ is monoidally equivalent to a full (strict) monoidal subcategory of $\mathcal{C}^{\mathrm{str}}$. This approach to proving the Strictification Theorem is modified from Theorem 2.8.5 of Etingof et al. [2015], which in turn was borrowed from Corollary 1.4 of Joyal and Street [1993].

Definition 3.27. For a monoidal category $(\mathcal{C}, \otimes, \mathbb{1}, a, \ell, r)$, define the monoidal category $\mathcal{C}^{\mathrm{str}}$ as follows.

(a) Objects: endofunctors $F : \mathcal{C} \to \mathcal{C}$ equipped with a natural isomorphism $u := \{u_{M,X} : F(M) \otimes X \xrightarrow{\sim} F(M \otimes X)\}_{M,X \in \mathcal{C}}$, such that the following pentagon axiom holds for all $M, X, Y \in \mathcal{C}$.

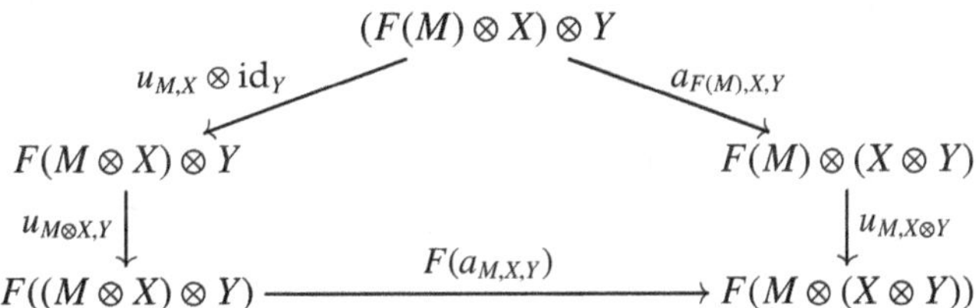

(b) Morphisms: $(F, u) \to (F', u')$ is a natural transformation $\theta : F \Rightarrow F'$ such that the diagram below commutes, for all $M, X \in \mathcal{C}$.

$$
\begin{array}{ccc}
F(M) \otimes X & \xrightarrow{u_{M,X}} & F(M \otimes X) \\
\downarrow{\scriptstyle \theta_M \otimes \mathrm{id}_X} & & \downarrow{\scriptstyle \theta_{M \otimes X}} \\
F'(M) \otimes X & \xrightarrow{u'_{M,X}} & F'(M \otimes X)
\end{array}
$$

(c) Monoidal product: $(F, u) \otimes^{\mathrm{str}} (F', u') := (FF', u'')$, with

$$u''_{M,X} := F(u'_{M,X}) \circ u_{F'(M),X} : FF'(M) \otimes X \to F(F'(M) \otimes X) \to FF'(M \otimes X).$$

(d) Monoidal unit: $\mathbb{1}^{\mathrm{str}} := (\mathrm{Id}_{\mathcal{C}}, \{\mathrm{id}_{M \otimes X}\}_{M,X \in \mathcal{C}}).$

Note that the natural isomorphism u is the inverse of the monoidal functor constraint, t^{-1}, for objects in Mod-$\mathcal{C}$. Moreover, verifying that $\mathcal{C}^{\mathrm{str}}$ is indeed strict monoidal is Exercise 3.15(a). Next, we relate a monoidal category $\mathcal{C}$ with $\mathcal{C}^{\mathrm{str}}$.

Theorem 3.28. *For any monoidal category $(\mathcal{C}, \otimes, \mathbb{1}, a, \ell, r)$, there exists a fully faithful, strong monoidal functor $\rho : \mathcal{C} \to \mathcal{C}^{\mathrm{str}}$.*

Proof. We define a functor $\rho : \mathcal{C} \to \mathcal{C}^{\mathrm{str}}$ on objects $Z \in \mathcal{C}$ by

$$\rho(Z) := (Z \otimes -, \ \{u_{M,X} := a_{Z,M,X}\}_{M,X \in \mathcal{C}}),$$

and on morphisms $f : Z \to Z' \in \mathcal{C}$ by

$$\rho(f : Z \to Z') := [\theta : (Z \otimes -) \Rightarrow (Z' \otimes -)], \quad \text{for } \theta_Y := f \otimes \mathrm{id}_Y,$$

for $Y \in \mathcal{C}$. Indeed, the compatibility condition in Definition 3.27(a) holds by the pentagon axiom for $\mathcal{C}$, and the compatibility condition in Definition 3.27(b) holds by the naturality of $a^{\mathcal{C}}$.

Claim 1. The functor ρ is fully faithful.

Proof of Claim 1. For fullness, take $\theta : (Z \otimes -) = \rho(Z) \Rightarrow \rho(Z') = (Z' \otimes -)$ in $\mathcal{C}^{\mathrm{str}}$. Consider the morphism $g := r_{Z'} \theta_{\mathbb{1}} r_Z^{-1} : Z \to Z'$. Then, $\rho(g) = \theta$ due to the commutative diagram below.

Indeed, the top region commutes by the naturality of θ at ℓ_Y; the bottom region commutes by definition; the left and right regions commute by the triangle axiom; and the middle region commutes by Definition 3.27(b). Thus, ρ is full.

For faithfulness, suppose that $\rho(f) = \rho(g)$, for some $f, g \in \mathrm{Hom}_{\mathcal{C}}(Z, Z')$. Then, taking the components at $\mathbb{1}$, we obtain that $f \otimes \mathrm{id}_{\mathbb{1}} = g \otimes \mathrm{id}_{\mathbb{1}}$ in $\mathrm{Hom}_{\mathcal{C}}(Z \otimes \mathbb{1}, Z' \otimes \mathbb{1})$. Now $f = g$ by the naturality of r, depicted by the commutative diagram below.

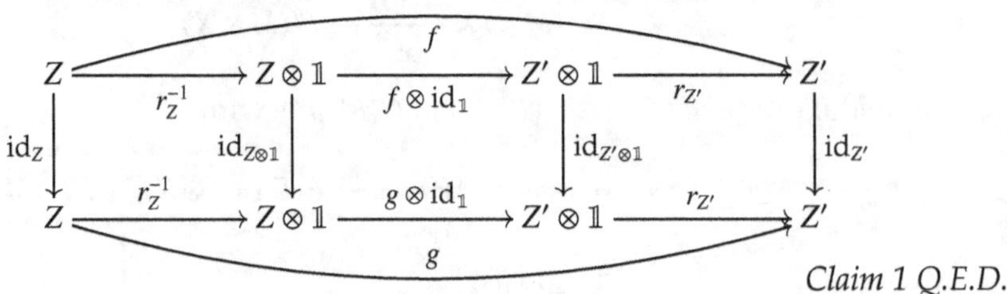

Claim 1 Q.E.D.

Claim 2. The functor ρ admits a strong monoidal structure.

Proof of Claim 2. Define the monoidal product constraint of ρ,

$$\rho^{(2)}_{W,Z} : [W \otimes (Z \otimes -), \{(\mathrm{id}_W \otimes a_{Z,M,X}) \circ a_{W,Z \otimes M,X}\}_{M,X \in \mathcal{C}}] = \rho(W) \otimes^{\mathrm{str}} \rho(Z)$$

$$\longrightarrow \rho(W \otimes Z) = [(W \otimes Z) \otimes -, \{a_{W \otimes Z,M,X}\}_{M,X \in \mathcal{C}}],$$

to be the natural transformation $a^{-1}_{W,Z,-}$. Indeed, by the pentagon axiom, the components of $\rho^{(2)}$ satisfy Definition 3.27(b) as depicted below.

$$
\begin{array}{ccccc}
(W \otimes (Z \otimes M)) \otimes X & \xrightarrow{\ a_{W,Z \otimes M,X}\ } & W \otimes ((Z \otimes M) \otimes X) & \xrightarrow{\ \mathrm{id}_W \otimes a_{Z,M,X}\ } & W \otimes (Z \otimes (M \otimes X)) \\
\downarrow{\scriptstyle a^{-1}_{W,Z,M} \otimes \mathrm{id}_X} & & & & \downarrow{\scriptstyle a^{-1}_{W,Z,M \otimes X}} \\
((W \otimes Z) \otimes M) \otimes X & & \xrightarrow{\ a_{W \otimes Z,M,X}\ } & & (W \otimes Z) \otimes (M \otimes X)
\end{array}
$$

We define the monoidal unit constraint of ρ as follows:

$$\rho^{(0)} : [\mathrm{Id}_{\mathcal{C}}, \{\mathrm{id}_{M \otimes X}\}_{M,X \in \mathcal{C}}] = \mathbb{1}^{\mathcal{C}^{\mathrm{str}}} \longrightarrow \rho(\mathbb{1}^{\mathcal{C}}) = \left[\mathbb{1}^{\mathcal{C}} \otimes -, \{a_{\mathbb{1}^{\mathcal{C}},M,X}\}_{M,X \in \mathcal{C}}\right],$$

to be the natural transformation ℓ^{-1}. By Exercise 3.1(a), the components of this natural transformation satisfy Definition 3.27(b) as depicted below.

$$
\begin{array}{ccc}
M \otimes X & \xrightarrow{\ \mathrm{id}_{M \otimes X}\ } & M \otimes X \\
\downarrow{\scriptstyle \ell^{-1}_M \otimes \mathrm{id}_X} & & \downarrow{\scriptstyle \ell^{-1}_{M \otimes X}} \\
(\mathbb{1}^{\mathcal{C}} \otimes M) \otimes X & \xrightarrow{\ a_{\mathbb{1}^{\mathcal{C}},M,X}\ } & \mathbb{1}^{\mathcal{C}} \otimes (M \otimes X)
\end{array}
$$

Exercise 3.15(b) asks us to verify that $\rho^{(2)}$ and $\rho^{(0)}$ satisfy the associativity and unit axioms making $(\rho, \rho^{(2)}, \rho^{(0)})$ a strong monoidal functor.   *Claim 2 Q.E.D.*

    This concludes the proof.   □

Now we complete the proof of the Strictification Theorem as follows.

Proof of the Strictification Theorem 3.26. We have that the essential image of the functor $\rho : \mathcal{C} \to \mathcal{C}^{\mathrm{str}}$ from the theorem above is a strict monoidal category, and is monoidally equivalent to $\mathcal{C}$. See Exercise 3.16 for details.   □

§3.4.2. Strictness for module categories

One can also consider strictness for module categories. Given a strict monoidal category $(\mathcal{C}, \otimes, \mathbb{1})$, a left $\mathcal{C}$-module category $(\mathcal{M}, \rhd, m, p)$ is **strict** if the components of the associativity constraint m and unitality constraint p are all identity maps.

Likewise, one can define strict right $\mathcal{C}$-module categories $(\mathcal{M}, \lhd, n, q)$, and strict $\mathcal{C}$-bimodule categories $(\mathcal{M}, \rhd, \lhd, m, n, p, q, b)$, where the components of the associativity constraints m, n, b and unitality constraints p, q are all identity maps.

Akin to Strictification Theorem (Theorem 3.26), we have the result below due to Theorem 1.3.8 of Greenough [2010], the proof of which follows closely to the proof of Theorem 3.26.

Theorem 3.29. *Any right C-module category is equivalent, as a right C-module category, to a strict right C-module category.* □

Likewise, the result is true for left C-module categories. We encourage the reader to explore if it is true for C-bimodule categories.

§3.4.3. Coherence

Given objects $X_1, \ldots, X_n$ in a monoidal category $(C, \otimes, \mathbb{1}, a, \ell, r)$, we call an ordered parenthesized monoidal product of $X_1, \ldots, X_n$, with possible $\mathbb{1}$'s inserted, an **expression** of $X_1 \otimes \cdots \otimes X_n$. For instance, $(X \otimes Y) \otimes (\mathbb{1} \otimes Z)$ and $\mathbb{1} \otimes [X \otimes ((Y \otimes \mathbb{1}) \otimes Z)]$ are both expressions of $X \otimes Y \otimes Z$. We are not only interested in expressions of objects, but we are also interested in how to move one expression to another.

If we have two expressions w and w' consisting of 3 non-unit objects, then we can get from w to w' in a unique way using the associativity and unitality isos. If w and w' consist of 4 non-unit objects, then we can still get from w to w' in a unique way using the associativity and unitality isos; the uniqueness is due to the pentagon and triangle axioms. The Strictification Theorem for monoidal categories also implies that we can get from w to w' in a unique way when w and w' consist of 5 or more non-unit objects. This is summarized in the theorem below, due to the work of Mac Lane [1998]; see Section VII.2 of that reference.

Theorem 3.30 (Coherence Theorem). *Let $w(X_1, \ldots, X_n)$ and $w'(X_1, \ldots, X_n)$ be two expressions of $X_1 \otimes \cdots \otimes X_n$ in a monoidal category C. Let*

$$f, g : w(X_1, \ldots, X_n) \to w'(X_1, \ldots, X_n)$$

be two isos given by compositions of associativity and unitality isos in C. Then, $f = g$.

Proof. Using Theorem 3.26, take a monoidal equivalence $\rho : C \to C'$, for C' a strict monoidal category. Let us write f as $f_1 \circ \cdots \circ f_r$, and g as $g_1 \circ \cdots \circ g_s$, where the f_i and g_j are associativity or unitality isos. Applying ρ to f and g yields the diagram below.

$$
\begin{array}{ccc}
\rho(w(X_1,\ldots,X_n)) = \rho(X_1 \otimes \cdots \otimes X_n) \xrightarrow[\rho(f_1)]{} \cdots \xrightarrow[\rho(f_r)]{} \rho(X_1 \otimes \cdots \otimes X_n) = \rho(w'(X_1,\ldots,X_n)) \\
\downarrow \qquad\qquad\qquad\qquad\qquad\qquad\qquad\qquad\qquad\qquad\qquad\qquad\qquad\qquad \downarrow \\
w(\rho(X_1),\ldots,\rho(X_n)) \xrightarrow{\ \mathrm{id}\ } \cdots \xrightarrow{\ \mathrm{id}\ } w'(\rho(X_1),\ldots,\rho(X_n)) \\
\downarrow \qquad\qquad\qquad\qquad\qquad\qquad\qquad\qquad\qquad\qquad\qquad\qquad\qquad\qquad \downarrow \\
\rho(w(X_1,\ldots,X_n)) = \rho(X_1 \otimes \cdots \otimes X_n) \xrightarrow{\rho(g_1)} \cdots \xrightarrow{\rho(g_s)} \rho(X_1 \otimes \cdots \otimes X_n) = \rho(w'(X_1,\ldots,X_n))
\end{array}
$$

with the top arc labeled $\rho(f)$ and the bottom arc labeled $\rho(g)$.

Here, the internal regions are broken up into sub-regions with vertical downward arrows being a combination of monoidal product or monoidal unit constraints, $\rho^{(2)}$ and $\rho^{(0)}$; each sub-region commutes due to the associativity or unitality axioms for the monoidal structure of ρ. Moreover, the middle row of morphisms consists of identity maps because $\mathcal{C}'$ is strict. Since ρ is strong, the vertical arrows are isos. Thus, $\rho(f) = \rho(g)$. Finally, $f = g$ as ρ is faithful. $\qquad\square$

§3.5. Graphical calculus

In this part, we introduce graphical notation to depict objects and morphisms in monoidal categories $\mathcal{C}$. This will enable us to compute compositions of morphisms in an elegant fashion when $\mathcal{C}$ is strict.

Standing hypothesis for graphical calculus computations. By the Strictification Theorem [Theorem 3.26], assume that monoidal categories $(\mathcal{C}, \otimes, \mathbb{1}, a, \ell, r)$ here are strict, that is, of the form $(\mathcal{C}, \otimes, \mathbb{1})$.

Consider the following conventions:

- An object X in $\mathcal{C}$ is depicted by a vertical string labeled by "X".

- A morphism $f : X \to Y$ in $\mathcal{C}$ is a vertical line with a box labeled by "f", with the string above the box labeled by X, and the string below the box labeled by Y.

- The composition of morphisms $gf := g \circ f : X \to Z$ in $\mathcal{C}$, for $f : X \to Y$ and $g : Y \to Z$ in $\mathcal{C}$, is drawn by connecting the string for g below the string for f. This process is called **vertical composition**.

- The monoidal product $X \otimes Y$ of objects of X and Y is drawn as two vertical strings side-by-side with the left (resp., right) string labeled by X (resp., Y). Likewise, the monoidal product of morphisms $f : X \to Y$ and $f' : X' \to Y'$ is drawn side-by-side. Both processes are called **horizontal composition**.

- The monoidal unit object is depicted with an empty or a dashed string.

This is illustrated in Figure 3.3. See Exercise 3.17 for practice.

Remark 3.31. A common convention for vertical composition is by stacking morphisms upward instead of connecting morphisms downward. Some refer to these, respectively, as the *optimist* and *pessimist* conventions. But the author views these, respectively, as the *idealist* and *realist* conventions as one realistically starts at the top of a sheet of paper and works their way downward to do computations (at least in many written languages). :)

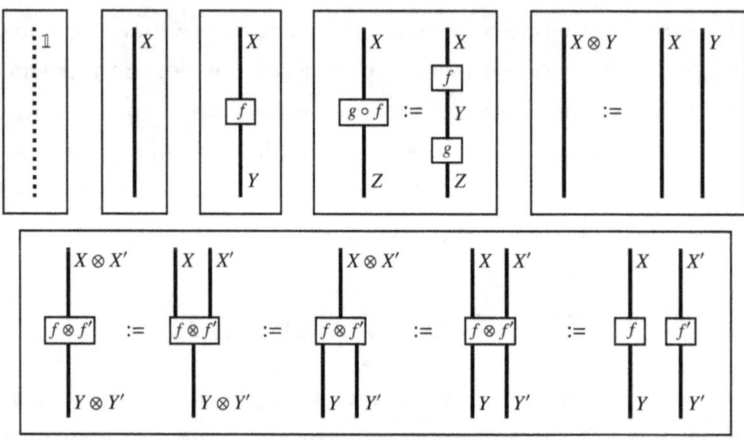

Figure 3.3: String diagrams for objects, morphisms in strict monoidal categories.

Various axioms in strict monoidal categories are encoded in string diagrams, including the associativity and unitality axioms for vertical compositions (see Figure 3.4), and for horizontal compositions (see Figure 3.5).

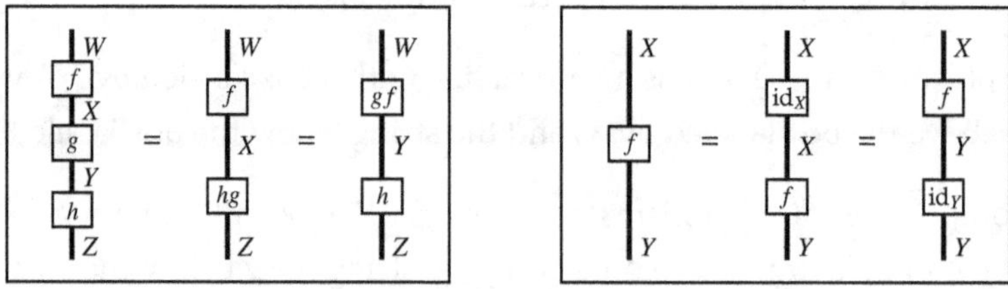

Figure 3.4: Diagrammatic associativity and unitality for vertical composition.

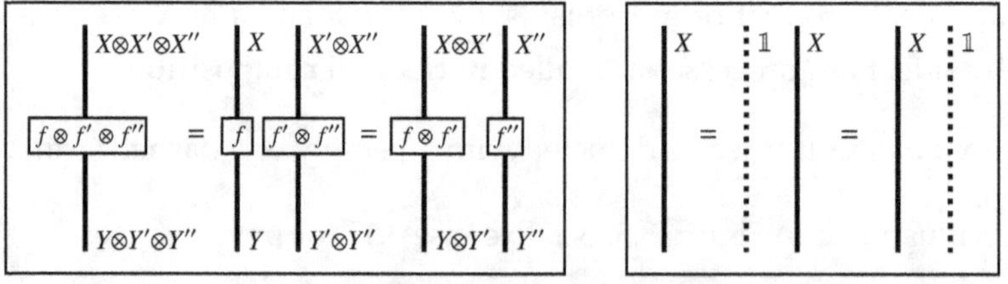

Figure 3.5: Diagrammatic associativity and unitality for horizontal composition.

Commutative diagrams can also be encoded in string diagrams. For instance, having morphisms in different slots is depicted by sliding morphisms up and down strings, as visualized in Figure 3.6. This is called **level exchange**.

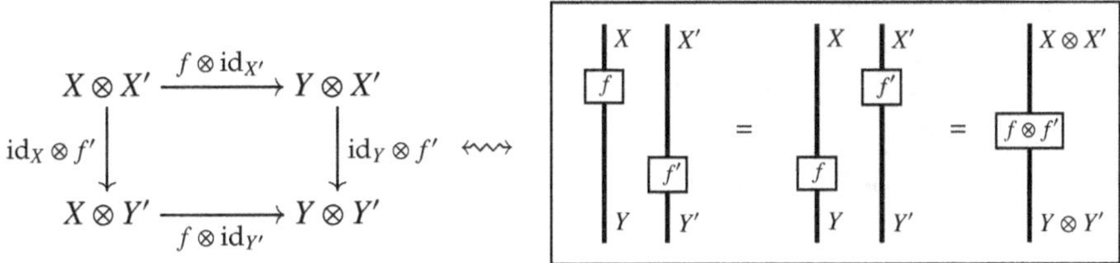

Figure 3.6: Level exchange: commutative diagram vs. string diagram.

§3.6. Rigid categories

Now we begin to discuss various features of monoidal categories that are used to get richer categorical models for applications. Here, we study when monoidal categories $\mathcal{C}$ contain *dual objects* and *dual morphisms*; in this case, $\mathcal{C}$ is called *rigid* or *autonomous*. We also examine module categories over such monoidal categories, namely, *dual module categories* over rigid monoidal categories.

§3.6.1. Rigid categories

Let us fix a monoidal category $(\mathcal{C}, \otimes, \mathbb{1}, a, \ell, r)$, and an object X in $\mathcal{C}$.

A **left dual of** X consists of the following data.

(a) An object $X^* \in \mathcal{C}$.

(b) A morphism $\mathrm{ev}_X^L : X^* \otimes X \to \mathbb{1}$ called **left evaluation**.

(c) A morphism $\mathrm{coev}_X^L : \mathbb{1} \to X \otimes X^*$ called **left coevaluation**.

This data must satisfy the **left rigidity axioms** below.

- The composition below is equal to id_X:

$$X \xrightarrow{\ell_X^{-1}} \mathbb{1} \otimes X \xrightarrow{\mathrm{coev}_X^L \otimes \mathrm{id}} (X \otimes X^*) \otimes X \xrightarrow{a_{X,X^*,X}} X \otimes (X^* \otimes X) \xrightarrow{\mathrm{id} \otimes \mathrm{ev}_X^L} X \otimes \mathbb{1} \xrightarrow{r_X} X.$$

- The composition below is equal to id_{X^*}:

$$X^* \xrightarrow{r_{X^*}^{-1}} X^* \otimes \mathbb{1} \xrightarrow{\mathrm{id} \otimes \mathrm{coev}_X^L} X^* \otimes (X \otimes X^*) \xrightarrow{a_{X^*,X,X^*}^{-1}} (X^* \otimes X) \otimes X^* \xrightarrow{\mathrm{ev}_X^L \otimes \mathrm{id}} \mathbb{1} \otimes X^* \xrightarrow{\ell_{X^*}} X^*.$$

Likewise, a **right dual of** X consists of the following data.

(a) An object $^*X \in \mathcal{C}$.

(b) A morphism $\mathrm{ev}_X^R : X \otimes {}^*X \to \mathbb{1}$ called **right evaluation**.

(c) A morphism $\mathrm{coev}_X^R : \mathbb{1} \to {}^*X \otimes X$ called **right coevaluation**.

This data must satisfy the **right rigidity axioms** below.

- The composition below is equal to id_X:

$$X \xrightarrow{r_X^{-1}} X \otimes \mathbb{1} \xrightarrow{\mathrm{id} \otimes \mathrm{coev}_X^R} X \otimes ({}^*X \otimes X) \xrightarrow{a_{X,{}^*X,X}^{-1}} (X \otimes {}^*X) \otimes X \xrightarrow{\mathrm{ev}_X^R \otimes \mathrm{id}} \mathbb{1} \otimes X \xrightarrow{\ell_X} X.$$

- The composition below is equal to $\mathrm{id}_{{}^*X}$:

$$ {}^*X \xrightarrow{\ell_{{}^*X}^{-1}} \mathbb{1} \otimes {}^*X \xrightarrow{\mathrm{coev}_X^R \otimes \mathrm{id}} ({}^*X \otimes X) \otimes {}^*X \xrightarrow{a_{{}^*X,X,{}^*X}} {}^*X \otimes (X \otimes {}^*X) \xrightarrow{\mathrm{id} \otimes \mathrm{ev}_X^R} {}^*X \otimes \mathbb{1} \xrightarrow{r_{{}^*X}} {}^*X.$$

We depict these structures in the strict case in Figure 3.7.

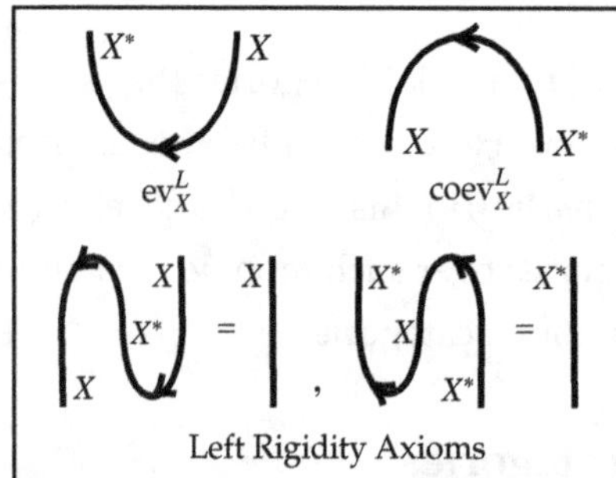

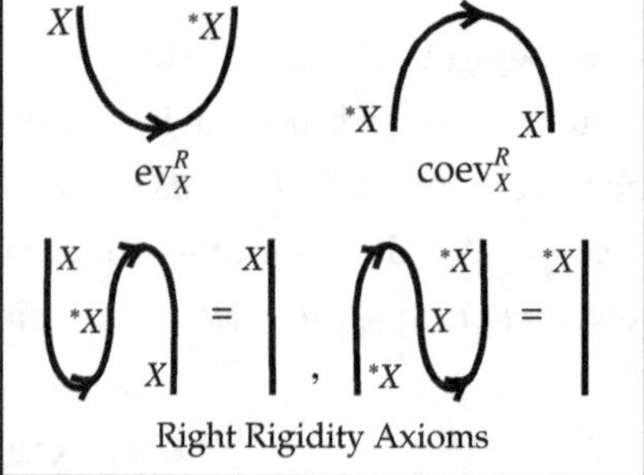

Figure 3.7: (Co)evaluation morphisms and rigidity axioms for left/right duality.

Also, consider the following terminology.

- If both a left dual and a right dual exists for X in $\mathcal{C}$, then X is called **rigid**.

- $\mathcal{C}$ is **left** (resp., **right) rigid** if each object in $\mathcal{C}$ has a left (resp., right) dual in $\mathcal{C}$.

- $\mathcal{C}$ is **rigid** if it is both left and right rigid (or, if each object is rigid).

One synonym for "rigid" is the term **autonomous**. Determining whether rigidity is a structure on or a property of a monoidal category is Exercise 3.18.

We now point out some features of dual objects in a rigid category.

Lemma 3.32. *In a rigid category $\mathcal{C}$, the following statements hold.*

(a) *Left dual objects are unique, up to unique iso compatible with the evaluation and coevaluation morphisms. Namely, if $X, Y_1, Y_2 \in \mathcal{C}$ are objects equipped with morphisms $e_i : Y_i \otimes X \to \mathbb{1}$ and $c_i : \mathbb{1} \to X \otimes Y_i$ for $i = 1, 2$, satisfying left rigidity axioms, then there is a unique iso $f : Y_1 \to Y_2$ in $\mathcal{C}$ such that*

$$e_1 = e_2 \circ (f \otimes \mathrm{id}_X) \quad and \quad c_2 = (\mathrm{id}_X \otimes f) \circ c_1.$$

(b) *Likewise, right dual objects are unique, up to unique iso compatible with the evaluation and coevaluation morphisms.*

(c) *If $X \in C$, then $^*(X^*) \cong X \cong (^*X)^*$ in C.*

(d) *We have that $\mathbb{1}^* = \mathbb{1} = {}^*\mathbb{1}$.*

(e) *The opposite monoidal categories C^{op}, $C^{\otimes\mathrm{op}}$, and C^{rev} are also rigid.*

Proof. Part (a) is Exercise 3.19, and part (b) holds likewise. Parts (c) and (d) comprise Exercise 3.20. Lastly, Exercise 3.21 is to verify part (e). □

Next, we present a vital result about endofunctors obtained by tensoring by an object, and by tensoring by its dual object.

Proposition 3.33. *In a monoidal category C, take $X, Y \in C$ for which its left and right dual objects exist in C. Then, we have the following adjunctions.*

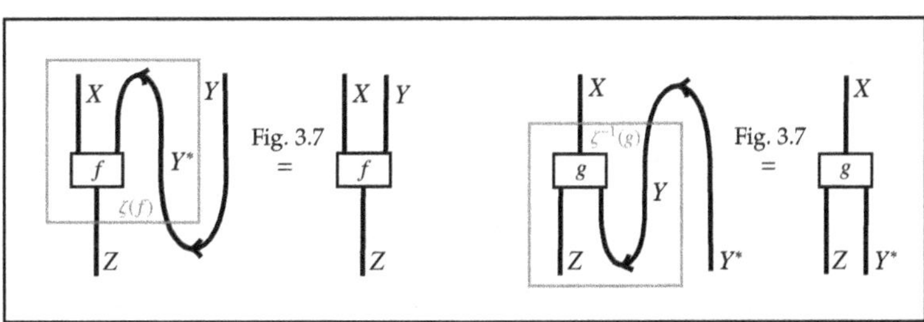

Proof. By Proposition 2.24, part (d) is equivalent to showing that there is a bijection of sets $\zeta_{X,Z} : \mathrm{Hom}_C(X \otimes Y, Z) \to \mathrm{Hom}_C(X, Z \otimes Y^*)$, that is natural in X and Z. Indeed, for $f : X \otimes Y \to Z$ and $g : X \to Z \otimes Y^*$, we can define

$$\zeta_{X,Z}(f) := (f \otimes \mathrm{id}_{Y^*}) \circ a^{-1}_{X,Y,Y^*} \circ (\mathrm{id}_X \otimes \mathrm{coev}^L_Y) \circ r^{-1}_X \quad : X \to Z \otimes Y^*$$

$$\zeta^{-1}_{X,Z}(g) := r_Z \circ (\mathrm{id}_Z \otimes \mathrm{ev}^L_Y) \circ a_{Z,Y^*,Y} \circ (g \otimes \mathrm{id}_Y) \qquad : X \otimes Y \to Z.$$

The fact that $(\zeta^{-1}_{X,Z} \circ \zeta_{X,Z})(f) = f$ and $(\zeta_{X,Z} \circ \zeta^{-1}_{X,Z})(g) = g$ follows from the rigidity axioms; see the depiction in the strict case in Figure 3.8. In the non-strict case, this can be verified with commutative diagrams. We leave it to the reader to examine naturality. Parts (a)-(c) follow similarly. □

Figure 3.8: Graphical calculus for part of the proof of Proposition 3.33.

§3.6.2. Examples of rigid categories

We now present several examples and non-examples of rigid monoidal categories, building on the examples of monoidal categories in §3.1.2. This is summarized in Table 3.1, which we discuss further below. Exercise 3.22 pertains to completing more of the details of Table 3.1.

Key: Rigid ($\checkmark$)   Not Rigid (X)   Sometimes Rigid ($*$)   Exercise (?)

FdVec	$\checkmark$	Vec_G	$\checkmark$	Cat	?	Top	?
Vec	X	Vec_G'	?	$\mathcal{A}_\sqcup$	?	Braid	?
G-FdMod	$\checkmark$	Vec_N	X	$\mathcal{A}_\sqcap$	?	Hilb	X
A-FdBimod	$*$	$\underline{N}$	?	$\mathrm{End}(\mathcal{A})$	X	FdHilb	$\checkmark$
A-Bimod	X	$\underline{G}$	$\checkmark$	$\mathrm{Aut}(\mathcal{A})$	?	Graph	?

Table 3.1: Examples and non-examples of rigid monoidal categories.

Example 3.34. The category of finite-dimensional $\Bbbk$-vector spaces, FdVec, is rigid. Namely, for $V \in$ FdVec, we have that $V^* = {}^*V = \mathrm{Hom}_\Bbbk(V, \Bbbk)$, with

$$\mathrm{ev}_V^L : V^* \otimes_\Bbbk V \to \Bbbk, \qquad \mathrm{coev}_V^L : \Bbbk \to V \otimes_\Bbbk V^*$$

$$f \otimes_\Bbbk v \mapsto f(v) \qquad\qquad 1_\Bbbk \mapsto \sum_{i=1}^{\dim_\Bbbk V} b_i \otimes_\Bbbk b_i^*,$$

where $\{b_i\}_i$ is a basis of V. One of the left rigidity axioms holds as follows:

$$V \xrightarrow{\ell_V^{-1}} \Bbbk \otimes_\Bbbk V \xrightarrow{\mathrm{coev}_V^L \otimes \mathrm{id}} (V \otimes_\Bbbk V^*) \otimes_\Bbbk V \xrightarrow{a_{V,V^*,V}} V \otimes_\Bbbk (V^* \otimes_\Bbbk V) \xrightarrow{\mathrm{id} \otimes \mathrm{ev}_V^L} V \otimes_\Bbbk \Bbbk \xrightarrow{r_V} V$$

$$v \longmapsto 1_\Bbbk \otimes_\Bbbk v \longmapsto \sum_i (b_i \otimes_\Bbbk b_i^*) \otimes_\Bbbk v \longmapsto \sum_i b_i \otimes_\Bbbk (b_i^* \otimes_\Bbbk v) \overset{**}{\longmapsto} v \otimes_\Bbbk 1_\Bbbk \longmapsto v.$$

For (**), assume that $v = \sum_j \lambda_j b_j$, for some $\lambda_j \in \Bbbk$. Then,

$$\sum_i b_i \otimes_\Bbbk b_i^*(v) = \sum_{i,j} \lambda_j b_i \otimes_\Bbbk b_i^*(b_j) = \sum_{i,j} \lambda_j b_i \otimes_\Bbbk \delta_{i,j} 1_\Bbbk = \sum_j \lambda_j b_j = v.$$

Likewise, the other left rigidity axiom holds. Moreover, there exist linear maps ev_V^R and coev_V^R defined in a similar way that make V right rigid.

Example 3.35. The category of arbitrary $\Bbbk$-vector spaces, Vec, is not rigid. Namely, for $V \in$ Vec, say there are $\Bbbk$-linear maps $e_V : W \otimes_\Bbbk V \to \Bbbk$ and $c_V : \Bbbk \to V \otimes_\Bbbk W$ satisfying the rigidity axioms. Then, $c_V(1_\Bbbk) = \sum_{i=1}^n b_i \otimes_\Bbbk w_i$ for some $w_i \in W$ and some finite n. Now for step (**) in Example 3.34, we get that

$$\sum_{i=1}^n b_i \otimes_\Bbbk w_i(v) = \sum_{i=1}^n \lambda_j b_i \otimes_\Bbbk e_V(w_i \otimes_\Bbbk b_j).$$

Since n is finite, there exists a basis element b_j of V not in the $\Bbbk$-span of $\{b_1, \ldots, b_n\}$. Therefore, the linear map $V \to V$ in Example 3.34 cannot be surjective, and a left rigidity axiom fails.

Example 3.36. The category G-FdMod is rigid. Take $(V, \triangleright) \in G$-FdMod, and define $(V, \triangleright)^* = {}^*(V, \triangleright) = (V^*, \blacktriangleright)$, with $(g \blacktriangleright f)(v) := f(g^{-1} \triangleright v)$ for $g \in G$, $f \in V^*$, and $v \in V$. The (co)evaluation morphisms are similar to that for Example 3.34.

By Exercise 3.35(f), FdVec$_G$, along with each of its 3-cocycle modifications, FdVec$_G^\omega$, are rigid.

Example 3.37. The category A-Bimod is not rigid, but the category A-FdBimod can be rigid though this is rarely the case. To see the latter, we mimic Example 3.34.

(a) For $M \in A$-FdBimod, take $M^* = {}^*M = \mathrm{Hom}_{A\text{-FdBimod}}(M, A)$. Also, let ev_M^L be the evaluation morphism defined by $\mathrm{ev}_M^L(f \otimes_A m) = f(m)$ for $f \in M^*$ and $m \in M$. But to define coev_M^L, we need a finite 'dual basis' of M, and this is equivalent to M being finitely generated and projective. See Section 2B of Lam [1999] for details. Hence, an A-bimodule M is rigid in this setting precisely when it is finitely generated and projective as both a left and a right A-module.

(b) On the other hand, we see that A-FdBimod is rigid when A is finite-dimensional and semisimple. Namely, projectivity holds for all A-bimodules M if and only if A is semisimple [Proposition 2.57]. Moreover, if A is finite-dimensional (as a $\Bbbk$-vector space), then M is finitely generated precisely when it is finite-dimensional (as a $\Bbbk$-vector space). Note that finite-dimensional, semisimple algebras are classified by the Artin-Wedderburn Theorem [Theorem 1.44].

Example 3.38. Continuing Example 3.37, take A to be the 2-dimensional commutative algebra $\Bbbk[x]/(x^2)$. Then, we will see below how A-FdBimod is not rigid.

(a) We have by Exercise 1.19 and Proposition 2.49(a) an example of a non-flat A-module V, that is, an example of a functor $(V \otimes_A -)$, which is not left exact. Now if A-FdBimod is rigid, then by Propositions 3.33(b) and 2.49(b), the functor $(V \otimes_A -)$ must be left exact, a contradiction.

(b) One can also use the discussion in Example 3.37, and the non-semisimplicity of A [Exercise 1.31(b)], to get that A-FdBimod is not rigid.

Example 3.39. The category $\underline{G}$ is rigid. For $g \in G$, take $g^* = {}^*g = g^{-1}$. Here, ev_g^L and ev_g^R are given by the group operation, that is, $\mathrm{ev}_g^L : g^{-1} \otimes g \to e$ and $\mathrm{ev}_g^R : g \otimes g^{-1} \to e$. Also, $\mathrm{coev}_g^L(e) := g \otimes g^{-1}$ and $\mathrm{coev}_g^R(e) := g^{-1} \otimes g$.

Example 3.40. The category Set $=$ Set$_\sqcap$ is not rigid. Take a set X of cardinality greater than 1. By way of contradiction, suppose that there exists a set Y and functions $e_X : Y \times X \to \{\cdot\}$ and $c_X : \{\cdot\} \to X \times Y$ that satisfy the rigidity axioms. Then, the image of c_X is equal to a singleton subset $\{(x, y)\}$ of $X \times Y$. Moreover, by the rigidity axioms, we have:

$$X \;=\; r_X \,(\mathrm{id}_X \times e_X)\, a_{X,Y,X}\, (c_X \times \mathrm{id}_X)\, \ell_X^{-1}(X)$$
$$=\; r_X \,(\mathrm{id}_X \times e_X)\, [\{x\} \times (\{y\} \times X)] \;=\; \{x\}, \qquad \text{(contradiction).}$$

Example 3.41. The category $\mathrm{End}(\mathcal{A})$ is not rigid in general, but there is an interesting characterization of when dual objects exist. Recall that $\mathrm{End}(\mathcal{A})$ is strict, and take $G \in \mathrm{End}(\mathcal{A})$. For $F \in \mathrm{End}(\mathcal{A})$ to serve as the left dual object of G, we require natural transformations

$$e := e_G : FG \Rightarrow \mathrm{Id}_{\mathcal{A}} \quad \text{and} \quad c := c_G : \mathrm{Id}_{\mathcal{A}} \Rightarrow GF,$$

such that the following left rigidity axioms hold:

$$(G * e) \circ^{\mathrm{ver}} (c * G) = \mathrm{Id}_G, \qquad (e * F) \circ^{\mathrm{ver}} (F * c) = \mathrm{Id}_F.$$

That is, F is the left dual object of G precisely when it is the left adjoint of G; see §2.5.1. Likewise, a right dual object of G is its right adjoint.

Therefore, $\mathrm{End}(\mathcal{A})$ is rigid precisely when left and right adjoints exist for all endofunctors of $\mathcal{A}$. This is quite a strong condition on $\mathcal{A}$, and to violate this, it suffices to cook up an endofunctor of $\mathcal{A}$ that fails to preserve some universal morphism in §2.2.1. See Proposition 2.26.

§3.6.3. Duality functors and functors preserving rigidity

We discussed dual objects in a rigid category $\mathcal{C}$ above. Now we define dual morphisms in $\mathcal{C}$; these are depicted in Figure 3.9 when $\mathcal{C}$ is strict.

- The **left dual of a morphism** $f : X \to Y$ is the composition of morphisms:

$$f^* := \ell_{X^*} \,(\mathrm{ev}_Y^L \otimes \mathrm{id}_{X^*})\,(\mathrm{id}_{Y^*} \otimes f \otimes \mathrm{id}_{X^*})\, a_{Y^*,X,X}^{-1}\, (\mathrm{id}_{Y^*} \otimes \mathrm{coev}_X^L)\, r_{Y^*}^{-1} : \; Y^* \to X^*.$$

- The **right dual of a morphism** $f : X \to Y$ is the composition:

$$^*f := r_X \,(\mathrm{id}_{^*X} \otimes \mathrm{ev}_Y^R)\,(\mathrm{id}_{^*X} \otimes f \otimes \mathrm{id}_{^*Y})\, a_{^*X,X,^*Y}\, (\mathrm{coev}_X^R \otimes \mathrm{id}_{^*Y})\, \ell_{^*Y}^{-1} : \; ^*Y \to {^*X}.$$

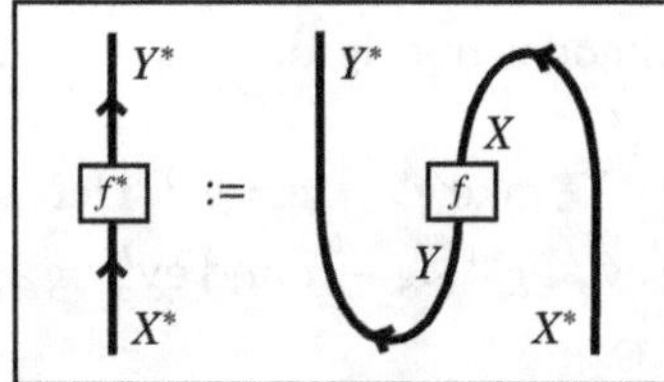
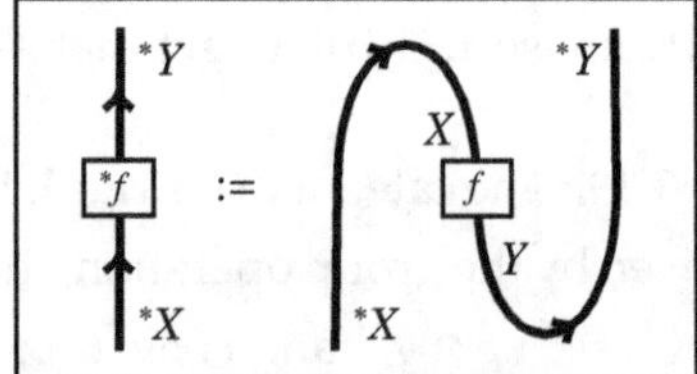

Figure 3.9: Duals of a morphism $f : X \to Y$ in a strict monoidal category.

With the duals of objects and morphisms defined above, we can now define the duality endofunctors for a rigid category $\mathcal{C}$.

- The **left duality functor** on a left rigid category C is given by

$$(-)^* : C \to C, \quad X \mapsto X^* \text{ (on objects)}, \quad f \mapsto f^* \text{ (on morphisms)}.$$

- The **right duality functor** on a right rigid category C is given by

$$^*(-) : C \to C, \quad X \mapsto {}^*X \text{ (on objects)}, \quad f \mapsto {}^*f \text{ (on morphisms)}.$$

Proposition 3.42. *Consider the left duality functor $(-)^*$ and right duality functor $^*(-)$ defined above for a rigid category C.*

(a) *These are contravariant functors; namely, $(g \circ f)^* = f^* \circ g^*$ and $^*(g \circ f) = {}^*f \circ {}^*g$.*

(b) *The left and right duality functors,*

$$(-)^*, \ ^*(-) : C^{\mathrm{rev}} \to C,$$

respectively, are strong monoidal functors. Here, Lemma 3.32(d) yields the monoidal unit. Moreover, the monoidal product constraint components,

$$d^L_{X,Y} := (-)^{*(2)}_{X,Y} : X^* \otimes Y^* \overset{\sim}{\to} (X \otimes^{\mathrm{op}} Y)^*,$$

$$d^R_{X,Y} := {}^*(-)^{(2)}_{X,Y} : {}^*X \otimes {}^*Y \overset{\sim}{\to} {}^*(X \otimes^{\mathrm{op}} Y),$$

are compositions of evaluation and coevaluation morphisms.

Proof. Complete Exercise 3.23 for the left duality case; the right duality case holds likewise. $\qquad\square$

Pushing this further, we also have **double duals of morphisms** in rigid categories, visualized in the strict case in Figure 3.10.

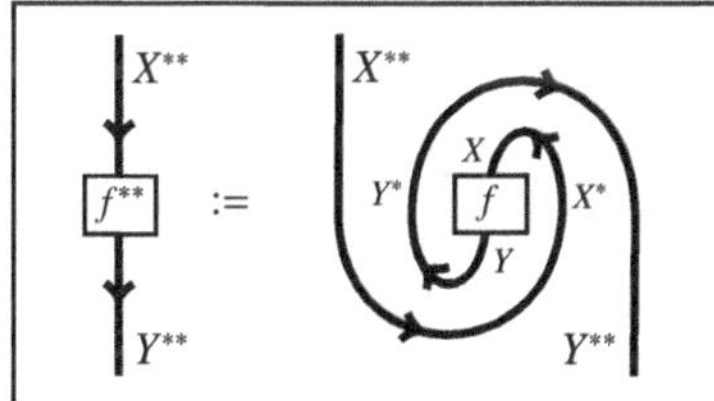

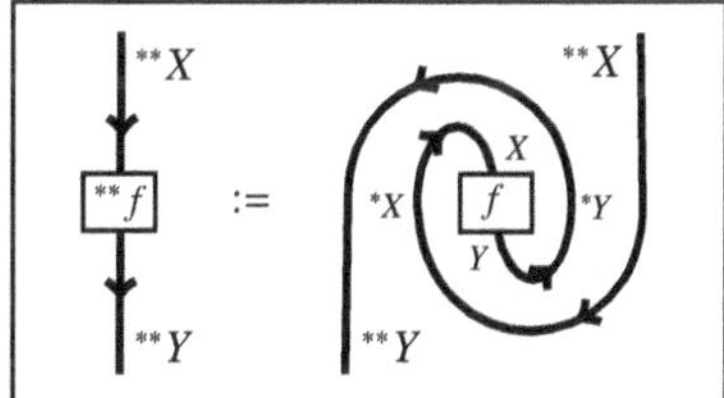

Figure 3.10: Double duals of a morphism $f : X \to Y$ in a strict monoidal category.

We can then form the **double duality functors**, which are covariant and strong monoidal functors:

$$(-)^{**} : C \to C, \quad X \mapsto X^{**}, \ f \mapsto f^{**}, \qquad ^{**}(-) : C \to C, \quad X \mapsto {}^{**}X, \ f \mapsto {}^{**}f.$$

Proving the double dual analogue to Proposition 3.42 is Exercise 3.24.

Next, we examine how rigidity is preserved under functors between monoidal categories. This is captured in the result below; the proof is Exercise 3.25.

Proposition 3.43. *Let C and D be monoidal categories, let $F : C \to D$ be a strong monoidal functor, and let X be an object in C.*

(a) *If there exists a left dual object $(X^*, \mathrm{ev}_X^L, \mathrm{coev}_X^L)$ in C, then $F(X)$ has a left dual object in D, where*

- $F(X)^* := F(X^*),$

- $\mathrm{ev}_{F(X)}^L : F(X^*) \otimes^D F(X) \xrightarrow{F_{X^*,X}^{(2)}} F(X^* \otimes^C X) \xrightarrow{F(\mathrm{ev}_X^L)} F(\mathbb{1}^C) \xrightarrow{F^{(-0)}} \mathbb{1}^D,$

- $\mathrm{coev}_{F(X)}^L : \mathbb{1}^D \xrightarrow{F^{(0)}} F(\mathbb{1}^C) \xrightarrow{F(\mathrm{coev}_X^L)} F(X \otimes^C X^*) \xrightarrow{F_{X,X^*}^{(-2)}} F(X) \otimes^D F(X^*).$

(b) *Likewise, F sends a right dual object in C to a right dual object in D.* $\qquad\square$

But the result above fails when F is not necessarily strong. To see this, use the forgetful functor from FdVec to Set, along with Examples 3.9, 3.34, and 3.40.

On a related note, we have the intriguing result below.

Proposition 3.44. *If C is a rigid category and D is a monoidal category, then any monoidal transformation $\phi : F \overset{\otimes}{\Rightarrow} F'$ between two strong monoidal functors $F, F' : C \to D$ is a monoidal isomorphism.*

Proof. For each $X \in C$, we need to display a morphism $\psi_X : F'(X) \to F(X)$ in D such that $\psi_X \circ \phi_X = \mathrm{id}_{F(X)}$ and $\phi_X \circ \psi_X = \mathrm{id}_{F'(X)}$. Assume that D is strict (via Theorem 3.26) for brevity, and define:

$$\psi_X : F'(X) \xrightarrow{\mathrm{coev}_{F(X)}^L \otimes\, \mathrm{id}} F(X) \otimes F(X^*) \otimes F'(X) \xrightarrow{\mathrm{id}\, \otimes\, \phi_{X^*}\, \otimes\, \mathrm{id}} F(X) \otimes F'(X^*) \otimes F'(X) \xrightarrow{\mathrm{id}\, \otimes\, \mathrm{ev}_{F'(X)}^L} F(X),$$

where $\mathrm{ev}_{F'(X)}^L$ and $\mathrm{coev}_{F(X)}^L$ are defined in Proposition 3.43. Verifying that ψ_X works is Exercise 3.26. $\qquad\square$

§3.6.4. Module categories over rigid categories

With the extra feature of a monoidal category C being rigid, we have an analogue of Proposition 3.33 for module categories. The proof is Exercise 3.27.

Proposition 3.45. *In a monoidal category C, take a rigid object $X \in C$. Then, for a left C-module category $(\mathcal{M}, \rhd, m, p)$, we have the following adjunctions.*

(a) $\quad \mathcal{M} \underset{{}^*X\, \rhd\, -}{\overset{X\, \rhd\, -}{\rightleftarrows}} \mathcal{M} \qquad\qquad$ (b) $\quad \mathcal{M} \underset{X\, \rhd\, -}{\overset{X^*\, \rhd\, -}{\rightleftarrows}} \mathcal{M} \qquad\qquad\square$

Next, with the rigidity of C, we can create new module categories over C from old module categories over C. We refer to the constructions in the result below as **(left, right) dual module categories**, and reserve the proof, along with further practice, for Exercise 3.28.

Proposition 3.46. *Let C be a rigid category, and take*

$$(\mathcal{M}, \triangleright, m, p) \in C\text{-Mod} \quad and \quad (\mathcal{N}, \triangleleft, n, q) \in \text{Mod-}C.$$

Recall the natural isomorphisms d^L and d^R in Proposition 3.42(b). Then, the following statements hold.

(a) *$\mathcal{M}^* := (\mathcal{M}^{\mathrm{op}}, \blacktriangleleft, \tilde{m}, \tilde{p}) \in \text{Mod-}C$, where we define*

$$M \blacktriangleleft X := X^* \triangleright M, \quad \tilde{m}_{M,X,Y} := m_{Y^*,X^*,M} \circ ((d^L_{Y,X})^{-1} \triangleright \mathrm{id}_M), \quad \tilde{p}_M = p_M.$$

*This is the **left dual of** $\mathcal{M}$.*

(b) *$\mathcal{N}^* := (\mathcal{N}^{\mathrm{op}}, \blacktriangleright, \tilde{n}, \tilde{q}) \in C\text{-Mod}$, where we define*

$$X \blacktriangleright N := N \triangleleft X^*, \quad \tilde{n}_{X,Y,N} := n_{N,Y^*,X^*} \circ (\mathrm{id}_N \triangleleft (d^R_{Y,X})^{-1}), \quad \tilde{q}_N = q_N.$$

*This is the **left dual of** $\mathcal{N}$.*

(c) *Likewise, there exists a **right dual of** $\mathcal{M}$ **and of** $\mathcal{N}$, namely $^*\mathcal{M} \in \text{Mod-}C$ where $M \blacktriangleleft X := {}^*X \triangleright M$, and $^*\mathcal{N} \in C\text{-Mod}$ where $X \blacktriangleright N := N \triangleleft {}^*X$.*

This is for all objects $M \in \mathcal{M}$, $N \in \mathcal{N}$, and $X, Y \in C$. $\qquad\qquad\square$

Recall that the dual of a finite-dimensional vector space V is defined as the collection of linear functionals $\mathrm{Hom}_{\Bbbk}(V, \Bbbk) = \mathrm{Hom}_{\mathsf{FdVec}}(V, \Bbbk)$. Moreover in Example 3.37, the dual of a finite-dimensional A-bimodule M is defined as $\mathrm{Hom}_{A\text{-}\mathsf{FdBimod}}(M, A)$ when M is finitely generated and projective (or when A is 'finite' and semisimple). Likewise, there are finiteness conditions on C, and on a C-module category $\mathcal{M}$ (e.g., see §2.9), for $\mathcal{M}^*$ and $^*\mathcal{M}$ to be equivalent to certain categories of C-module functors from $\mathcal{M}$ to C. This leads us to the result below, which is Proposition 2.4.9 in Section 2.4.3 of Douglas et al. [2020].

Proposition 3.47. *If C is a finite rigid category, $\mathcal{M}$ is a finite left C-module category, and $\mathcal{N}$ is a finite right C-module category, then the following statements hold.*

(a) *$^*\mathcal{M} \simeq \mathrm{Rex}_{C\text{-Mod}}(\mathcal{M}, C_{\mathrm{reg}})$ as right C-module categories.*

(b) *$\mathcal{N}^* \simeq \mathrm{Rex}_{\mathsf{Mod-}C}(\mathcal{N}, C_{\mathrm{reg}})$ as left C-module categories.*

Here, $\mathrm{Rex}_-(-, -)$ is the subcategory of $\mathsf{Fun}_-(-, -)$ consisting of right exact C-module functors, and C_{reg} is the regular (right, left) C-module category. $\qquad\square$

Here, $\mathrm{Rex}_{\mathsf{Mod-}C}(\mathcal{N}, C_{\mathrm{reg}})$ is a left C-module category via Example 3.23(a), and by Example 3.18 applied to Exercise 3.11. Likewise, $\mathrm{Rex}_{C\text{-Mod}}(\mathcal{M}, C_{\mathrm{reg}})$ is right C-module category. The finiteness conditions are used to handle exactness for C-module functors; compare to §2.8.

§3.7. Pivotal categories

Here, we discuss a structure imposed on rigid categories $\mathcal{C}$ that identifies objects with their double left dual, or equivalently, identifies an object's left and right dual objects. This is useful for applications, especially since it provides a way of measuring objects in $\mathcal{C}$, e.g., via their *quantum dimension*.

Standing hypothesis. $\mathcal{C} := (\mathcal{C}, \otimes, \mathbb{1}, a, \ell, r, (-)^*, {}^*(-))$ is a rigid category.

---

§3.7.1. Pivotal categories and pivotal functors

We say that $\mathcal{C}$ is **pivotal** if there exists a monoidal natural isomorphism:

$$j : \mathrm{Id}_{\mathcal{C}} \overset{\otimes}{\underset{\Rightarrow}{\cong}} (-)^{**}.$$

Namely, a **pivotal structure** on $\mathcal{C}$ is a collection of isos $j := \{j_X : X \overset{\sim}{\to} X^{**}\}_{X \in \mathcal{C}}$ in $\mathcal{C}$, natural in X, such that

$$j_{X \otimes Y} = ((d^L_{Y,X})^*)^{-1} \circ d^L_{X^*,Y^*} \circ (j_X \otimes j_Y) \tag{3.48}$$

as morphisms from $X \otimes Y$ to $(X \otimes Y)^{**}$ in $\mathcal{C}$. The natural isomorphism d^L is from Proposition 3.42(b). Moreover, the compositions

$$((d^L_{Y,X})^*)^{-1} \circ d^L_{X^*,Y^*} : X^{**} \otimes Y^{**} \overset{\sim}{\to} (Y^* \otimes X^*)^* \overset{\sim}{\to} (X \otimes Y)^{**}$$

are the monoidal product components of the double dual version of Proposition 3.42 [Exercise 3.24]. The monoidal unit is preserved by j [Lemma 3.32(d)].

Note that a synonym for "pivotal" is the term **sovereign.**

We can characterize pivotal structures on rigid categories as the rigid categories for which the left and right duality monoidal functors (from Proposition 3.42) coincide. See the result below.

Proposition 3.49. *A rigid category $\mathcal{C}$ has a pivotal structure j if and only if we have a monoidal natural isomorphism,*

$$\hat{j} : (-)^* \overset{\otimes}{\underset{\Rightarrow}{\cong}} {}^*(-),$$

for the duality functors introduced in §3.6.3.

By this result, the monoidal natural isomorphism $\hat{j}$ is also referred to as a **pivotal structure** on $\mathcal{C}$.

Proof of Proposition 3.49. Given the components $\{\hat{j}_X : X^* \overset{\sim}{\to} {}^*X\}_{X \in \mathcal{C}}$ of a monoidal natural isomorphism between $(-)^*$ and ${}^*(-)$, by using Lemma 3.32(c), define morphisms:

$$j_X := \hat{j}_X^* : X \cong ({}^*X)^* \to X^{**}.$$

The morphism j_X is an iso because functors, including $(-)^*$, preserve isos [Exercise 2.19(a)]. We will show in Figure 3.11 that (3.48) holds. Here, the assumption that $\hat{j}_{X\otimes Y} = d^R_{Y,X} \circ (\hat{j}_Y \otimes \hat{j}_X) \circ (d^L_{Y,X})^{-1}$ is denoted by (†).

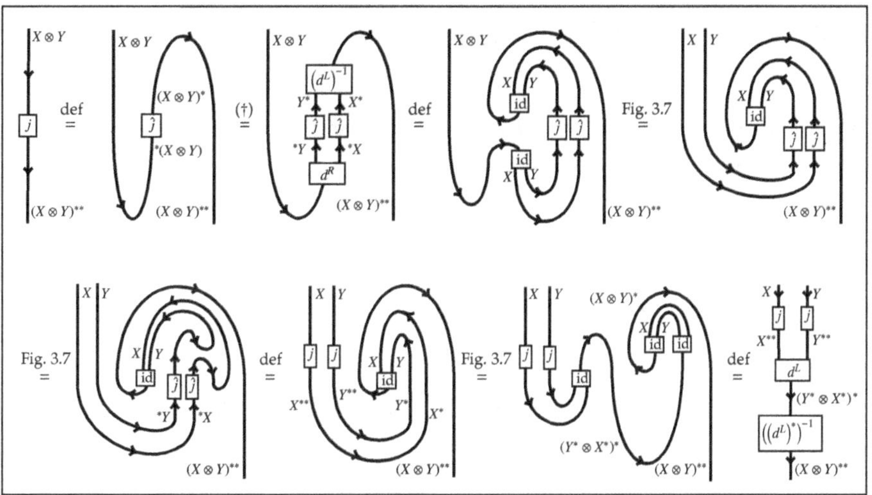

Figure 3.11: Part of the proof of Proposition 3.49.

Conversely, given a pivotal structure $j = \{j_X : X \xrightarrow{\sim} X^{**}\}_{X\in\mathcal{C}}$ on $\mathcal{C}$, by using Lemma 3.32(c), define morphisms $\hat{j}_X := {}^*j_X : {}^*(X^{**}) \cong X^* \to {}^*X$. With a similar argument, j_X are isos, and we also get that $\hat{j}_{X\otimes Y} = d^R_{Y,X} \circ (\hat{j}_Y \otimes \hat{j}_X) \circ (d^L_{Y,X})^{-1}$ due to j being a monoidal natural transformation. $\square$

Since we can identify the left and right duality structures in pivotal categories $\mathcal{C}$, we can also simplify the graphical calculus introduced in Figure 3.7; see Figure 3.12. For an object $X \in \mathcal{C}$, use the following notation.

- $X^\vee := X^*$, which is isomorphic to ${}^*X$ via $\hat{j}_X$.

- $\mathrm{ev}_X := \mathrm{ev}^L_X : X^\vee \otimes X \to \mathbb{1}$, which is identified with $\mathrm{ev}^R_{X^\vee} : X^\vee \otimes X^{\vee\vee} \to \mathbb{1}$ by precomposing ev^L_X with $\mathrm{id}_{X^\vee} \otimes j_X^{-1}$.

- $\mathrm{coev}_X := \mathrm{coev}^L_X : \mathbb{1} \to X \otimes X^\vee$, identified with $\mathrm{coev}^R_{X^\vee} : \mathbb{1} \to X^{\vee\vee} \otimes X^\vee$ by composing coev^L_X with $j_X \otimes \mathrm{id}_{X^\vee}$.

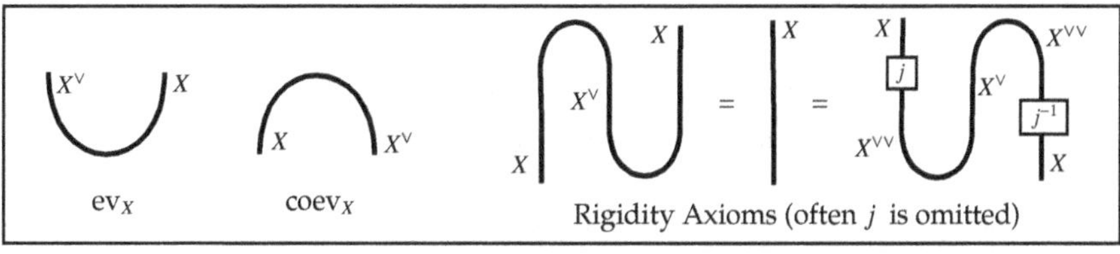

Figure 3.12: (Co)evaluation morphisms and rigidity in pivotal categories.

These identifications enable us to get a right (resp., left) dual structure from left (resp., right) dual structure in a pivotal category. Likewise, we write:

- $f^\vee$ for the left or right dual of a morphism f in $\mathcal{C}$.

Next, we relate two pivotal categories via functors that respect pivotality. Let $(\mathcal{C}, j)$ and $(\mathcal{C}', j')$ be two pivotal categories.

A **pivotal functor** $(\mathcal{C}, j) \to (\mathcal{C}', j')$ is a strong monoidal functor $(F, F^{(2)}, F^{(0)})$ from $\mathcal{C}$ to $\mathcal{C}'$, such that the morphisms below from $F(X^\vee)$ to $F(X)^\vee$ in $\mathcal{C}'$ are equal, for each $X \in \mathcal{C}$. We suppress all subscripts here:

$$F(X^\vee) \xrightarrow{(\mathrm{id}\, \otimes'\, \mathrm{coev})r'^{-1}} F(X^\vee) \otimes' (F(X) \otimes' F(X)^\vee) \xrightarrow{a'^{-1}} (F(X^\vee) \otimes' F(X)) \otimes' F(X)^\vee$$

$$\Big\downarrow F^{(2)} \otimes' \mathrm{id}$$

$$F(X)^\vee \xleftarrow{\ell'(F^{(-0)}\, \otimes'\, \mathrm{id})} F(\mathbb{1}) \otimes' F(X)^\vee \xleftarrow{F(\mathrm{ev})\, \otimes'\, \mathrm{id}} F(X^\vee \otimes X) \otimes' F(X)^\vee$$

$$F(X^\vee) \xrightarrow{(\mathrm{coev}\, \otimes'\, \mathrm{id})\ell'^{-1}} (F(X)^\vee \otimes' F(X)^{\vee\vee}) \otimes' F(X^\vee) \xrightarrow{(\mathrm{id}\, \otimes'\, j'^{-1}\, \otimes'\, \mathrm{id})\, a'} F(X)^\vee \otimes' (F(X) \otimes' F(X^\vee))$$

$$\Big\downarrow \mathrm{id} \otimes' F^{(2)}$$

$$F(X)^\vee \xleftarrow{r'(\mathrm{id}\, \otimes'\, F^{(-0)})} F(X)^\vee \otimes' F(\mathbb{1}) \xleftarrow{\mathrm{id}\, \otimes'\, F(\mathrm{ev}(j\, \otimes\, \mathrm{id}))} F(X)^\vee \otimes' F(X \otimes X^\vee)$$

We also say that $(\mathcal{C}, j)$ and $(\mathcal{C}', j')$ are **equivalent** (resp., **isomorphic**) **as pivotal categories** if there exists a pivotal functor $(F, F^{(2)}, F^{(0)}) : (\mathcal{C}, j) \to (\mathcal{C}', j')$ such that F is an equivalence (resp., isomorphism) of the underlying categories. In this case, we write $(\mathcal{C}, j) \overset{\mathrm{piv.}\otimes}{\simeq} (\mathcal{C}', j')$ (resp., $(\mathcal{C}, j) \overset{\mathrm{piv.}\otimes}{\cong} (\mathcal{C}', j')$).

§3.7.2. Examples of pivotal categories

We now present several examples of pivotal structures on rigid categories, building on the examples of rigid categories in §3.6.2.

- FdVec: The pivotal structure is given by $j_V : V \to V^{**}$ for $V \in$ FdVec, where we define $j_V(v) := [f \mapsto f(v)] \in V^{**}$, for $f \in V^*$. See Exercises 1.4(d) and 2.23. Equivalently, the monoidal natural isomorphism $\hat{j}$ in Proposition 3.49 is given by $\hat{j}_V = \mathrm{id}_{\mathrm{Hom}_\Bbbk(V,\Bbbk)}$ as $V^* = {}^*V = \mathrm{Hom}_\Bbbk(V, \Bbbk)$; see Example 3.34.

- G-FdMod: Likewise, the pivotal structure can be given by Proposition 3.49 with $\hat{j}_{(V,\triangleright)}$ being the identity morphism; see Example 3.36.

- FdVec$_G$, FdVec$_G^\omega$: These are pivotal categories by Exercise 3.35(g).

- $\underline{G}$: The pivotal structure is given by $j_g := \mathrm{id}_g : g \to g^{**}\ (= g)$, for $g \in \underline{G}$.

So far, we presented pivotal categories in which the left and right dual objects of a given object are equal; in this case, the pivotal structure $\hat{j}$ is the identity monoidal natural isomorphism. We will see more interesting pivotal structures later in a future volume. Namely, categories of modules over *semisimple Hopf algebras* are pivotal. These have the underlying structure of a finite semisimple category (see §§2.7.3, 2.9).

In fact, there is a research direction studying the pivotality of certain semisimple rigid categories called *fusion categories* (see §3.9.1 later). This is prompted by Conjecture 2.8 of Etingof et al. [2005], also stated as Question 4.8.3 of Etingof et al. [2015], which, in general, is the following inquiry.

Research Question 3.50. Are all finite, semisimple, rigid categories pivotal?

§3.7.3. Pivotal trace and pivotal dimension

With a pivotal structure j on a rigid category $\mathcal{C}$, let us attach invariants to morphisms and objects in $\mathcal{C}$ as follows. This is modeled on traces of linear endomorphisms and dimensions of vector spaces in linear algebra.

Let $f : X \to X$ be a morphism in a pivotal category $(\mathcal{C}, j)$. The **left pivotal trace** and **right pivotal trace** of f are defined, respectively, as:

$$\mathrm{tr}_j^L(f) : \mathbb{1} \xrightarrow{\mathrm{coev}_{X^\vee}} X^\vee \otimes X^{\vee\vee} \xrightarrow{\mathrm{id}_{X^\vee} \otimes j_X^{-1}} X^\vee \otimes X \xrightarrow{\mathrm{id}_{X^\vee} \otimes f} X^\vee \otimes X \xrightarrow{\mathrm{ev}_X} \mathbb{1},$$

$$\mathrm{tr}_j^R(f) : \mathbb{1} \xrightarrow{\mathrm{coev}_X} X \otimes X^\vee \xrightarrow{f \otimes \mathrm{id}_{X^\vee}} X \otimes X^\vee \xrightarrow{j_X \otimes \mathrm{id}_{X^\vee}} X^{\vee\vee} \otimes X^\vee \xrightarrow{\mathrm{ev}_{X^\vee}} \mathbb{1}.$$

Pivotal trace is also known as **categorical trace**, as **quantum trace**, or simply as **trace** in the literature.

Moreover, we define the **(left) pivotal dimension** of an object X in $\mathcal{C}$ as follows:

$$\dim_j(X) := \mathrm{tr}_j^L(\mathrm{id}_X).$$

Likewise, pivotal dimension is called **categorical dimension** or **quantum dimension**. Here, we do not use the right trace to define dimension since

$$\mathrm{tr}_j^R(\mathrm{id}_X) = \mathrm{tr}_j^L(\mathrm{id}_X^\vee) = \mathrm{tr}_j^L(\mathrm{id}_{X^\vee}) = \dim_j(X^\vee). \tag{3.51}$$

See Exercise 3.30. When $\mathcal{C}$ is strict, we depict these invariants in Figure 3.13.

For instance, if $\mathcal{C}$ is the pivotal category FdVec (as in §3.7.2), then the pivotal trace is the usual trace of a linear endomorphism, and the pivotal dimension is vector space dimension; see Exercise 3.29. See also Exercise 3.31 on how pivotal trace is preserved under monoidal product.

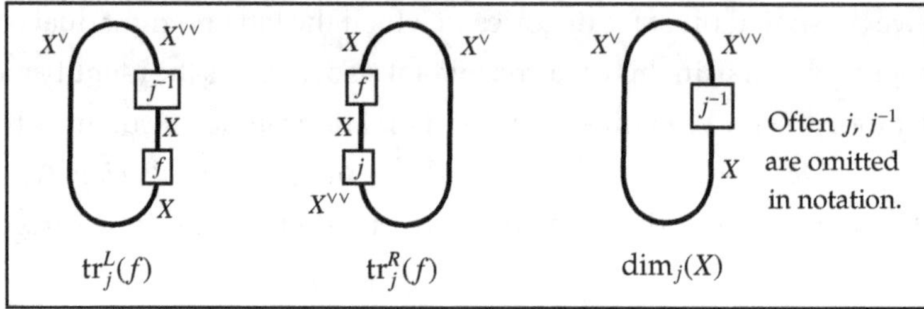

Figure 3.13: Pivotal trace and pivotal dimension

§3.7.4. Module categories over pivotal categories

With the extra structure of a monoidal category $\mathcal{C}$ being pivotal, one may want to construct module categories over $\mathcal{C}$ that have extra structure accompanying the pivotality of $\mathcal{C}$. This has been achieved when $\mathcal{C}$ is a finite *tensor category* (see §3.10.1 later). This is due to the work of K. Shimizu, building on the work of Schaumann [2015]. After reading the rest of the chapter, we encourage the reader to check out Shimizu [2020] (especially Section 5.2) and Shimizu [2019] (especially Section 3.6) for details. In particular, Theorem 3.13 of Shimizu [2019] states that given a pivotal finite tensor category $\mathcal{C}$, along with a *pivotal module category* $\mathcal{M}$ over $\mathcal{C}$, one obtains that $(\mathrm{Rex}_{\mathcal{C}\text{-Mod}}(\mathcal{M}, \mathcal{M}))^{\otimes\mathrm{op}}$ is a pivotal monoidal category.

§3.8. Spherical categories

Recall from §3.7.3 that in a pivotal category $(\mathcal{C}, j)$, the pivotal dimension $\dim_j(X)$ of an object X in $\mathcal{C}$ is defined using the left trace of the morphism id_X. Moreover, the right trace of id_X turns out to be $\dim_j(X^\vee)$ (see (3.51)). Next, we will impose a property that will imply that $\dim_j(X^\vee) = \dim_j(X)$, for each object X in $\mathcal{C}$.

Standing hypothesis.   Take $(\mathcal{C}, j)$ to be a pivotal category here.

---

§3.8.1. Trace-spherical categories

A pivotal category $(\mathcal{C}, j)$ is said to be **trace-spherical** if the left and right pivotal traces of any morphism $f : X \to X$ in $\mathcal{C}$ are equal. Often these are called **spherical** categories, for short. In this case, we denote

$$\mathrm{tr}_j(f) := \mathrm{tr}_j^L(f) = \mathrm{tr}_j^R(f).$$

In this case, we have that $\dim_j(X^\vee) = \dim_j(X)$, for each object X in $\mathcal{C}$; see (3.51). See Figure 3.14 for a visualization of the pivotal trace in this case.

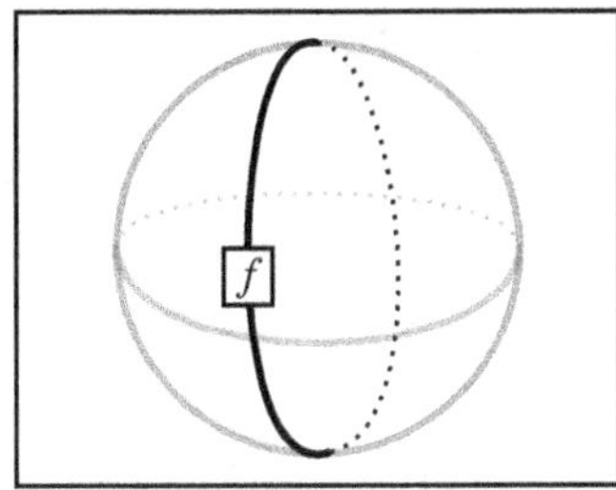

Figure 3.14: Pivotal trace "on" a sphere in a trace-spherical category (cf. Fig. 3.13)

Since sphericality is a property of the pivotal structure on a rigid category, a functor between spherical categories is simply a pivotal functor. Moreover, two spherical categories are **equivalent** if they are equivalent as pivotal categories (see §3.7.1). See Exercises 3.31 and 3.32 for practice.

Examples of spherical categories include the pivotal categories discussed in §3.7.2: FdVec, G-FdVec, FdVec$_G$, FdVec$_G^\omega$ (see Exercise 3.35(g)), $\underline{G}$, and also categories of modules over *semisimple Hopf algebras* (discussed in a future volume).

§3.8.2. DSPS-spherical categories

Recall that Research Question 3.50 inquires whether all finite, semisimple, rigid categories are pivotal. This question can be extended to trace-sphericality as there are no counterexamples to date.

On the other hand, in the nonsemisimple setting, sphericality has been generalized by the work of Douglas–Schommer-Pries–Snyder. See Section 3.5.2 of Douglas et al. [2020] for their notion of sphericality for finite tensor categories (see §3.10.1 here); we refer to these as *DSPS-spherical categories*. In fact, it is shown in Section 3.5.3 of Douglas et al. [2020] that:

- Semisimple DSPS-spherical categories are trace-spherical.

After reading about Hopf algebras in a future volume, one can appreciate the examples of (non-)spherical categories presented in Section 3.5.3 of Douglas et al. [2020] that illustrate the following statements.

- There exist nonsemisimple DSPS-spherical categories, that are not trace-spherical; see Example 3.5.5 of Douglas et al. [2020].

- There exist nonsemisimple trace-spherical categories, that are not DSPS-spherical; see Example 3.5.6 of Douglas et al. [2020].

§3.8.3. Module categories over spherical categories

As for rigidity and pivotality (cf. §§3.6.4, 3.7.4), one may want to have module categories over spherical categories C that have extra features that accompany the spherical condition of C. But we have a couple of choices for sphericality: trace-sphericality or DSPS-sphericality. Recent results in the literature employ the latter notion as it extends well to the nonsemisimple setting. See the work of Fuchs–Galindo–Jaklitsch–Schweigert, especially Section 5.6 of Fuchs et al. [2022], for details about *spherical module categories*.

§3.9. Fusion categories

Next, we focus on a certain class of $\Bbbk$-linear monoidal categories whose behavior is largely understood in terms of its simple objects. Such categories are used quite often in applications. Consider the terminology below, which was established in the article by Etingof et al. [2005]; see also the textbook by Etingof et al. [2015].

§3.9.1. Fusion categories and fusion rules

i. Fusion categories

A monoidal category $C := (C, \otimes, \mathbb{1}, a, \ell, r)$ is **fusion** if the items below hold.

(a) C is abelian (see §§2.2.2iii, 3.1.3).

(b) C is $\Bbbk$-linear (see §§2.2.2i, 3.1.3). We get that it is enriched over Vec (see §3.11 later). We also obtain that the monoidal product $\otimes$ is $\Bbbk$-bilinear on morphisms: for $f, f' : X \to Y$ and $g : W \to Z$ in C and scalars $\lambda, \lambda' \in \Bbbk$, we have that

$$(\lambda f + \lambda' f') \otimes g = \lambda(f \otimes g) + \lambda'(f' \otimes g), \qquad g \otimes (\lambda f + \lambda' f') = \lambda(g \otimes f) + \lambda'(g \otimes f').$$

(c) C is locally finite (see §2.9), which implies that it is enriched over FdVec.

(d) The unit object $\mathbb{1}$ is absolutely simple, that is, $\mathrm{End}_C(\mathbb{1}) \cong \Bbbk$ (see §2.9).

(e) C is rigid (see §3.6.1).

(f) C is semisimple (see §2.7.3).

(g) C is finite (see §2.9).

If we omit the condition (d) above, then we have a **multifusion** category.

By item (a), a fusion category C is **decomposable** if it is equivalent to a product of nonzero fusion categories (see §§2.1.3, 2.2.2ii). Else, C is **indecomposable**.

We have the following result pertaining to item (d).

Lemma 3.52. *Assume that a rigid category C is abelian and linear (i.e., satisfying items (a,b,e) above). If $\mathbb{1}$ is absolutely simple, then $\mathbb{1}$ is a simple object in C. Conversely, with local finiteness (i.e., item (c)), $\mathbb{1}$ is absolutely simple when $\mathbb{1}$ is simple in C.*

Proof. This argument is modified from Theorem 4.3.8 of Etingof et al. [2015]. For the first statement, take a nonzero subobject $(X,\ \iota : X \to \mathbb{1})$ of $\mathbb{1}$. We aim to show that $X = \mathbb{1}$. Without loss of generality, assume that X is simple. Moreover, it suffices to show that ι is epic. To proceed, consider the short exact sequence:

$$0 \longrightarrow X \xrightarrow{\ \iota\ } \mathbb{1} \longrightarrow \operatorname{coker}(\iota) \longrightarrow 0.$$

Since $(-)^*$ is exact by Propositions 3.33(b,d) and 2.49(b), we get the following short exact sequence:

$$0 \longrightarrow \operatorname{coker}(\iota)^* \longrightarrow \mathbb{1} \xrightarrow{\ \iota^*\ } X^* \longrightarrow 0.$$

In particular, $\mathbb{1}^* \cong \mathbb{1}$. Next, by Propositions 3.33(a,b) and 2.49(b), $X \otimes -$ is exact, which yields the short exact sequence below:

$$0 \longrightarrow X \otimes \operatorname{coker}(\iota)^* \longrightarrow X \otimes \mathbb{1} \cong X \xrightarrow{\ \operatorname{id} \otimes \iota^*\ } X \otimes X^* \longrightarrow 0.$$

Note that $\operatorname{coev}_X^L : \mathbb{1} \to X \otimes X^*$ is a nonzero morphism, so $X \otimes X^*$ is a nonzero object. Now X is simple, so we obtain that $\operatorname{id}_X \otimes \iota^*$ is monic. The morphism $\operatorname{id}_X \otimes \iota^*$ is already epic, thus $\operatorname{id}_X \otimes \iota^*$ is an iso. Therefore, we get a nonzero morphism

$$\phi := (\operatorname{id}_X \otimes \iota^*)^{-1} \operatorname{coev}_X^L : \mathbb{1} \to X.$$

Further, $\iota\phi : \mathbb{1} \to \mathbb{1}$ is also a nonzero morphism. Since $\mathbb{1}$ is absolutely simple, we get that $\iota\phi$ a nonzero scalar multiple of $\operatorname{id}_{\mathbb{1}}$. Hence, $\iota\phi$ is an iso, and thus epic. This implies that ι is epic, as desired.

The converse direction holds essentially by Schur's Lemma [Proposition 2.35]. Namely, Corollary 2.36 implies that $\operatorname{End}_C(\mathbb{1})$ is a division algebra over $\Bbbk$. Then, item (c) and Lemma 1.36 implies that $\operatorname{End}_C(\mathbb{1}) \cong \Bbbk$. $\square$

Two fusion categories C and D are said to be **equivalent** (resp., **isomorphic**) if there exists a $\Bbbk$-linear functor $F : C \to D$ that is an equivalence (resp., isomorphism) of monoidal categories. In this case, we still write $C \overset{\otimes}{\simeq} D$ (resp., $C \overset{\otimes}{\cong} D$).

Example 3.53. Examples of the rigid categories from §3.6.2 that are fusion include: FdVec; G-FdMod; FdVec$_G$; along with the categories FdVec$_G^\omega$ defined in Exercise 3.35, for G a finite group. See Exercises 3.33 and 3.35(h).

One nice feature of (not-necessarily-finite) pivotal fusion categories is that nonzero simple objects have nonzero pivotal dimension.

Proposition 3.54. *Let C be a monoidal category satisfying conditions (a)-(f) above, and assume that C is pivotal. Take a nonzero simple object X in C. Then, $\dim_j(X) \in \mathrm{End}_C(\mathbb{1})$ is nonzero (or, equivalently, $\dim_j(X) \in \mathrm{End}_C(\mathbb{1})$ is invertible).*

Proof. Since X is simple, we have by Corollary 2.36 and Lemma 1.36 that $\dim_{\Bbbk}\mathrm{Hom}_C(X, X) = 1$. So, $\dim_{\Bbbk}\mathrm{Hom}_C(\mathbb{1}, X^\vee \otimes X^{\vee\vee}) = 1$ by Proposition 3.33(b,d).

Next, take Y to be the image of $\mathrm{coev}_{X^\vee} : \mathbb{1} \to X^\vee \otimes X^{\vee\vee}$. Then, there exists an object $Z \in C$ such that

$$X^\vee \otimes X^{\vee\vee} \cong Y \sqcup Z \; (\cong Y \sqcap Z)$$

in C by Proposition 2.40. Since $\mathrm{coev}_{X^\vee}$ is monic, $\mathbb{1} \cong Y$. Consider

$$\mathbb{1} \cong Y \xrightarrow{\;\;\alpha_Y\;\;} Y \sqcup Z \cong X^\vee \otimes X^{\vee\vee},$$

for the morphism α_Y from the universal property of the coproduct. Using Exercise 2.8, we then have:

$$1 = \dim_{\Bbbk}\mathrm{Hom}_C(\mathbb{1}, X^\vee \otimes X^{\vee\vee}) \geq \dim_{\Bbbk}\mathrm{Hom}_C(\mathbb{1}, Y) + \dim_{\Bbbk}\mathrm{Hom}_C(\mathbb{1}, Z).$$

As a result, $\dim_{\Bbbk}\mathrm{Hom}_C(\mathbb{1}, Y) = 1$ and $\dim_{\Bbbk}\mathrm{Hom}_C(\mathbb{1}, Z) = 0$. Therefore, $\mathrm{coev}_{X^\vee}$ is a nonzero scalar multiple of α_Y.

We also claim that $\mathrm{ev}_X \, (\mathrm{id}_{X^\vee} \otimes j_X^{-1}) : X^\vee \otimes X^{\vee\vee} \to \mathbb{1}$ is a nonzero scalar multiple of $\alpha'_Y : Y \sqcap Z \to Y$, the morphism arising from the universal property of the product. By semisimplicity and Lemma 2.41, we have that $\dim_{\Bbbk}\mathrm{Hom}_C(Y, \mathbb{1}) = 1$, $\dim_{\Bbbk}\mathrm{Hom}_C(Z, \mathbb{1}) = 0$, and $\dim_{\Bbbk}\mathrm{Hom}_C(X^\vee \otimes X^{\vee\vee}, \mathbb{1}) = 1$. So, the claim holds by using the morphism,

$$X^\vee \otimes X^{\vee\vee} \cong Y \sqcap Z \xrightarrow{\;\;\alpha'_Y\;\;} Y \cong \mathbb{1}.$$

Now we obtain that $\dim_j(X) := \mathrm{ev}_X \, (\mathrm{id}_{X^\vee} \otimes j_X^{-1}) \, \mathrm{coev}_{X^\vee}$ is a nonzero scalar multiple of $\alpha'_Y \alpha_Y = \mathrm{id}_Y$ [§2.2.2ii], as desired. For the parenthetical statement in the result, note that nonzero pivotal dimension is equivalent to invertible pivotal dimension by the simplicity of $\mathbb{1}$ and by Schur's Lemma [Proposition 2.35]. $\qquad\square$

ii. Fusion rules and rank

Next, let us delve into the internal structure of fusion categories. Take:

- $\mathrm{Irr}(C)$: the finite set of isoclasses of simple objects of C;

- $i \in \mathrm{Irr}(C)$ corresponding to a simple object X_i of C, or to an isoclass $[X_i]$ in C.

The cardinality of the finite set $\mathrm{Irr}(C)$ is called the **rank** of C.

Let $\{X_i\}_{i \in \mathrm{Irr}(C)}$ be a collection of representatives of simple objects of C. Then,

$$X_i \otimes X_j \cong \coprod_{k \in \mathrm{Irr}(C)} X_k^{\sqcup N_k^{i,j}}, \tag{3.55}$$

for some $N_k^{i,j} \in \mathbb{Z}_{\geq 0}$ by the semisimplicity of C. These integers $\{N_k^{i,j}\}_{i,j,k \in \mathrm{Irr}(C)}$ are called the **fusion rules** of C, also referred to as the **fusion rule** of C.

By local finiteness, each object in C has finite length (see §2.7.2). We then obtain for $i, j, k \in \mathrm{Irr}(C)$ that

$$N_k^{i,j} = [X_i \otimes X_j : X_k], \quad \text{the multiplicity of } X_k \text{ in } X_i \otimes X_j. \tag{3.56}$$

We will see in Example 3.66 below that simple objects and fusion rules are not enough to fully determine a fusion category, up to equivalence. That is, there exist inequivalent fusion categories with the same fusion rule.

On the other hand, we do have a powerful finiteness result for fusion categories that appeared as Theorem 2.27 in the article by Etingof et al. [2005], which was inspired by unpublished work of A. Ocneanu.

Theorem 3.57 (Ocneanu rigidity). *There are only finitely many equivalence classes of fusion categories that have a given fusion rule.* □

Returning to the notion of rank, Ostrik [2003a] inquired if a similar finiteness result holds for fusion categories, as follows.

Research Question 3.58 (Rank finiteness for fusion categories). Are there only finitely many equivalence classes of fusion categories of a given rank?

The only fusion category of rank 1 is FdVec, up to equivalence; see Exercise 3.34. For rank 2, there are precisely four equivalence classes of fusion categories, by the work of Ostrik [2003a]. So, the question is settled for rank ≤ 2. Research Question 3.58 is settled in higher ranks in special cases, such as for *modular* fusion categories (discussed in a future volume) by Bruillard et al. [2016].

§3.9.2. Frobenius-Perron (FP-)dimension

We can study the size of a fusion category C, and the size of objects in C, via its fusion rules. To do so, consider the result below due to the work of Perron [1907] and of Frobenius [1912].

Theorem 3.59 (Frobenius-Perron Theorem). *Take a square matrix M with entries in $\mathbb{R}_{\geq 0}$. Then, M has a non-negative real eigenvalue $\mathrm{FP}(M)$ that is greater than or equal to the absolute value of all other eigenvalues of M.* □

i. FP-dimension

Now take an object X in C, and consider the square matrix

$$N_X := \left([X \otimes X_j : X_k] \right)_{j,k \in \mathrm{Irr}(C)}. \tag{3.60}$$

See also (3.55) and (3.56) for the case when X is a simple object of C. We define the **Frobenius-Perron (FP-)dimension of $X \in C$** to then be

$$\mathrm{FPdim}_C(X) := \mathrm{FP}(N_X) \in \mathbb{R}.$$

Moreover, the **Frobenius-Perron (FP-)dimension of** C is defined as:

$$\text{FPdim}(C) := \sum_{i \in \text{Irr}(C)} \text{FPdim}_C(X_i)^2.$$

Frobenius-Perron dimension was introduced by Etingof–Nikshych–Ostrik in Etingof et al. [2005], and we refer the reader to that article and the book by Etingof et al. [2015] for a full account of this measure. But here are some useful facts.

The isoclasses $[X_i]_{i \in \text{Irr}(C)}$ form an unital ring with

$$[X_i] + [X_j] := [Y], \text{ for which } \exists \text{ a short exact sequence } 0 \to X_i \to Y \to X_j \to 0,$$

$$[X_i][X_j] := [X_i \otimes X_j] = \sum_{k \in \text{Irr}(C)} [X_i \otimes X_j : X_k] \, X_k \tag{3.61}$$

This is called the **Grothendieck ring** of C, is denoted by $\text{Gr}(C)$, and its unit is $[\mathbb{1}]$. Objects of $\text{Gr}(C)_{\Bbbk} := \text{Gr}(C) \otimes_{\mathbb{Z}} \Bbbk$, that are not necessarily objects of C, are referred to as **virtual objects** of C.

Proposition 3.62. *Let C be a fusion category. Then, the following items hold.*

(a) *If X_i is a simple object of C, then (along with being a real number) $\text{FPdim}_C(X_i)$ is an algebraic integer (e.g., a root of a monic polynomial in $\mathbb{Z}[x]$).*

(b) *If X_i is a simple object of C, we also have that $\text{FPdim}_C(X_i) \geq 1$.*

(c) *There exists a virtual object R_C of C, unique up to rescaling, that satisfies the following conditions in $\text{Gr}(C)_{\Bbbk}$, for each object $X \in C$:*

$$X \otimes R_C \cong \text{FPdim}_C(X) \, R_C \cong R_C \otimes X.$$

*Moreover, $R_C \cong \coprod_{i \in \text{Irr}(C)} \text{FPdim}_C(X_i) \, X_i$. Here, R_C is called the **regular object** of C.*

(d) *For any pair of objects $X, Y \in C$, we have that*

$$\text{FPdim}_C(X \sqcup Y) \;=\; \text{FPdim}_C(X) + \text{FPdim}_C(Y),$$

$$\text{FPdim}_C(X \otimes Y) \;=\; \text{FPdim}_C(X) \, \text{FPdim}_C(Y).$$

(e) *If D is another fusion category, and $F : C \to D$ is an exact, faithful, linear, monoidal functor, then $\text{FPdim}_C(X) = \text{FPdim}_D(F(X))$ for any $X \in C$.*

Proof. The proofs of parts (a)–(c) are given in Propositions 3.3.4 and 3.3.6 of Etingof et al. [2015] in the language of fusion rings, and we recommend trying out the proof for yourself before consulting this reference. Now part (d) holds by the computations in $\mathrm{Gr}(\mathcal{C})_{\Bbbk}$ below:

$$\mathrm{FPdim}_{\mathcal{C}}(X \sqcup Y)\, R_{\mathcal{C}} \;\overset{(c)}{\cong}\; (X \sqcup Y) \otimes R_{\mathcal{C}} \;\overset{\mathrm{Lem.\,3.4}}{\cong}\; (X \otimes R_{\mathcal{C}}) \sqcup (Y \otimes R_{\mathcal{C}})$$
$$\overset{(c)}{\cong}\; (\mathrm{FPdim}_{\mathcal{C}}(X) + \mathrm{FPdim}_{\mathcal{C}}(Y))\, R_{\mathcal{C}},$$

$$\mathrm{FPdim}_{\mathcal{C}}(X \otimes Y)\, R_{\mathcal{C}} \;\overset{(c)}{\cong}\; (X \otimes Y) \otimes R_{\mathcal{C}} \;\cong\; X \otimes (Y \otimes R_{\mathcal{C}})$$
$$\overset{(c)}{\cong}\; X \otimes (\mathrm{FPdim}_{\mathcal{C}}(Y)\, R_{\mathcal{C}}) \;\overset{\mathrm{Lem.\,3.4}}{\cong}\; \mathrm{FPdim}_{\mathcal{C}}(Y)\, (X \otimes R_{\mathcal{C}})$$
$$\overset{(c)}{\cong}\; (\mathrm{FPdim}_{\mathcal{C}}(X)\, \mathrm{FPdim}_{\mathcal{C}}(Y))\, R_{\mathcal{C}}.$$

For part (e), see Proposition 4.5.7 of Etingof et al. [2015]. $\qquad\square$

ii. Integrality

Next, we turn our attention to integrality. A fusion category $\mathcal{C}$ is said to be **integral** if $\mathrm{FPdim}_{\mathcal{C}}(X) \in \mathbb{Z}$ for all $X \in \mathcal{C}$, and $\mathcal{C}$ is called **weakly integral** if $\mathrm{FPdim}(\mathcal{C}) \in \mathbb{Z}$.

Many curious results imply integrality for fusion categories. For instance, a fusion category of odd Frobenius-Perron dimension must be integral, not just weakly integral. See Corollary 3.5.8 of Etingof et al. [2015], which again is given in the language of fusion rings.

Moreover, a fusion category is integral if and only if it is equivalent to the category of finite-dimensional modules over a semisimple *quasi-Hopf algebra*. See Proposition 6.1.14 of Etingof et al. [2015] for a general result after reading §3.10.

Each of the examples of fusion categories presented in §3.9.1i are integral fusion categories. We now discuss the integrality of FdVec and G-FdMod as follows, and leave the discussion of FdVec_G and $\mathsf{FdVec}_G^{\omega}$ to the reader as Exercise 3.35(i).

Example 3.63. (a) We have that $\mathrm{Irr}(\mathsf{FdVec}) = \{[\Bbbk]\}$ and $|\mathrm{Irr}(\mathsf{FdVec})| = 1$. Now for $V \in \mathsf{FdVec}$, we get that

$$[V \otimes_{\Bbbk} \Bbbk : \Bbbk] \;=\; \dim_{\Bbbk}\mathrm{Hom}_{\mathsf{Vec}}(V \otimes_{\Bbbk} \Bbbk, \Bbbk) \;=\; \dim_{\Bbbk} V;$$

see (2.38). Thus, $\mathrm{FPdim}_{\mathsf{FdVec}}(V) = \dim_{\Bbbk} V$, and $\mathrm{FPdim}(\mathsf{FdVec}) = \dim_{\Bbbk}\Bbbk = 1$.

(b) Take a finite group G. Now $\Bbbk G$ is a semisimple algebra by Maschke's Theorem [Theorem 1.47], and $\Bbbk G$ is also finite-dimensional. So, the Artin-Wedderburn Theorem [Theorem 1.44] can be applied to get that

$$\Bbbk G \;\cong\; \textstyle\prod_{i=1}^{r} \mathrm{Mat}_{n_i}(\Bbbk),$$

as $\Bbbk$-algebras. So, $|\mathrm{Irr}(\mathsf{FdVec})| = r$, and $\mathrm{Irr}(G\text{-}\mathsf{FdMod}) = \{[V_1], \ldots, [V_r]\}$, with $\dim_{\Bbbk} V_i = n_i$ by Proposition 1.50. In $G\text{-}\mathsf{FdMod} \cong \Bbbk G\text{-}\mathsf{FdMod}$, consider the regular module $(\Bbbk G)_{\mathrm{reg}}$ and modules $V, W \in \Bbbk G\text{-FdMod}$ to get

$$\dim_\Bbbk \mathrm{Hom}_{\Bbbk G\text{-Mod}}(V \otimes_\Bbbk (\Bbbk G)_{\mathrm{reg}}, W)$$

$$\overset{\text{Prop.3.33(a)}}{=} \dim_\Bbbk \mathrm{Hom}_{\Bbbk G\text{-Mod}}((\Bbbk G)_{\mathrm{reg}}, {}^*V \otimes_\Bbbk W)$$

$$\overset{\text{Free-Forget adj.}}{=} \dim_\Bbbk \mathrm{Hom}_{\mathrm{Vec}}(\Bbbk, {}^*V \otimes_\Bbbk W)$$

$$= \dim_\Bbbk ({}^*V \otimes_\Bbbk W) \;\; = \;\; (\dim_\Bbbk V)(\dim_\Bbbk W).$$

Therefore, we get that in $\mathrm{Gr}(\mathcal{C})_\Bbbk$:

$$[V][(\Bbbk G)_{\mathrm{reg}}] \;\; \overset{(3.61)}{=} \;\; \sum_{i \in \mathrm{Irr}(\Bbbk G\text{-Mod})} [V \otimes_\Bbbk (\Bbbk G)_{\mathrm{reg}} : V_i] \, [V_i]$$

$$\overset{(2.38)}{=} \sum_{i \in \mathrm{Irr}(\Bbbk G\text{-Mod})} (\dim_\Bbbk V)(\dim_\Bbbk V_i) \, [V_i]$$

$$= (\dim_\Bbbk V) \sum_{i \in \mathrm{Irr}(\Bbbk G\text{-Mod})} (\dim_\Bbbk V_i) \, [V_i] \;\; = \;\; (\dim_\Bbbk V)[(\Bbbk G)_{\mathrm{reg}}].$$

So, by Proposition 3.62(c), $(\Bbbk G)_{\mathrm{reg}}$ is the regular object of $\Bbbk G\text{-FdMod}$, and

$$\mathrm{FPdim}_{G\text{-FdMod}}(V) = \dim_\Bbbk V.$$

In particular, $\mathrm{FPdim}_{G\text{-FdMod}}(V_i) = n_i$. Therefore,

$$\mathrm{FPdim}(G\text{-FdMod}) \;\; = \;\; \sum_{i=1}^r n_i^2 \;\; = \;\; \dim_\Bbbk \left(\prod_{i=1}^r \mathrm{Mat}_{n_i}(\Bbbk) \right) \;\; = \;\; \dim_\Bbbk \Bbbk G \;\; = \;\; |G|.$$

See Exercise 3.36 for the case when G is the dihedral group D_8 of order 8.

(c) As a special case of part (b), suppose that G is a finite abelian group. Then, $\Bbbk G$ is a finite-dimensional, semisimple, commutative $\Bbbk$-algebra, and thus is isomorphic to $\Bbbk^{\oplus |G|}$ as $\Bbbk$-algebra; see Figure 1.2. In this case, $\Bbbk G\text{-FdMod}$ has rank $|G|$, and by Exercise 1.30(a), each simple $\Bbbk G$-module is 1-dimensional as a $\Bbbk$-vector space (each forming a singleton isoclass of $\Bbbk G$-modules).

§3.9.3. More examples of fusion categories

Previously we discussed some **numerical invariants** of fusion categories, that is, numerical data that equivalent fusion categories share. This included rank, fusion rules, and the FP-dimension of objects. However, this data is not enough to fully determine a fusion category; i.e., such invariants are not **complete**. We will see this below as we discuss more examples of fusion categories.

Example 3.64. Consider the **Fibonacci fusion category**, denoted by Fib, with simple objects $\mathbb{1}$ and X, and with fusion rules given by:

$$\mathbb{1} \otimes \mathbb{1} \cong \mathbb{1}, \qquad \mathbb{1} \otimes X \cong X \cong X \otimes \mathbb{1}, \qquad X \otimes X \cong \mathbb{1} \sqcup X.$$

This is also known as the **Yang-Lee fusion category**. To determine the Frobenius-Perron dimension of objects, we have the following matrices from (3.60):

$$N_\mathbb{1} = \begin{pmatrix} 1 & 0 \\ 0 & 1 \end{pmatrix} \quad \text{and} \quad N_X = \begin{pmatrix} 0 & 1 \\ 1 & 1 \end{pmatrix}.$$

Now $\mathrm{FPdim}_{\mathsf{Fib}}(\mathbb{1}) = \mathrm{FP}(N_\mathbb{1}) = 1$ and $\mathrm{FPdim}_{\mathsf{Fib}}(X) = \mathrm{FP}(N_X) = \frac{1+\sqrt{5}}{2}$. The latter value is the *golden ratio* for the Fibonacci sequence of integers, prompting the first name above for this fusion category. Moreover, $\mathrm{FPdim}(\mathsf{Fib}) = \frac{5+\sqrt{5}}{2}$. So, Fib is neither integral, nor weakly integral.

The example above shows explicitly that rank is not a complete invariant for fusion categories. Both $\mathbb{Z}_2$-Mod and Fib have rank 2, but the FP-dimensions of simple objects are, respectively, $\{1, 1\}$ and $\{1, \frac{1+\sqrt{5}}{2}\}$. (See Example 3.63(c).) Therefore, $\mathbb{Z}_2$-Mod and Fib are inequivalent as fusion categories.

Example 3.65. Consider the **Ising fusion category**, denoted by Ising, with simple objects $\mathbb{1}$, σ, and X, and with fusion rules given by:

$$\mathbb{1} \otimes \mathbb{1} \cong \mathbb{1}, \qquad \mathbb{1} \otimes \sigma \cong \sigma \cong \sigma \otimes \mathbb{1}, \qquad \mathbb{1} \otimes X \cong X \cong X \otimes \mathbb{1},$$

$$\sigma \otimes \sigma \cong \mathbb{1}, \qquad \sigma \otimes X \cong X \cong X \otimes \sigma, \qquad X \otimes X \cong \mathbb{1} \sqcup \sigma.$$

This fusion category is weakly integral, but not integral; see Exercise 3.37. A detailed study of Ising fusion categories can be found in the article by Appendix B of Drinfeld et al. [2010] (check this reference only after doing Exercise 3.37).

The details about the associativity and unitality constraints (a, ℓ, r) of Fib and of Ising can be found in the work of Ostrik [2003a] and of Drinfeld et al. [2010], respectively.

Next, we present more examples of weakly integral fusion categories, that are sometimes integral. The construction is provided in the work of Tambara and Yamagami [1998], and we recommend checking out this reference for details.

Example 3.66. For a finite abelian group G, let us consider a **Tambara-Yamagami (TY-)fusion category**, with simple objects X and $\{\sigma_g\}_{g \in G}$, where $\mathbb{1} = \sigma_e$. Here, for $g, h \in G$, the fusion rules are given by:

$$\sigma_g \otimes \sigma_h \cong \sigma_{gh}, \qquad \sigma_g \otimes X \cong X \cong X \otimes \sigma_g, \qquad X \otimes X \cong \coprod_{g \in G} \sigma_g.$$

The associativity and unitality constraints (a, ℓ, r) of these categories are parameterized by a square root τ of $|G|$, along with a *bicharacter* of χ of G. We denote such categories by $\mathrm{TY}(G, \chi, \tau)$.

This extra data (χ, τ) is used to determine equivalence classes of the TY-fusion categories. For instance, there are finite abelian groups G for which there exist pairs

(χ, τ) and (χ', τ') corresponding to inequivalent TY-fusion categories $\mathrm{TY}(G, \chi, \tau)$ and $\mathrm{TY}(G, \chi', \tau')$. Also, no matter what the pairs (χ, τ) are, we get that:

$$\mathrm{FPdim}_{\mathrm{TY}(G,\chi,\tau)}(\sigma_g) = 1, \qquad \mathrm{FPdim}_{\mathrm{TY}(G,\chi,\tau)}(X) = \sqrt{|G|}.$$

So, $\mathrm{FPdim}(\mathrm{TY}(G, \chi, \tau)) = 2|G|$.

TY-fusion categories also capture many interesting examples of fusion categories, such as D_8-FdMod from Exercise 3.36, and Ising from Example 3.65. Confirming the details here, and exploring other facts about TY-fusion categories, is the open-ended Exercise 3.38.

§3.9.4. Module categories over fusion categories

Recall that a fusion category $\mathcal{C}$ is an abelian, $\Bbbk$-linear, semisimple, rigid category with additional finiteness properties (see §3.9.1). To have a rich theory of module categories $\mathcal{M}$ over such $\mathcal{C}$, it is common to impose conditions on $\mathcal{M}$ that reflect the behavior of $\mathcal{C}$.

Namely, a **left module category over a (multi)fusion category** $\mathcal{C}$ is a left $\mathcal{C}$-module category $(\mathcal{M}, \triangleright, m, p)$ as in §3.3.1 that satisfies the conditions below.

(a) $\mathcal{M}$ is abelian (see §§2.2.2iii, 3.3.4).

(b) $\mathcal{M}$ is $\Bbbk$-linear (see §§2.2.2i, 3.3.4). In particular, the left action bifunctor $\triangleright$ is $\Bbbk$-bilinear on morphisms.

(c) $\mathcal{M}$ is locally finite (see §2.9).

(d) $\mathcal{M}$ is semisimple (see §2.7.3).

(e) $\mathcal{M}$ is finite (see §2.9).

Likewise, we can define **right module categories** (resp., **bimodule categories**) **over (multi)fusion categories**. Moreover, two such (left, right, bi) module categories are **equivalent** (resp. **isomorphic**) when they are equivalent (resp., isomorphic) in the sense of §3.3.1 via $\Bbbk$-linear functors.

The conditions (a)–(e) are useful for examining module categories over fusion categories for several reasons, including classification results, and for constructing well-behaved *higher categorical* structures with such gadgets. See the material in §§4.10.2 and 4.10.3 later for a preview.

Pertaining to classification, note that equivalence classes of module categories over the fusion categories Vec_G^ω of Exercise 3.35, which are indecomposable in the sense of §3.3.4, have been characterized by explicit group-theoretical data. This is due to the work of Ostrik [2003b] and of Natale [2017].

On the other hand, classification results for module categories over fusion categories would be far more difficult if the conditions (a)–(e) were removed. For instance, left module categories over FdVec (just as an abelian monoidal category, as in §3.3.4) include all of the categories, A-FdMod, for a $\Bbbk$-algebra A.

§3.10. Tensor categories

Now we discuss generalizations of fusion categories, where neither semisimplicity nor finiteness are required. See Chapter 4 of Etingof et al. [2015] for more context.

§3.10.1. Tensor categories

A monoidal category $C := (C, \otimes, \mathbb{1}, a, \ell, r)$ is **tensor** if the items below hold.

(a) C is abelian (see §§2.2.2iii, 3.1.3).

(b) C is $\Bbbk$-linear (see §§2.2.2i, 3.1.3). Namely, it is enriched over Vec (see §3.11 later). In particular, the monoidal product $\otimes$ is $\Bbbk$-bilinear on morphisms.

(c) C is locally finite (see §2.9), which implies that it is enriched over FdVec.

(d) The unit object $\mathbb{1}$ is absolutely simple, i.e., $\mathrm{End}_C(\mathbb{1}) \cong \Bbbk$.

(e) C is rigid (see §3.6.1).

If we omit the condition (d) above, then we have a **multitensor** category.

See also Lemma 3.52 pertaining to condition (d).

Notice that a tensor category C that is also semisimple and finite is, by definition, a fusion category (see §3.9.1i).

Example 3.67. (a) Examples of nonsemisimple, non-finite tensor categories include G-FdMod, for G an infinite group. See Exercise 1.32(a,b) for the case when $G = \mathbb{Z}$ and the ground field is $\mathbb{C}$.

(b) Examples of nonsemisimple, finite tensor categories include $\mathbb{F}G$-FdMod, for G a finite group, and $\mathbb{F}$ an algebraically closed field whose characteristic divides the order of G. Use Exercise 1.32(c,d) for the case when G is the cyclic group C_2 of order 2, and the ground field is the finite field $\mathbb{F}_2$ of order 2.

Two tensor categories C and D are **equivalent** (resp., **isomorphic**) if there exists a $\Bbbk$-linear functor $F : C \to D$ that is an equivalence (resp., isomorphism) of monoidal categories. In this case, we still write $C \overset{\otimes}{\simeq} D$ (resp., $C \overset{\otimes}{\cong} D$).

§3.10.2. Exactness, projectivity, and FP-dimension

In losing semisimplicity when moving from fusion categories to tensor categories, we must now pay attention to exactness. For instance, if $\mathcal{C}$ is semisimple, the functors $\mathrm{Hom}_{\mathcal{C}}(P, -)$ and $\mathrm{Hom}_{\mathcal{C}}(-, Q)$ are exact, for all objects $P, Q \in \mathcal{C}$. Without semisimplicity, we need P (resp., Q) to be a projective (resp., an injective) object of $\mathcal{C}$ for right exactness to occur. See Propositions 2.52, 2.53, 2.54 and Corollary 2.56.

Here, we briefly examine exactness and the role of projective objects for both the monoidal product of tensor categories and in defining FP-dimension for finite tensor categories. The details are in Sections 4.5 and 6.1 of Etingof et al. [2015].

Proposition 3.68. *Let $(\mathcal{C}, \otimes)$ be an abelian rigid category. Then, the endofunctors $(X \otimes -)$ and $(- \otimes X)$ of $\mathcal{C}$ are exact, for all $X \in \mathcal{C}$. Here, we say that $\otimes$ is* **biexact.**

Proof. This follows from Propositions 2.49(b) and 3.33. $\qquad\square$

Next, we examine how projectivity is preserved under the monoidal product.

Proposition 3.69. *Let $(\mathcal{C}, \otimes, \mathbb{1}, a, \ell, r, (-)^*, {}^*(-))$ be an abelian rigid category. If $P \in \mathcal{C}$ is projective, then $P \otimes X$ and $X \otimes P$ are projective objects in $\mathcal{C}$, for all $X \in \mathcal{C}$.*

Proof. By Proposition 3.33(d), we have that as functors:

$$\mathrm{Hom}_{\mathcal{C}}(P \otimes X, -) \cong \mathrm{Hom}_{\mathcal{C}}(P, - \otimes X^*).$$

The functor $(- \otimes X^*)$ has a right adjoint, namely $(- \otimes X^{**})$ by Proposition 3.33(d). So, $(- \otimes X^*)$ is right exact by Proposition 2.49(b). Moreover, $\mathrm{Hom}_{\mathcal{C}}(P, -)$ is right exact by Proposition 2.53. Since the composition of right exact functors is right exact, $\mathrm{Hom}_{\mathcal{C}}(P \otimes X, -)$ is right exact. Thus, $P \otimes X$ is projective by Proposition 2.53. We leave it to the reader to verify that $X \otimes P$ is projective. $\qquad\square$

Now take a finite tensor category $\mathcal{C}$. Then, the **FP-dimension of** $X \in \mathcal{C}$ is defined as in the fusion case (see §3.9.2). But the **FP-dimension of** $\mathcal{C}$ is defined as

$$\mathrm{FPdim}(\mathcal{C}) := \textstyle\sum_{i \in \mathrm{Irr}(\mathcal{C})} \mathrm{FPdim}_{\mathcal{C}}(X_i)\, \mathrm{FPdim}_{\mathcal{C}}(P(X_i)),$$

where $P(X_i)$ is the projective cover of X_i (see §2.8.3). Indeed, in the fusion case, we obtain that $P(X_i) = X_i$; see Corollary 2.56.

§3.10.3. Module categories over tensor categories

Similar to the fusion case (see §3.9.4), to have a rich theory of module categories $\mathcal{M}$ over tensor categories $\mathcal{C}$, we impose conditions on $\mathcal{M}$ that reflect the behavior of $\mathcal{C}$. As mentioned in §3.10.2, we must take care when handling exactness when $\mathcal{C}$ is tensor.

In particular, for a left C-module category $\mathcal{M}$ (as in §3.3.1), the bifunctor $\triangleright : C \times \mathcal{M} \to \mathcal{M}$ satisfies the condition below:

(*) $(X \triangleright -) : \mathcal{M} \to \mathcal{M}$ is exact, for all $X \in C$,

by Propositions 2.49(b) and 3.45.

We define a **left module category over a (multi)tensor category** C to be a left C-module category $(\mathcal{M}, \triangleright, m, p)$ as in §3.3.1 that satisfies the items below.

(a) $\mathcal{M}$ is abelian (see §§2.2.2iii, 3.3.4).

(b) $\mathcal{M}$ is $\Bbbk$-linear (see §§2.2.2i, 3.3.4). In particular, the left action bifunctor $\triangleright$ is $\Bbbk$-bilinear on morphisms.

(c) $\mathcal{M}$ is locally finite (see §2.9).

(d) The functor $(- \triangleright M) : C \to \mathcal{M}$ is exact, for all $M \in \mathcal{M}$.

Moreover, if C is finite, then we require:

(e) $\mathcal{M}$ is finite (see §2.9).

Likewise, we can define **right module categories** (resp., **bimodule categories**) **over (finite) multitensor categories**. Moreover, two such (left, right, bi) module categories are **equivalent** (resp. **isomorphic**) when they are equivalent (resp., isomorphic) in the sense of §3.3.1 via $\Bbbk$-linear functors.

In particular, if $\mathcal{M}$ is the regular left C-module category C_{reg} with $\triangleright := \otimes$ [Example 3.17], then the items (*) and (d) above are consistent with Proposition 3.68.

§3.10.4. Exact module categories

Next, we consider an additional condition on module categories over tensor categories to be consistent with Proposition 3.69 for the regular module category.

First, one can check that the following condition holds by using Proposition 3.45(a) and techniques similar to the proof of Proposition 3.69.

- For any object $X \in C$ and any projective object $P \in \mathcal{M}$, we get that $X \triangleright P$ is a projective object of $\mathcal{M}$.

This prompts the following notion.

- A left module category $(\mathcal{M}, \triangleright, m, p)$ over a multitensor category C with enough projectives is said to be **exact** if, for any projective object $P \in C$ and any object $M \in \mathcal{M}$, we get that $P \triangleright M$ is a projective object of $\mathcal{M}$.

Likewise, a right module category $(\mathcal{M}, \triangleleft, n, q)$ over a multitensor category $\mathcal{C}$ with enough projectives is said to be **exact** if, for any projective object $P \in \mathcal{C}$ and any object $M \in \mathcal{M}$, we get that $M \triangleleft P$ is a projective object of $\mathcal{M}$.

Exact module categories were introduced by Etingof and Ostrik [2004]. They satisfy many useful properties, such as having enough projectives, and having admitting a Krull-Schmidt decomposition (into a coproduct of indecomposable module categories). See Section 7.6 of Etingof et al. [2015] for details.

§3.11. Enriched categories

We discussed previously certain categories that are *enriched* over other categories. This is also related to monoidal categories for which the monoidal product, with a variable fixed, has a right adjoint (i.e., to the notion of *closure*). We define this terminology here, and refer the reader to Chapter 1 of Kelly [2005] for more details.

Standing hypothesis.   Fix a monoidal category $\mathcal{V} := (V, \otimes^{\mathcal{V}}, \mathbb{1}^{\mathcal{V}}, a^{\mathcal{V}}, \ell^{\mathcal{V}}, r^{\mathcal{V}})$.

---

§3.11.1. Enriched categories

A $\mathcal{V}$-**category** $\mathcal{A}$ consists of the following data.

(a) A collection of objects, $\mathrm{Ob}(\mathcal{A})$, of $\mathcal{A}$. Here, we write $X \in \mathcal{A}$ for $X \in \mathrm{Ob}(\mathcal{A})$.

(b) For every pair of objects $X, Y \in \mathcal{A}$, a **Hom object** $\mathcal{A}(X, Y)$ in $\mathcal{V}$.

(c) For all triples of objects $X, Y, Z \in \mathcal{A}$, a **composition morphism** in $\mathcal{V}$:

$$\gamma_{X,Y,Z} : \mathcal{A}(Y, Z) \otimes^{\mathcal{V}} \mathcal{A}(X, Y) \longrightarrow \mathcal{A}(X, Z).$$

(d) For each object $X \in \mathcal{A}$, a **unit morphism**, $v_X : \mathbb{1}^{\mathcal{V}} \longrightarrow \mathcal{A}(X, X)$, in $\mathcal{V}$.

This data must satisfy the axioms below.

- **(associativity)**   We have the following equality of morphisms in $\mathcal{V}$ from the object $[\mathcal{A}(Y, Z) \otimes^{\mathcal{V}} \mathcal{A}(X, Y)] \otimes^{\mathcal{V}} \mathcal{A}(W, X)$ to the object $\mathcal{A}(W, Z)$:

$$\gamma_{W,Y,Z} \circ (\mathrm{id}_{\mathcal{A}(Y,Z)} \otimes^{\mathcal{V}} \gamma_{W,X,Y}) \circ a^{\mathcal{V}}_{\mathcal{A}(Y,Z),\mathcal{A}(X,Y),\mathcal{A}(W,X)} = \gamma_{W,X,Z} \circ (\gamma_{X,Y,Z} \otimes^{\mathcal{V}} \mathrm{id}_{\mathcal{A}(W,X)}).$$

- **(unitality)**   We have the following equalities of morphisms in $\mathcal{V}$:

$$\ell^{\mathcal{V}}_{\mathcal{A}(X,Y)} = \gamma_{X,Y,Y} \circ (v_Y \otimes^{\mathcal{V}} \mathrm{id}_{\mathcal{A}(X,Y)}),$$
$$r^{\mathcal{V}}_{\mathcal{A}(X,Y)} = \gamma_{X,X,Y} \circ (\mathrm{id}_{\mathcal{A}(X,Y)} \otimes^{\mathcal{V}} v_X).$$

In this case, $\mathcal{A}$ is also said to be **enriched over** $\mathcal{V}$.

We have seen examples of enriched categories, including the ones below.

- Locally small categories [§2.1.1] are enriched over $(\mathsf{Set}, \times, \{*\})$.

- Preadditive categories [§2.2.2i] are enriched over $(\mathsf{Ab}, \otimes_{\mathbb{Z}}, \mathbb{Z})$.

- $\Bbbk$-linear categories [§2.2.2i] are enriched over $(\mathsf{Vec}, \otimes_{\Bbbk}, \Bbbk)$.

- Locally finite $\Bbbk$-linear categories [§2.9] are enriched over $(\mathsf{FdVec}, \otimes_{\Bbbk}, \Bbbk)$.

Moreover, rigid categories are enriched over themselves; see Exercise 3.39.

Note that a $\mathcal{V}$-category $\mathcal{A}$ is not necessarily an ordinary category, i.e., we may not have a collection of morphisms $\mathrm{Hom}_{\mathcal{A}}(X, Y)$ between objects $X, Y \in \mathcal{A}$. To remedy this, consider the **underlying category** $\mathcal{A}_0$ of $\mathcal{A}$ defined by the data below.

(a) The same class of objects $X, Y, \ldots$ as $\mathcal{A}$.

(b) A morphism $f : X \to Y$ in $\mathcal{A}_0$, for each morphism $\mathbb{1}^{\mathcal{V}} \to \mathcal{A}(X, Y)$ in $\mathcal{V}$.

(c) The composition of morphisms $f : X \to Y$ and $g : Y \to Z$ corresponding to:

$$\mathbb{1}^{\mathcal{V}} \xrightarrow{\;\cong\;} \mathbb{1}^{\mathcal{V}} \otimes^{\mathcal{V}} \mathbb{1}^{\mathcal{V}} \xrightarrow{\;g \otimes f\;} \mathcal{A}(Y, Z) \otimes^{\mathcal{V}} \mathcal{A}(X, Y) \xrightarrow{\;\gamma_{X,Y,Z}\;} \mathcal{A}(X, Z).$$

(d) The identity morphism of X corresponding to $\upsilon_X : \mathbb{1}^{\mathcal{V}} \to \mathcal{A}(X, X)$.

We leave it to the reader to verify that $\mathcal{A}_0$ is indeed a category.

§3.11.2. Enriched functors and enriched equivalence

To address when two enriched categories are the same, we need to introduce $\mathcal{V}$-functors. Let $\mathcal{A}$ and $\mathcal{A}'$ be two $\mathcal{V}$-categories.

A **$\mathcal{V}$-functor** $F : \mathcal{A} \to \mathcal{A}'$ consists of the following data.

(a) An object $F(X)$ in $\mathcal{A}'$, for each $X \in \mathcal{A}$.

(b) For every pair of objects $X, Y \in \mathcal{A}$, a morphism in $\mathcal{V}$:

$$F_{X,Y} : \mathcal{A}(X, Y) \longrightarrow \mathcal{A}'(F(X), F(Y)).$$

We must also have the equalities of morphisms in $\mathcal{V}$ below, for all $X, Y, Z \in \mathcal{A}$.

- (**respects composition**) From $\mathcal{A}(Y, Z) \otimes^{\mathcal{V}} \mathcal{A}(X, Y)$ to $\mathcal{A}'(F(X), F(Z))$:

$$F_{X,Z} \circ \gamma_{X,Y,Z} \;=\; \gamma'_{F(X),F(Y),F(Z)} \circ (F_{Y,Z} \otimes^{\mathcal{V}} F_{X,Y}).$$

- (**respects unitality**) $F_{X,X} \circ \upsilon_X = \upsilon'_{F(X)}$ from $\mathbb{1}^{\mathcal{V}}$ to $\mathcal{A}'(F(X), F(X))$.

Now two $\mathcal{V}$-categories $\mathcal{A}$ and $\mathcal{A}'$ are said to be $\mathcal{V}$-**equivalent**, written as $\mathcal{A} \simeq^{\mathcal{V}} \mathcal{A}'$, if there exists a $\mathcal{V}$-functor $F : \mathcal{A} \to \mathcal{A}'$ satisfying the two conditions below.

- For each pair of objects $X, Y \in \mathcal{A}$, we get that $F_{X,Y}$ is an iso in $\mathcal{V}$. Here, we say that F is $\mathcal{V}$-**fully faithful**.

- Every object $X' \in \mathcal{A}'_0$ is isomorphic to $F(X)$ in $\mathcal{A}'_0$, for some object $X \in \mathcal{A}_0$. Here, we say that F is $\mathcal{V}$-**essentially surjective**.

§3.11.3. Closed monoidal categories

Next, we will consider monoidal categories, where the monoidal product with a variable fixed has a right adjoint. These provide examples of enriched categories.

A monoidal category $\mathcal{C} := (\mathcal{C}, \otimes, \mathbb{1}, a, \ell, r)$ is said to be **left closed monoidal** if the functor $(X \otimes -) : \mathcal{C} \to \mathcal{C}$ has a right adjoint

$$(X \otimes -) \dashv \underline{\mathrm{Hom}}(X, -) : \mathcal{C} \to \mathcal{C},$$

for each $X \in \mathcal{C}$. That is,

$$\mathrm{Hom}_{\mathcal{C}}(X \otimes Z, Y) \cong \mathrm{Hom}_{\mathcal{C}}(Z, \underline{\mathrm{Hom}}(X, Y))$$

for $Y, Z \in \mathcal{C}$. The object $\underline{\mathrm{Hom}}(X, Y)$ of $\mathcal{C}$ is called the **left internal Hom** of X and Y.

A monoidal category $\mathcal{C}$ is said to be **right closed monoidal** if the functor $(- \otimes Y) : \mathcal{C} \to \mathcal{C}$ has a right adjoint

$$(- \otimes Y) \dashv \underline{\mathrm{Hom}}(Y, -) : \mathcal{C} \to \mathcal{C},$$

for each $Y \in \mathcal{C}$. That is,

$$\mathrm{Hom}_{\mathcal{C}}(X \otimes Y, Z) \cong \mathrm{Hom}_{\mathcal{C}}(X, \underline{\mathrm{Hom}}(Y, Z))$$

for all $X, Z \in \mathcal{C}$. The object $\underline{\mathrm{Hom}}(Y, Z)$ of $\mathcal{C}$ is the **right internal Hom** of Y and Z.

For example, $\mathcal{C}$ is right closed monoidal when it is left rigid by taking $\underline{\mathrm{Hom}}(Y, -) := - \otimes Y^*$. Also, $\mathcal{C}$ is left closed monoidal when it is right rigid. See Exercise 3.40.

Moreover, left (or right) closed monoidal categories $\mathcal{C}$ are enriched over themselves with $\mathcal{C}(X, Y) := \underline{\mathrm{Hom}}(X, Y)$; verifying this fact is Exercise 3.41.

Examples of closed monoidal categories are provided in Exercise 3.42.

§3.11.4. Internal Homs of module categories

Broadening the setting of regular module categories arising via the last section to arbitrary module categories, we consider the following terminology.

A left $\mathcal{C}$-module category $(\mathcal{M}, \rhd)$ is said to be **closed** if, for each $M \in \mathcal{M}$, the functor $(- \rhd M) : \mathcal{C} \to \mathcal{M}$ has a right adjoint

$$\mathcal{C} \underset{\underline{\mathrm{Hom}}(M,-)}{\overset{- \rhd M}{\rightleftarrows}} \mathcal{M}$$

That is, for all $Z \in \mathcal{C}$ and $N \in \mathcal{M}$:

$$\mathrm{Hom}_{\mathcal{M}}(Z \rhd M, N) \cong \mathrm{Hom}_{\mathcal{C}}(Z, \underline{\mathrm{Hom}}(M, N)).$$

A right $\mathcal{C}$-module category $(\mathcal{M}, \lhd)$ is said to be **closed** if, for each $M \in \mathcal{M}$, the functor $(M \lhd -) : \mathcal{C} \to \mathcal{M}$ has a right adjoint

$$\mathcal{C} \underset{\underline{\mathrm{Hom}}(M,-)}{\overset{M \lhd -}{\rightleftarrows}} \mathcal{M}$$

That is, for all $Z \in \mathcal{C}$ and $N \in \mathcal{M}$:

$$\mathrm{Hom}_{\mathcal{M}}(M \lhd Z, N) \cong \mathrm{Hom}_{\mathcal{C}}(Z, \underline{\mathrm{Hom}}(M, N)).$$

In either case, the object $\underline{\mathrm{Hom}}(M, N)$ of $\mathcal{C}$ is the **internal Hom** of M and N.

Example 3.70. If $\mathcal{C} := (\mathcal{C}, \otimes, \mathbb{1}, a, \ell, r)$ is right closed monoidal, then the regular left $\mathcal{C}$-module category $\mathcal{C}_{\mathrm{reg}}$ (from Example 3.17) is closed. Here, $\underline{\mathrm{Hom}}(-, -)$ is the internal Hom from the right closed monoidal condition on $\mathcal{C}$.

We also have the result below; the proof holds by Corollary 2.62.

Proposition 3.71. *Suppose that $\mathcal{C}$ is a finite, abelian, linear, monoidal category, and $\mathcal{M}$ is a left $\mathcal{C}$-module category. If $(- \rhd M) : \mathcal{C} \to \mathcal{M}$ is right exact for all $M \in \mathcal{M}$, then $\mathcal{M}$ is closed.* $\qquad\square$

Now the following consequence holds by definition (as in §3.10.3). .

Corollary 3.72. *Suppose that $\mathcal{C}$ is a finite (multi)tensor category, and $\mathcal{M}$ is a left $\mathcal{C}$-module category. Then, $\mathcal{M}$ is closed.* $\qquad\square$

Similar statements hold for right $\mathcal{C}$-module categories.

We will use internal Homs of module categories in Chapter 4 to build *algebras in monoidal categories*. They will also serve as representatives of Morita equivalence classes of algebras in this context.

§3.12. Summary

The focus of this chapter was on monoidal categories, which are categories that come equipped with an operation ($\otimes$) and an object ($\mathbb{1}$) that mimic the structure of a monoid. This structure endows a category with a rich structure that allows us to combine objects and morphisms in a systematic way. For instance, the category FdVec of finite-dimensional $\mathbb{k}$-vector spaces and $\mathbb{k}$-linear maps is monoidal with $\otimes := \otimes_{\mathbb{k}}$ and $\mathbb{1} := \mathbb{k}$.

Just as FdVec has dual structures (e.g., the dual space $V^* := \mathrm{Hom}_{\mathbb{k}}(V, \mathbb{k}) \in$ FdVec, for $V \in$ FdVec), we discussed when objects of monoidal categories are paired with dual objects. This led to the concept of a rigid category. One can also ask when double duality formed in a rigid category is essentially the same as taking no duals, and this prompted the notion of a pivotal structure on a monoidal category. With pivotal structure, we can then measure objects via their pivotal dimension, which recovers vector space dimension for FdVec. Computing pivotal dimension involves taking a (one-sided) pivotal trace of a morphism, and we defined spherical categories to cover the case when the left and right pivotal traces of morphisms are equal. It is an open question of whether finite, semisimple, rigid categories are automatically pivotal (or spherical).

Next, we imposed various conditions on categories discussed in previous chapter to endow monoidal categories with richer structure. Recall that abelian categories were convenient for the plethora of universal constructions that they contain. This condition, along with linearity (resp., along with semisimplicity and finiteness), is imposed on rigid categories to form tensor categories (resp., fusion categories). In particular, there exists another useful measure for objects in fusion categories and in finite tensor categories: Frobenius-Perron dimension.

We also discussed how to represent the various monoidal categories above with certain categories that reflect their behavior. This led to the notion of a module category, which is modeled on modules over algebras presented in the first chapter. The more complex the monoidal category, the more conditions we require for its module categories. This, in turn, allows one to have a fruitful representation theory for monoidal categories, including classification results.

We ended by discussing enriched categories, which include categories whose collections of morphisms admit the structure of a monoidal category. These encompass many categories covered in this and in the last chapter. See Figure 3.15 for a summary of the various monoidal categories examined here.

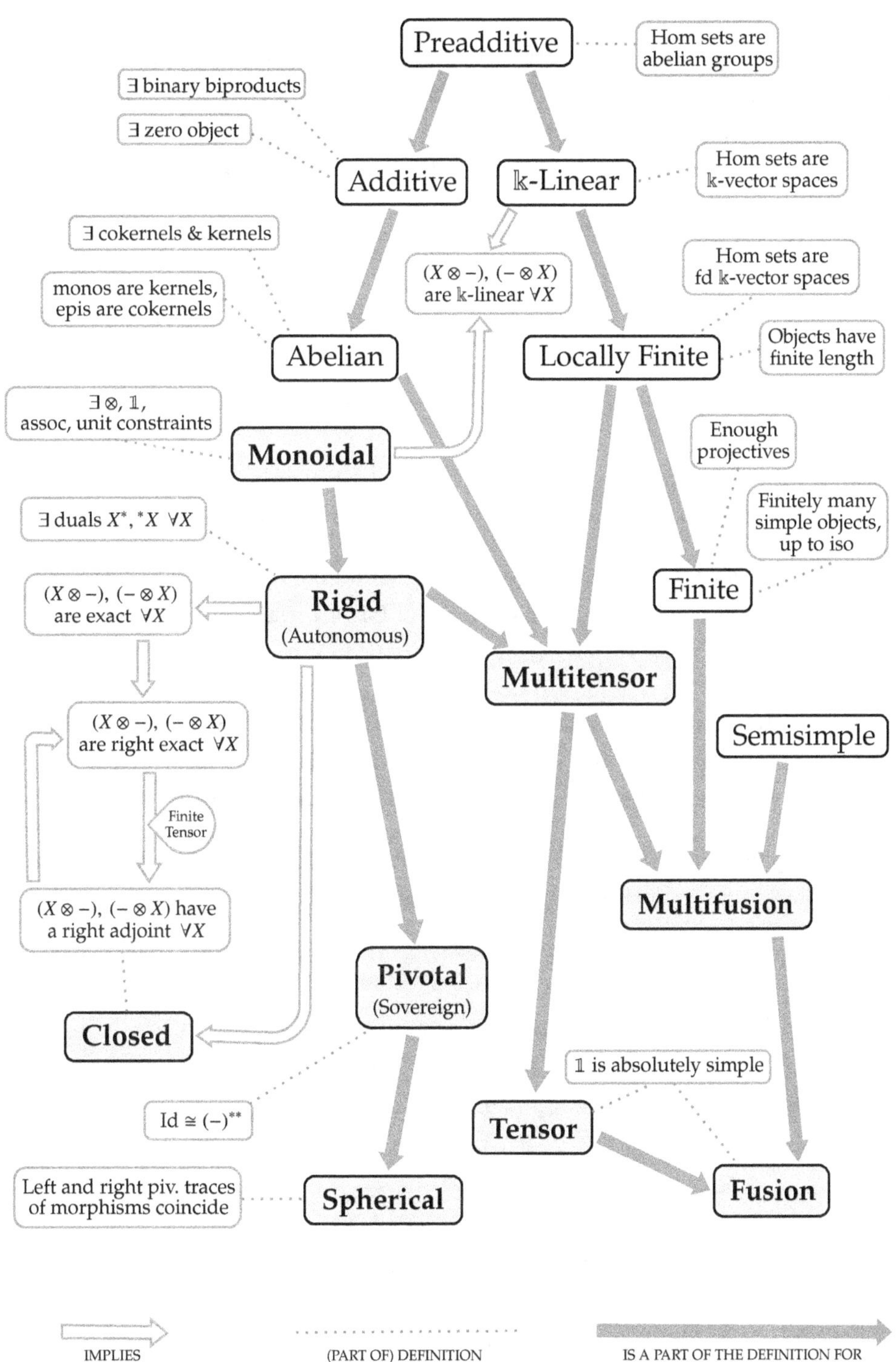

Figure 3.15: Various monoidal categories.

§3.13. Modern applications

We now illustrate how various notions that were introduced in this chapter on monoidal categories are used in modern mathematics. A full understanding of the resources here is not expected. Instead, we aim to put the chapter's material into context by offering videos and content to casually explore.

A fantastic overview of the program to **classify fusion categories** via tools presented in this chapter is provided in the video below.

Julia Plavnik's 2021 Institute for Pure and Applied Mathematics' lecture on
"Classifying small fusion categories"
`https://youtu.be/Vin1VJzw1ZM`

An intriguing talk on suitable structures on the **category of Hilbert spaces** is given in the video below. Those who like functional analysis will enjoy this.

Chris Heunen's 2021 Online Worldwide Seminar on Logic and Semantics talk on
"Axioms for the category of Hilbert spaces"
`https://youtu.be/c3iOH4pOiko`

A friendly invitation to **2-dimensional topological quantum field theories (2-TQFTs)** is presented in the videos below. Monoidal categories, and algebraic structures within them (as a preview of Chapter 4), play a key role.

Nils Carqueville's 2019 Young Researchers Integrability School and Workshop
lectures on "Boundaries and defects"
– Part 1 begins with motivation and background on categories and functors.
`https://youtu.be/kIIaBHXZB7w`
– Part 2 defines monoidal categories, monoidal functors, and 2-TQFTs.
`https://youtu.be/JGfB1M_dxIU`
– Part 3 goes through examples of 2-TQFTs and discusses state sum models.
`https://youtu.be/UWLY1mXWDzo`
– Part 4 branches out to open/ closed 2-TQFTs, ending with examples.
`https://youtu.be/V1CB-n-eZhw`

An insightful talk that highlights the role of monoidal categories and their module categories in studying **conformal field theories** is provided below.

Christoph Schweigert's 2020 lecture for the online workshop on Topological
Orders and Higher Structures: "Bulk fields in conformal field theory"
`https://youtu.be/ZbASleujoAU`

§3.14. References for further exploration

- The landmark article by Joyal and Street [1993] is a must-read. Even though the focus is on *braided* monoidal categories (studied in a future volume), Section 1 of this work pertains to the material covered in the first half of this chapter. There, a monoidal category is referred to as a "tensor category". Be aware of this difference in terminology throughout the literature.

- The textbook by Etingof et al. [2015] is an essential reference for those interested in the monoidal categories appearing in the last half of this chapter. We adopt many of the conventions from the textbook here, and this (and the previous) chapter serves as a starter guide for their textbook. One of the goals of their book is to provide additional context for the foundational article by Etingof et al. [2005] on fusion categories.

- The textbook by Turaev and Virelizier [2017] is a comprehensive resource to learn about monoidal categories and their role in Quantum Topology. The book begins with background material on monoidal categories, and proceeds to examine various *state sum 3-TQFTs*. Braided monoidal categories, along with spherical and fusion categories, are all used to build the monoidal categories that go hand-in-hand with 3-TQFTs: *modular* fusion categories.

- Another great textbook is by Heunen and Vicary [2019], which is a friendly introduction to the role of monoidal categories in Quantum Information Theory. It does not need prerequisites beyond what was used to begin Chapters 2 and 3 here. Highlights include connections to quantum computational notions, such as *entanglement, quantum teleportation,* and *ZX-calculus.*

§3.15. Exercises

3.1   In a monoidal category $(\mathcal{C}, \otimes, \mathbb{1}, a, \ell, r)$, verify the identities below, for $X, Y \in \mathcal{C}$:

    (a) $\ell_X \otimes \mathrm{id}_Y = \ell_{X \otimes Y}\, a_{\mathbb{1}, X, Y}$;   (b) $r_{X \otimes Y} = (\mathrm{id}_X \otimes r_Y)\, a_{X, Y, \mathbb{1}}$;   (c) $\ell_{\mathbb{1}} = r_{\mathbb{1}}$.

Hint.   For part (a), it suffices to show that the perimeter of the following diagram commutes. The bottom region commutes by dividing it into quadrangles for naturality, and triangles for the triangle axiom.

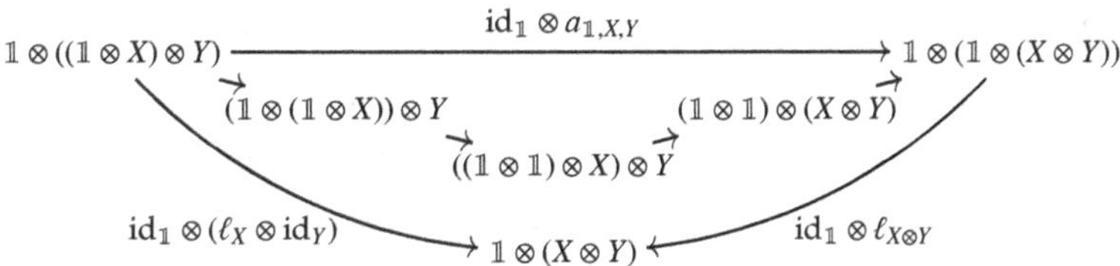

3.2 Recall the discussion of opposite monoidal categories from §3.1.1.

    (a) Verify that C^{op}, $C^{\otimes op}$, and C^{rev} are monoidal categories.

    (b) Consider the monoidal category operations $(-)^{op}$, $(-)^{\otimes op}$, and $(-)^{rev}$. Can one operation be obtained by composing the other two?

3.3 [Open-ended] Recall the monoidal categories from §3.1.2.

    (a) Pick your favorite three examples in §3.1.2, and verify in detail that they are indeed monoidal categories, and discuss whether they are strict.

    (b) Derive more examples of monoidal categories in addition to what is in §3.1.2; try to cook up at least one of each mathematical type.

3.4 Recall monoidal functors from §3.2.1. Given two monoidal functors

$$(F, F^{(2)}, F^{(0)}) : C \to D \quad \text{and} \quad (G, G^{(2)}, G^{(0)}) : D \to \mathcal{E},$$

show that $GF : C \to \mathcal{E}$ is monoidal. Namely, describe $(GF)^{(2)}$ and $(GF)^{(0)}$ in terms of $F^{(2)}$, $G^{(2)}$, $F^{(0)}$, and $G^{(0)}$.

3.5 Prove Proposition 3.7 on monoidal isomorphism and monoidal equivalence.

3.6 Recall the notion of monoidal functors and monoidally isomorphic categories in §§3.2.1, 3.2.2. Take a group G and recall §1.3.4.

    (a) Verify Example 3.8, showing that Forg : G-Mod $\to$ Vec is strong monoidal.

    (b) Give the category, Rep(G), of representations of G the structure of a monoidal category, such that G-Mod $\stackrel{\otimes}{\cong}$ Rep(G).

    (c) Expand on part (b) to then show that $\Bbbk G$-Mod from Exercise 2.30 is monoidally isomorphic to G-Mod.

    (d) For a group embedding $\iota : H \to G$, show that the restriction functor below can be given the structure of a strong monoidal functor.

$$\mathrm{Res}^G_H : \mathrm{Rep}(G) \longrightarrow \mathrm{Rep}(H), \quad (\rho : G \to GL(V)) \mapsto (\rho\iota : H \to GL(V)).$$

    (e) Show that part (d) generalizes part (a).

3.7 Recall the actions of monoidal categories on categories from §3.2.3. Show that a left action, $\rho : C \to \mathrm{End}(A)$, of a monoidal category C on a category A is precisely A equipped with a bifunctor $* : C \times A \to A$ with two additional structure natural isomorphisms (corresponding to $\rho^{(2)}$ and $\rho^{(0)}$).

3.8 Complete the details of Example 3.13 in determining the monoidal product and monoidal unit constraints, $\rho^{(2)}$ and $\rho^{(0)}$, that define an action of a monoid N on Vec_N, and define an action of a group G on Vec_G.

3.9 [Open-ended]   Recall the examples of monoidal functors and monoidal equivalence in §3.2.3. Derive additional examples of such notions. One may start by modifying the examples in §3.2.3.

3.10 Recall the discussion about structure versus 'structure category' in §3.3.1. Before reading §3.3.3, for monoidal categories $\mathcal{C}$ and $\mathcal{D}$, write down the definitions of a $(\mathcal{C}, \mathcal{D})$-bimodule category and a $(\mathcal{C}, \mathcal{D})$-bimodule functor.

3.11 Akin to Exercise 1.26(a) for modules over algebras, for a monoidal category $\mathcal{C}$, and for the regular module category $\mathcal{C}_{\mathrm{reg}}$ in Example 3.17 of §3.3.2, show that there exist strong monoidal functors:

$$\rho_1 : \mathcal{C} \longrightarrow \mathrm{End}_{\mathrm{Mod}\text{-}\mathcal{C}}(\mathcal{C}_{\mathrm{reg}}), \qquad \rho_2 : \mathcal{C}^{\otimes \mathrm{op}} \longrightarrow \mathrm{End}_{\mathcal{C}\text{-}\mathrm{Mod}}(\mathcal{C}_{\mathrm{reg}}).$$

3.12 Provide the details of Example 3.18, along with the details of two additional examples from Examples 3.17–3.24 in §3.3.2, including describing explicitly the module associativity and module unitality constraints, and verifying the pentagon and triangle axioms.

Hint. For the triangle axiom for Example 3.18, proceed as follows: (i) fill in the objects and morphisms of the diagram below; (ii) justify why each internal region commutes; and then (iii) conclude why the perimeter commutes.

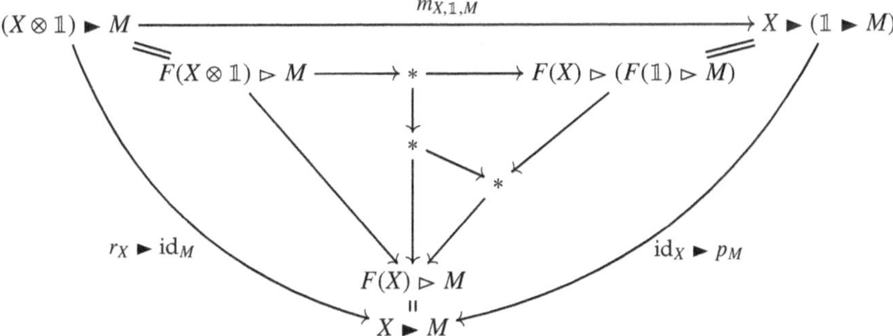

For step (ii), the outer regions should commute by definition. Now justification is needed for the commutativity of the right quadrant, the left quadrant, and the bottom triangle in the interior. Moreover, step (iii) in these types of arguments does not need a written justification when the diagram has all paths flowing from one perimeter object to another (e.g., as drawn above).

3.13 [Open-ended]   This pertains to §3.3.3. Derive examples of bimodule categories over monoidal categories. They could be built from some of the one-sided module categories examined in Exercise 3.12.

3.14 This pertains to the material in §3.3.4. Let A be a $\Bbbk$-algebra.

(a) Show that A-FdMod is a right FdVec-module category.

(b) Verify that if A-FdMod is an indecomposable right FdVec-module category, then A is indecomposable as a $\Bbbk$-algebra (see §1.4.1).

3.15 Consider the proof of the Strictification Theorem in §3.4.1.

(a) Verify that C^{str} from Definition 3.27 is strict monoidal.

(b) Complete the proof of Claim 2 in Theorem 3.26 by showing that $\rho^{(2)}$ and $\rho^{(0)}$ defined there satisfy the associativity and unitality axioms making $(\rho, \rho^{(2)}, \rho^{(0)})$ a strong monoidal functor.

3.16 Recall the discussion about monoidal functors and monoidally equivalent categories in §§3.2.1, 3.2.2. Here, we build on Exercise 2.20.

(a) Show that if $F : C \to D$ is a strong monoidal functor, then its essential image $\mathrm{Im}^{\mathrm{ess}}(F)$ is a monoidal subcategory of D.

(b) Show that if, further, D is strict, then $\mathrm{Im}^{\mathrm{ess}}(F)$ must be strict.

(c) Prove that, further, when F is fully faithful, then $\mathrm{Im}^{\mathrm{ess}}(F) \overset{\otimes}{\cong} C$.

(d) Do any of the parts above fail if F is not strong?

3.17 Take a strict monoidal category $(C, \otimes, \mathbb{1})$ as in §3.4 with morphisms $f : X \to Y$, $g : Y \to Z$, $f' : X' \to Y'$, $g' : Y' \to Z'$ in C.

(a) Verify that $(g \circ f) \otimes (g' \circ f') = (g \otimes g') \circ (f \otimes f')$.

(b) Depict this identity using the graphical calculus discussed in §3.5.

Hint. For part (a), recall that $\otimes$ is a functor and revisit this definition.

3.18 Recall the notion of rigidity in §3.6.1, and discuss the following items.

(a) Whether rigidity is a structure on or a property of a monoidal category.

(b) The notion of a monoidal functor being rigid.

(c) The notion of an equivalence of rigid categories.

3.19 Prove part (a) of Lemma 3.32 in §3.6.1 about the uniqueness of left dual objects in (left) rigid categories. (You may assume that $\mathcal{C}$ is strict to use graphical calculus.)

3.20 Prove parts (c) and (d) of Lemma 3.32 in §3.6.1 about dual objects in rigid categories $\mathcal{C}$. Namely, show that $^*(X^*) \cong X \cong (^*X)^*$ for each $X \in \mathcal{C}$, and that $\mathbb{1}^* = \mathbb{1} = {}^*\mathbb{1}$ in $\mathcal{C}$. (You may assume that $\mathcal{C}$ is strict to use graphical calculus.)

3.21 Prove part (e) of Lemma 3.32 in §3.6.1 to show that the opposite monoidal categories $\mathcal{C}^{\mathrm{op}}$, $\mathcal{C}^{\otimes\mathrm{op}}$, and $\mathcal{C}^{\mathrm{rev}}$ of a rigid category $\mathcal{C}$ (from §3.1.1) are rigid.

3.22 [Open-ended]   Complete and elaborate on some of the missing details in Table 3.1 in §3.6.2 for your favorite (non-)examples of rigid monoidal categories. Also explore other (non-)examples, building on Exercise 3.3.

3.23 Take a left rigid category $\mathcal{C}$ as in §3.6.1, and establish the following statements about the left duality functor $(-)^*$ on $\mathcal{C}$; see Proposition 3.42. (Feel free to assume $\mathcal{C}$ is strict to use graphical calculus.)

    (a) For morphisms $f : X \to Y$ and $g : Y \to Z$ in $\mathcal{C}$, verify that $(g \circ f)^*$ is equal to $f^* \circ g^*$ as morphisms from Z^* to X^* in $\mathcal{C}$.

    (b) For objects $X, Y \in \mathcal{C}$, show that the left dual of $X \otimes Y$ is $Y^* \otimes X^*$.

    (c) Complete the proof of Proposition 3.42(b) to show that $(-)^*$ is a strong monoidal functor.

3.24 This builds on Exercise 3.23. Take a rigid category $\mathcal{C}$ as in §3.6.1, and write down and prove the double left dual analogue of Proposition 3.42. (Feel free to assume that $\mathcal{C}$ is strict to use graphical calculus.)

3.25 Prove Proposition 3.43 in §3.6.3 on how rigidity is preserved under strong monoidal functors between monoidal categories.

3.26 Complete the proof of Proposition 3.44 in §3.6.3: If $\mathcal{C}$ is a rigid category and $\mathcal{D}$ is a monoidal category, then any monoidal transformation between two strong monoidal functors $\mathcal{C} \to \mathcal{D}$ is a monoidal isomorphism.

3.27 Complete the proof of Proposition 3.45 in §3.6.4 on proving that the functor derived from the action bifunctor for a left module category over a rigid category, by having the first slot fixed, has a left adjoint and a right adjoint.

3.28 Recall the discussion of dual module categories in §3.6.4.

   (a) Prove Proposition 3.46 in §3.6.4.

   (b) Verify for a rigid category C and for $M \in C\text{-Mod}$ that $^*(M^*) \simeq M$ as left C-module categories. Likewise, show that $(^*M)^* \simeq M$ in C-Mod.

   (c) Is M^{**} equivalent to M as left C-module categories?

3.29 Recall pivotal trace and pivotal dimension in §3.7.3. Take a finite-dimensional $\Bbbk$-vector space V with basis elements $v_1, \ldots, v_n$, and take a $\Bbbk$-linear endomorphism f of V given by $v_k \mapsto \sum_{\ell=1}^n \lambda_{k,\ell} v_\ell$, for some $\lambda_{k,\ell} \in \Bbbk$.

   (a) Compute $\text{tr}_j^L(f)$ and $\text{tr}_j^R(f)$, with j for FdVec given in §3.7.2.

   (b) Compute $\dim_j(V)$.

3.30 Recall the pivotal trace from §3.7.3. For a morphism $f : X \to X$ in a pivotal category (C, j), show that $\text{tr}_j^R(f) = \text{tr}_j^L(f^\vee)$ and $\text{tr}_j^L(f) = \text{tr}_j^R(f^\vee)$.

3.31 Recall the pivotal trace from §3.7.3 for a pivotal category (C, j). Consider the action of $\text{End}_C(\mathbb{1})$ on $\text{Hom}_C(X, Y)$ given as follows. For any morphism $\phi : \mathbb{1} \to \mathbb{1}$, and any morphism $h : X \to Y$ in C, define

$$\phi \triangleright h := \ell_Y \, (\phi \otimes h) \, \ell_X^{-1}, \qquad h \triangleleft \phi := r_Y \, (h \otimes \phi) \, r_X^{-1}.$$

(This is defined without pivotality; see Section 1.3.2 of Turaev and Virelizier [2017] for details.) Take morphisms $f : X \to X$ and $g : Y \to Y$ in C.

   (a) Show that $\text{tr}_j^L(f \otimes g) = \text{tr}_j^L\left(\text{tr}_j^L(f) \triangleright g\right)$ as morphisms from $\mathbb{1}$ to $\mathbb{1}$.

   (b) Show that $\text{tr}_j^R(f \otimes g) = \text{tr}_j^R\left(f \triangleleft \text{tr}_j^R(g)\right)$ as morphisms from $\mathbb{1}$ to $\mathbb{1}$.

   (c) Now show that in a trace-spherical category (C, j), as in §3.8.1, pivotal trace is $\otimes$-multiplicative: namely, that $\text{tr}_j(f \otimes g) = \text{tr}_j(f) \circ \text{tr}_j(g)$.

3.32 Recall the discussion about pivotal and trace-spherical categories from §§3.7.1, 3.8.1. Show that if (C, j) and (C', j') are equivalent pivotal categories, then trace-sphericality must transfer from one pivotal category to the other.

3.33 Recall fusion categories from §3.9.1i. Verify the following statements.

   (a) FdVec is a fusion category.

   (b) G-FdMod is a fusion category if and only if G is a finite group.

3.34 Recall the discussion about fusion categories and rank from §3.9.1. Verify that the only fusion category of rank 1 is FdVec, up to equivalence.

3.35 This is an extended exercise to modify the monoidal category Vec_G in §3.1.2i and study its properties and additional structure. See §§3.1.1, 3.1.2 for parts (a)–(d), §2.7.3 for part (e), §3.6.1 for part (f), §§3.7.1, 3.8.1 for part (g), §3.9.1i for part (h), and §3.9.2 for part (i).

First, let us summarize some basic terminology from **Group Cohomology**. Take a group G, and as in Exercise 1.14, a left G-module on an abelian group:

$$M := (M, +_M, 0_M, \triangleright : G \times M \to M).$$

- An **n-cochain of G with value in M** is a function from $G^n \to M$, where $G^n := G \times G \times \cdots \times G$ (n-times). The collection of such functions is denoted by $C^n(G, M)$. Here, $C^0(G, M) \cong M$, after identifying G^0 with $\langle e_G \rangle$.

- We have that $C^n(G, M)$ is an abelian group via

$$(\phi + \phi')(g_1, \ldots, g_n) := \phi(g_1, \ldots, g_n) +_M \phi'(g_1, \ldots, g_n),$$

and with the identity element being the zero map $\underline{0} : G^n \to M$, defined by $(g_1, \ldots, g_n) \mapsto 0_M$, for $g_1, \ldots, g_n \in G$.

- The **n-th coboundary map** is $d^{n+1} : C^n(G, M) \to C^{n+1}(G, M)$ given by

$$
\begin{aligned}
(d^{n+1}\phi)(g_1, \ldots, g_{n+1}) := \quad & g_1 \triangleright \phi(g_2, \ldots, g_{n+1}) \\
& + \sum_{i=1}^{n}(-1)^i \phi(g_1, \ldots, g_{i-1}, g_i g_{i+1}, g_{i+2}, \ldots, g_{n+1}) \\
& + (-1)^{n+1}\phi(g_1, \ldots, g_n).
\end{aligned}
$$

Here, $+$ is the operation $+_M$ of M. For example,

$$
\begin{aligned}
(d^2\phi)(g_1, g_2) &= g_1 \triangleright \phi(g_2) - \phi(g_1 g_2) + \phi(g_1), \\
(d^3\phi)(g_1, g_2, g_3) &= g_1 \triangleright \phi(g_2, g_3) - \phi(g_1 g_2, g_3) + \phi(g_1, g_2 g_3) - \phi(g_1, g_2), \\
(d^4\phi)(g_1, g_2, g_3, g_4) &= g_1 \triangleright \phi(g_2, g_3, g_4) - \phi(g_1 g_2, g_3, g_4) + \phi(g_1, g_2 g_3, g_4) \\
& \qquad - \phi(g_1, g_2, g_3 g_4) + \phi(g_1, g_2, g_3).
\end{aligned}
$$

(It is common to denote the coboundary maps simply by "d", which we will not do here.)

- Moreover, we obtain that $d^{n+1} \circ d^n$ is the zero map, so the cochain groups and their coboundary maps above form a *cochain complex* for G.

- We denote the kernel of the n-th coboundary map d^{n+1} by $Z^n(G, M)$ and refer to its elements as **n-cocycles of G with value in M**.

- We denote the image of d^n by $B^n(G, M)$ and refer to its elements as **n-coboundaries of G with value in M**.

- Since $d^{n+1} \circ d^n$ is the zero map, $B^n(G, M)$ is a subgroup of $Z^n(G, M)$. The corresponding quotient group

$$H^n(G, M) := Z^n(G, M)/B^n(G, M)$$

is referred to as the n-**cohomology group of G with values in M**.

- We say that $\phi, \phi' \in Z^n(G, M)$ are **cohomologous** if $\phi = \phi'$ in $H^n(G, M)$.

- We call an n-cocycle $\phi \in Z^n(G, M)$ **cohomologically trivial** if ϕ is cohomologous to the trivial n-cocycle, that is, if ϕ is equal to an n-coboundary.

To proceed with the exercise, take M to be the multiplicative abelian group $\Bbbk^\times$, on which G acts trivially, that is, $g \triangleright \lambda = \lambda$ for all $g \in G$, $\lambda \in \Bbbk$. Now take a $\Bbbk^\times$**-valued 3-cocycle ω on G**, which is a function $\omega : G \times G \times G \to \Bbbk^\times$ that satisfies the following condition:

$$
\begin{aligned}
&\omega(g_1 g_2, g_3, g_4)\, \omega(g_1, g_2, g_3 g_4) \\
&= \omega(g_1, g_2, g_3)\, \omega(g_1, g_2 g_3, g_4)\, \omega(g_2, g_3, g_4),
\end{aligned}
\tag{3.73}
$$

for all $g_1, g_2, g_3, g_4 \in G$.

(a) Take the monoidal category Vec_G. Verify that the objects of Vec_G are isomorphic to a direct sum of simple objects

$$\delta_g = \bigoplus_{k \in G} (\delta_g)_k,$$

with $(\delta_g)_g = \Bbbk$, and where $(\delta_g)_k$ is the zero vector space for $k \neq g$.

(b) Describe the Hom sets of Vec_G in terms of the simple objects in part (a). Namely, verify that Vec_G is $\Bbbk$-linear (see §2.2.2iii).

(c) Describe the monoidal product and the unit object of Vec_G in terms of the simple objects in part (a).

(d) With the monoidal product and the unit object in part (c), define the following linear maps (which are linear isomorphisms).

$$a^\omega_{g,h,k} := \omega(g, h, k)\, \mathrm{id}_{\delta_{ghk}}, \qquad \ell_{\delta_g} := \omega(e, e, g)^{-1}\, \mathrm{id}_{\delta_g}, \qquad r_{\delta_g} := \omega(g, e, e)\, \mathrm{id}_{\delta_g}.$$

Show that these maps form the components of associativity and unitality constraints of a monoidal structure on the category Vec_G.

- We denote this new monoidal category by Vec_G^ω.

(e) Show that Vec_G^ω is a semisimple category.

(f) Verify that FdVec_G^ω is rigid with $\delta_g^* = {}^*\delta_g = \delta_{g^{-1}}$, along with:

$$
\begin{aligned}
\mathrm{ev}^L_{\delta_g} &:= \mathrm{id}_{\delta_e} : \delta_g^* \otimes \delta_g \to \mathbb{1}, &\qquad \mathrm{coev}^L_{\delta_g} &:= v_g^{-1}\, \mathrm{id}_{\delta_e} : \mathbb{1} \to \delta_g \otimes \delta_g^* \\
\mathrm{ev}^R_{\delta_g} &:= v_g\, \mathrm{id}_{\delta_e} : \delta_g \otimes {}^*\delta_g \to \mathbb{1}, &\qquad \mathrm{coev}^R_{\delta_g} &:= \mathrm{id}_{\delta_e} : \mathbb{1} \to {}^*\delta_g \otimes \delta_g.
\end{aligned}
$$

Here, $v_g := \omega(e, e, g)\, \omega(g, g^{-1}, g)\, \omega(g, e, e)$.

(g) Show that FdVec_G^ω is pivotal, and further, is trace-spherical.

(h) Prove that FdVec_G^ω is fusion if and only if G is a finite group.

(i) For a finite group G, show that FdVec_G^ω is integral, and compute its Frobenius-Perron dimension.

3.36 Recall the material on fusion categories from §§3.9.1, 3.9.2. Take the dihedral group D_8 of order 8 generated by r and s, with relations $r^4 = s^2 = srsr = e$. Consider the following D_8-modules.

$$
\begin{aligned}
V_1 &:= \Bbbk v_1, && \text{with} & r \triangleright v_1 &= v_1, & s \triangleright v_1 &= v_1 \\
V_2 &:= \Bbbk v_2, && \text{with} & r \triangleright v_2 &= v_2, & s \triangleright v_2 &= -v_2 \\
V_3 &:= \Bbbk v_3, && \text{with} & r \triangleright v_3 &= -v_3, & s \triangleright v_3 &= v_3 \\
V_4 &:= \Bbbk v_4, && \text{with} & r \triangleright v_4 &= -v_4, & s \triangleright v_4 &= -v_4 \\
V_5 &:= \Bbbk v_5 \oplus \Bbbk v_5', && \text{with} & r \triangleright v_5 &= -v_5', & s \triangleright v_5 &= v_5 \\
&&&& r \triangleright v_5' &= v_5, & s \triangleright v_5' &= -v_5'
\end{aligned}
$$

(a) Verify that these finite-dimensional D_8-modules are pairwise not isomorphic, and that each module is simple.

(b) Determine the Artin-Wedderburn parameters for the corresponding group algebra $\Bbbk D_8$; see §1.6.1 and Exercise 3.6(c).

(c) Recall the monoidal product $\otimes = \otimes_\Bbbk$ and unit $\mathbb{1}$ of D_8-FdMod in §3.1.2. For each pair of simple modules $(V, \triangleright)$ and $(V', \triangleright')$ above, compute the module $(V, \triangleright) \otimes_\Bbbk (V', \triangleright')$ in D_8-FdMod. Also, identify the module $\mathbb{1}^{D_8\text{-FdMod}}$.

(d) For each monoidal product $(V, \triangleright) \otimes_\Bbbk (V', \triangleright')$ in part (c), decompose it into a direct sum of simple D_8-modules, up to isomorphism.

(e) Use part (d) to compute the FP-dimensions of the (isoclasses of) simple D_8-modules, and of the category D_8-FdMod.

3.37 Show that the Ising fusion category, Ising, from Example 3.65 in §3.9.3 is weakly integral, but not integral. In particular, compute:

$$
\mathrm{FPdim}_{\mathsf{Ising}}(\mathbb{1}), \quad \mathrm{FPdim}_{\mathsf{Ising}}(\tau), \quad \mathrm{FPdim}_{\mathsf{Ising}}(\sigma), \quad \mathrm{FPdim}(\mathsf{Ising}).
$$

3.38 [Open-ended] Explore details about the Tambara-Yamagami (TY-)fusion categories from Example 3.66 in §3.9.3, using the article by Tambara and Yamagami [1998].

(a) Write down their associativity and unitality constraints (a, ℓ, r), and their (co)evaluation morphisms, making it a rigid category.

(b) Discuss when two TY-fusion categories are equivalent.

(c) Verify the FP-dimension of objects, and of the category itself, is as given in Example 3.66.

(d) Verify that D_8-FdMod from Exercise 3.36, and Ising from Example 3.65, are each TY-fusion categories.

(e) Explore the literature for uses of the TY-fusion categories.

3.39 Recall rigid categories from §3.6.1 and enriched categories from §3.11.1.

(a) Show that a left rigid category $C := (C, \otimes, \mathbb{1}, a, \ell, r, (-)^*)$ is enriched over itself, with Hom objects $C(X, Y) := Y \otimes X^*$, for all $X, Y \in C$.

(b) Likewise, verify that a right rigid category is enriched over itself.

(c) Explore part (a) in detail for $C = $ FdVec. Namely, relate the Hom objects with $\Bbbk$-linear maps between finite-dimensional $\Bbbk$-vector spaces.

3.40 Recall rigid categories from §3.6.1, and closed monoidal categories from §3.11.3. Take $C := (C, \otimes, \mathbb{1}, a, \ell, r)$ to be a monoidal category.

(a) Show that if C is left rigid, then it is right closed monoidal, where we have that $\underline{\mathrm{Hom}}(Y, -) := - \otimes Y^*$.

(b) Likewise, show that if C is right rigid, then it is left closed monoidal.

3.41 Recall the enriched categories from §3.11.1, and closed monoidal categories from §3.11.3.

(a) Prove that if C is a right closed monoidal category, then C is enriched over itself with $C(Y, Z) := \underline{\mathrm{Hom}}(Y, Z)$.

(b) Prove a similar statement for left closed monoidal categories.

3.42 Recall the notion of a closed monoidal category from §3.11.3. Establish the following (counter-)examples of closed monoidal categories.

(a) Show that $(\mathrm{Set}, \times, \{\cdot\})$ is closed monoidal.

(b) For an algebra A, prove $(A\text{-Bimod}, \otimes_A, A)$ is closed monoidal.

(c) Prove that the monoidal category $(\mathrm{Ring}, \otimes_{\mathbb{Z}}, \mathbb{Z})$ is not closed monoidal.

$\cdot$ C H A P T E R $4 \cdot$

Algebras in monoidal categories

History

An algebra in a monoidal category $(\mathcal{C}, \otimes, \mathbb{1})$ is an object A in $\mathcal{C}$, equipped with morphisms $m : A \otimes A \to A$ and $u : \mathbb{1} \to A$ in $\mathcal{C}$, such that (A, m, u) mimics the structure of a monoid. They were introduced by Bénabou [1964] as *monoïdes à unité*, and were formalized in Section VII.3 of MacLane [1971]. Modules over algebras in this setting were introduced in Section VII.4 of MacLane [1971], and were also studied in detail by Pareigis [1977a]; they were called, resp., *objects with action* and *A-objects*. Key examples of such algebras are in the monoidal category of endofunctors: *monads*; they were introduced by Godement [1958] and formalized by Huber [1961] as *standard constructions*. More recently, the theory of algebras in fusion categories was advanced by the work of Ostrik [2003c], and properties of and further structure on algebras in monoidal categories were used by Fuchs et al. [2002] to model rational conformal field theories in mathematical physics.

Overview

An introduction to algebras in monoidal categories is covered in §4.1, and subalgebras and quotient algebras are covered in §4.2. Constructions of algebras in monoidal categories via adjunction, including monads, are discussed in §4.3. Modules and bimodules in monoidal categories, including those over monads, are presented in §4.4. Various operations of algebras and (bi)modules are examined in §4.5. Graded algebraic structures in monoidal categories are introduced in §4.6. Then, capstone results on the Morita equivalence of algebras, namely, the Generalized Eilenberg-Watts Theorem and the Generalized Morita's Theorem, are presented in §4.7. Another capstone result, Ostrik's Theorem, on representing categories of modules in fusion categories by internal End algebras is discussed in §4.8. Various properties of algebras in monoidal categories are studied in §4.9; some are intrinsic to algebraic structure, while others are module-theoretic. Finally, in §4.10, a detailed study of bimodules is presented; this includes a discussion of the monoidal category of bimodules, using bimodules for a notion of Morita equivalence for tensor categories, and a discussion of a higher categorical structure on bimodules. The chapter ends with a summarizing diagram in §4.11, modern applications in §4.12, references in §4.13, and several exercises.

Standing hypotheses. Linear structures are over an algebraically closed field $\Bbbk$ of characteristic 0, and $\Bbbk$-algebras are associative and unital. Here, $\mathcal{A}$ is an ordinary category (not necessarily monoidal), and $\mathcal{C} := (\mathcal{C}, \otimes, \mathbb{1}, a, \ell, r)$ is a monoidal category (with additional features as specified below).

---

§4.1. Algebras in monoidal categories

In this part, we introduce algebras in monoidal categories, which will be analogous to the material in §1.1.5 on $\Bbbk$-algebras. Next, we turn our attention to the *Eckmann-Hilton Principle*, which relates operations of algebras. Then, we discuss endomorphism algebras in the enriched setting (see §3.11), which generalizes the material in §1.2.1 on matrix algebras and endomorphism algebras over $\Bbbk$.

§4.1.1. Algebras

An **algebra** (or an **algebra object**) in a monoidal category $\mathcal{C}$ consists of:

(a) An object A in $\mathcal{C}$;

(b) **(multiplication morphism)** A morphism $m := m_A : A \otimes A \to A$ in $\mathcal{C}$;

(c) **(unit morphism)** A morphism $u := u_A : \mathbb{1} \to A$ in $\mathcal{C}$.

This data must satisfy the commutative diagrams below.

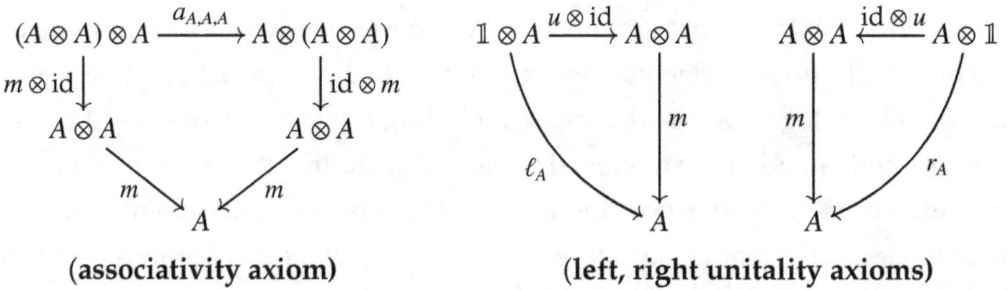

$$\text{(associativity axiom)} \qquad\qquad \text{(left, right unitality axioms)}$$

We sometimes write A_{obj} to denote the underlying object A of (A, m, u). The string diagrams for algebras in $\mathcal{C}$ (when $\mathcal{C}$ is strict) are given in Figure 4.1.

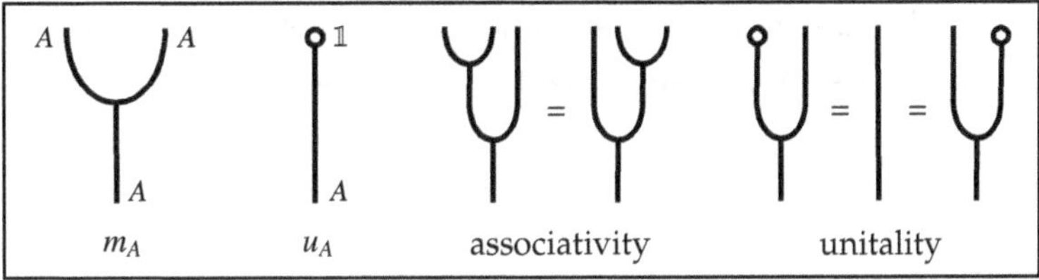

Figure 4.1: String diagrams for algebras in (strict) monoidal categories.

Some examples of algebras in monoidal categories include the following.

- Algebras in $(\mathrm{Vec}, \otimes_{\Bbbk}, \Bbbk)$ from §3.1.2i are $\Bbbk$-algebras (as in §1.1.5).

- Algebras in $(\mathrm{Set}, \times, \{\cdot\})$ from §3.1.2ii are monoids (as in §1.1.1).

- Algebras in $(\mathrm{End}(\mathcal{A}), \circ, \mathrm{Id}_{\mathcal{A}})$ from §3.1.2ii are *monads* (see §4.3.2 later).

Verifying the first two examples, along with others, is Exercise 4.1. Moreover, algebras in the monoidal categories, G-Mod, Vec_G, and Vec_G^{ω} are examined respectively in Exercises 4.2, 4.3, and 4.4; see also §3.1.2i and Exercise 3.35.

"Algebra" versus "monoid" in a monoidal category. "Monoid" is the original name for the construction above; see Section VII.3 of MacLane [1971]. However the term "algebra" is now frequently used, especially when the monoidal category has an addition, an additive identity, additive inverses, and a scalar multiplication, e.g., when objects have the underlying structure of a vector space. We choose to use the term "algebra" in this textbook due to our main sources of examples, but "monoid" is equally as descriptive.

Consider the trivial examples of algebras below.

Example 4.1. (a) The unit object $\mathbb{1}$ of $\mathcal{C}$ admits the structure of an algebra in $\mathcal{C}$ with $m_{\mathbb{1}} := \ell_{\mathbb{1}}$ $(= r_{\mathbb{1}})$ and $u_{\mathbb{1}} = \mathrm{id}_{\mathbb{1}}$. We refer to this as the **unit algebra** of $\mathcal{C}$.

(b) If $\mathcal{C}$ has a zero object 0 (see §2.2.1i), then the **zero algebra** is 0 equipped with multiplication morphism $_{0\otimes 0}\vec{0}: 0 \otimes 0 \to 0$ and unit morphism $_{\mathbb{1}}\vec{0}: \mathbb{1} \to 0$.

Verifying that the associativity and unitality conditions hold is Exercise 4.5.

We also have a useful result below.

Lemma 4.2. *The multiplication morphism m_A of an algebra A in $\mathcal{C}$ is epic.*

Proof. Suppose that $g, g' : A \to B$ are morphisms in $\mathcal{C}$ such that $g\, m_A = g'\, m_A$. Then, by the left unitality axiom, we get that:

$$g \;=\; g\, m_A\, (u_A \otimes \mathrm{id}_A)\, \ell_A^{-1} \;=\; g'\, m_A\, (u_A \otimes \mathrm{id}_A)\, \ell_A^{-1} \;=\; g'. \qquad \square$$

Next, an **algebra morphism** between two algebras (A, m, u) and (A', m', u') in C is a morphism $\phi : A \to A'$ in C such that the following diagrams commute.

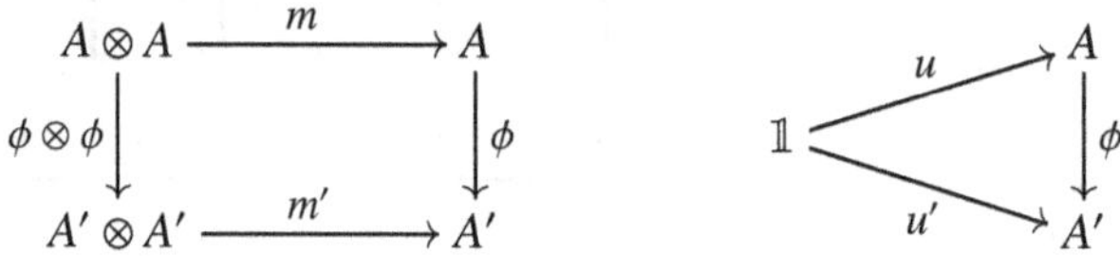

Now we can form the **category of algebras in a monoidal category** C:

- $\mathsf{Alg}(C)$: objects are algebras in C and morphisms are algebra morphisms in C.

The collection of algebra morphisms $(A, m, u) \to (A', m', u')$ is denoted by $\mathrm{Hom}_{\mathsf{Alg}(C)}(A, A')$, which is a subcollection of $\mathrm{Hom}_C(A_{\mathrm{obj}}, A'_{\mathrm{obj}})$. See Exercise 4.6.

A **monic** (resp., **epic, iso-) morphism of algebras** in C is an algebra morphism that is a mono (resp., an epi, an iso) on the underlying objects in C.

Algebras in the monoidal setting are also preserved under monoidal functors. Namely, if C and C' are monoidal categories, then a monoidal functor $C \to C'$ yields a functor $\mathsf{Alg}(C) \to \mathsf{Alg}(C')$, seen as follows. The proof is Exercise 4.7.

Proposition 4.3. *Recall §§3.2.1, 3.2.2. Take a monoidal functor $(F, F^{(2)}, F^{(0)}) : C \to C'$.*

(a) *For $(A, m_A, u_A) \in \mathsf{Alg}(C)$, we have that*

$$\left(F(A), \quad m_{F(A)} := F(m_A) \circ F^{(2)}_{A,A}, \quad u_{F(A)} := F(u_A) \circ F^{(0)}\right) \in \mathsf{Alg}(C').$$

(b) *The assignment in part (a) yields a functor $\mathsf{Alg}(F) : \mathsf{Alg}(C) \to \mathsf{Alg}(C')$, which is an isomorphism of categories when F is an isomorphism of monoidal categories.* □

See Exercises 4.8 and 4.9 for examples and an open-ended task, respectively.

§4.1.2. The Eckmann-Hilton Principle

We examine a way to relate various algebra operations on a given object in a monoidal category. This was derived from Theorem 4.17 of Eckmann and Hilton [1962], and it has been applied to many mathematical structures. In this section, we proceed in the strict case for ease via the Strictification Theorem [Theorem 3.26].

Theorem 4.4 (Eckmann-Hilton (EH-)Principle). *Take an object A in a strict monoidal category $(C, \otimes, \mathbb{1})$, and assume that there is a morphism in C, $\mathrm{flip} : A \otimes A \to A \otimes A$, that swaps the A's across $\otimes$, such that the identities in Figure 4.2 hold.*

Next, suppose that there exist morphisms in C,

$$m_1, m_2 : A \otimes A \to A \quad and \quad u_1, u_2 : \mathbb{1} \to A,$$

such that m_1 and m_2 are multiplication morphisms for A (that are not necessarily associative), which are unital via u_1 and u_2, respectively. Now if we have that

$$m_2 (m_1 \otimes m_1) = m_1 (m_2 \otimes m_2)(\mathrm{id}_A \otimes \mathrm{flip} \otimes \mathrm{id}_A) \qquad (4.5)$$

as morphisms $A \otimes A \otimes A \otimes A \to A$ in C, then the following statements hold.

(a) $u_1 = u_2$ (=: u_0)   *and* $m_1 = m_2$ (=: m_0).

(b) m_0 *is associative, that is, (A, m_0, u_0) is an algebra in C.*

(c) $m_0 = m_0 \circ \mathrm{flip}$, *that is, (A, m_0, u_0) is, in a sense, a commutative algebra in C.*

We will define commutative algebras in *symmetric* monoidal categories in a future volume. We call (4.5) the **Eckmann-Hilton (EH-)condition.**

Proof. By employing graphical calculus for this proof, parts (a), (b), and (c) are verified in Figures 4.3, 4.4, and 4.5, respectively. $\qquad\qquad\square$

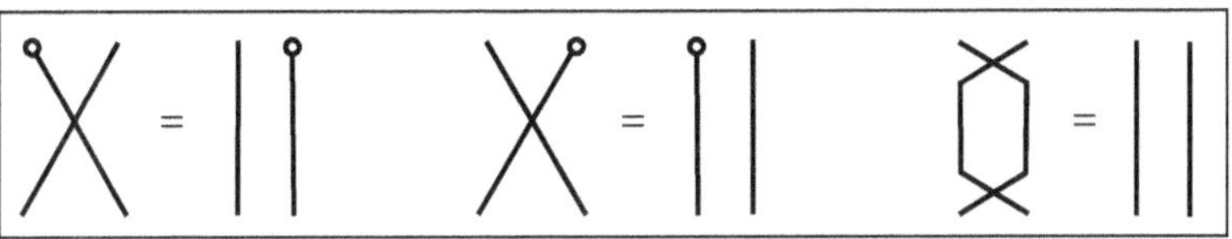

Figure 4.2: Flip identities.

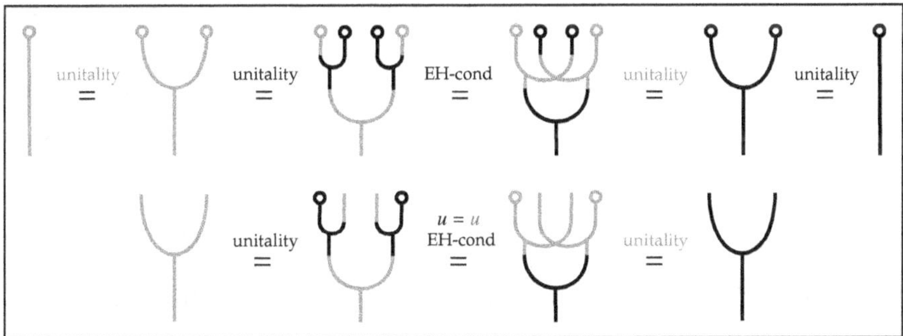

Figure 4.3: Proof of operation identification in the EH-Principle [Theorem 4.4(a)]. Here, m_1 and u_1 are depicted in black, and m_2 and u_2 are depicted in gray.

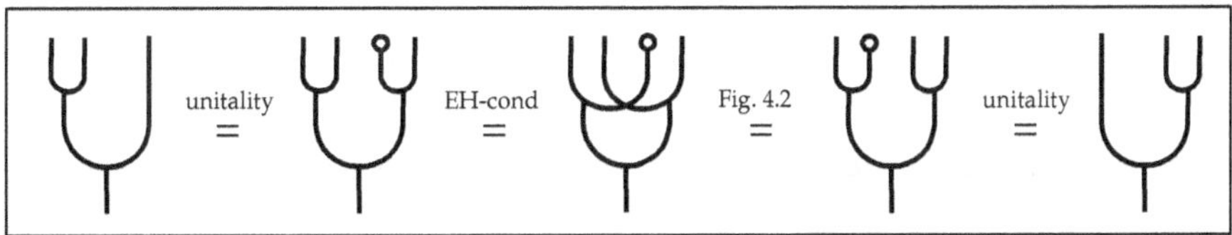

Figure 4.4: Proof of associativity in the EH-Principle [Theorem 4.4(b)].

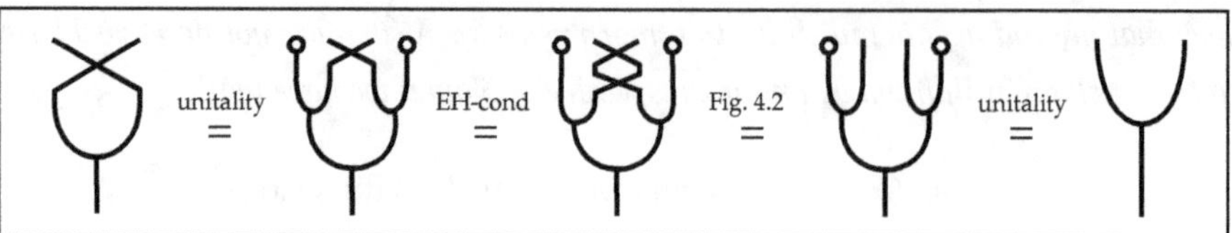

Figure 4.5: Proof of commutativity in the EH-Principle [Theorem 4.4(c)].

Example 4.6. With the EH-Principle, one gets that each algebra in the monoidal category of monoids, $(\mathsf{Monoid}, \times, \{e\})$, is commutative. That is,

$$\mathsf{Alg}(\mathsf{Monoid}) \cong \mathsf{ComMonoid},$$

where the latter is the category of commutative monoids. Namely, take a monoid $N = (N_{\mathsf{set}}, \star, e)$, and equip it with a multiplication morphism $m : N \times N \to N$ and a unit morphism $u : \{e\} \to N$ such that $(N, m, u) \in \mathsf{Alg}(\mathsf{Monoid})$. Note that $m : N \times N \to N$ is a morphism in Monoid. So, $N \times N$ being a monoid with product $(\star \times \star)(\mathrm{id} \times \mathrm{flip} \times \mathrm{id})$ yields the following commutative diagram.

$$
\begin{array}{ccc}
N \times N \times N \times N & \xrightarrow{\ (\star \times \star)(\mathrm{id} \times \mathrm{flip} \times \mathrm{id})\ } & N \times N \\
{\scriptstyle m \times m}\big\downarrow & & \big\downarrow{\scriptstyle m} \\
N \times N & \xrightarrow{\qquad\qquad \star \qquad\qquad} & N
\end{array}
$$

So, the Eckmann-Hilton condition (4.5) holds for $(N_{\mathsf{set}}, \star, e, m, u) \in \mathsf{Alg}(\mathsf{Monoid})$ with $m_1 = m$ and $m_2 = \star$. Thus, we achieve the desired result by Theorem 4.4.

Another example involving *group objects* in categories is introduced in Exercise 4.10. This pertains to the original work of Eckmann and Hilton [1962], which in turn, has applications to topology and geometry.

Next, we discuss an application of the EH-Principle in the enriched setting.

Example 4.7. Take a monoidal category $(\mathcal{C}, \otimes, \mathbb{1})$ that is enriched over a monoidal category $(\mathcal{V}, \otimes^{\mathcal{V}}, \mathbb{1}^{\mathcal{V}})$ as in §3.11.1. Then, the algebra $\mathsf{End}_{\mathcal{C}}(\mathbb{1}) \in \mathcal{V}$ satisfies the Eckmann-Hilton condition (4.5) with m_1 being the composition $\circ$ of endomorphisms in $\mathsf{End}_{\mathcal{C}}(\mathbb{1})$, and with m_2 being the monoidal product $\otimes$ of $\mathcal{C}$; see Exercise 3.17. Both u_1 and u_2 are the identity morphism $\mathrm{id}_{\mathbb{1}}$. We call

$$\mathsf{End}_{\mathcal{C}}(\mathbb{1}) \in \mathsf{Alg}(\mathcal{V}),$$

the **ground algebra** of $(\mathcal{C}, \otimes, \mathbb{1})$. In particular, we have the following results.

(a) If $\mathcal{V} = (\mathsf{Set}, \times, \{*\})$ (e.g., if $\mathcal{C}$ is locally small), then $\mathsf{End}_{\mathcal{C}}(\mathbb{1}) \in \mathsf{ComMonoid}$.

(b) If $\mathcal{V} = (\mathsf{Vec}, \otimes_{\Bbbk}, \Bbbk)$ (e.g., if $\mathcal{C}$ is $\Bbbk$-linear), then $\mathsf{End}_{\mathcal{C}}(\mathbb{1}) \in \mathsf{ComAlg}$.

Finally, recall the two operations of natural transformations of functors, vertical and horizontal composition, discussed in §2.3.5.

Example 4.8. Observe that the interchange law (2.9) between vertical composition $\circ^{\text{ver}}$ and horizontal composition $\circ^{\text{hor}}$ of natural transformations satisfies the Eckmann-Hilton condition (4.5). In fact, we can consider the collection of natural transformations from an identity functor, $\text{Id}_{\mathcal{A}}$, on a category $\mathcal{A}$ to itself, obtaining that $\text{End}(\text{Id}_{\mathcal{A}})$ is a commutative algebra (cf. Exercise 2.38). This generalizes the previous example, but we need higher category theory introduced in a future volume (see also §4.10.3 later) to fully capture the setting of this example.

§4.1.3. Endomorphism algebras

Here, we examine algebras that generalize the endomorphism algebras $\text{End}_{\Bbbk}(V)$ from §1.2.1. Here, V is a $\Bbbk$-vector space, and $\text{End}_{\Bbbk}(V)$ is a $\Bbbk$-algebra with multiplication $\circ$, and with unit id_V. But given an object X in a category $\mathcal{A}$, we have that $\text{End}_{\mathcal{A}}(X)$ is just a collection of morphisms without further structure if we do not impose conditions on $\mathcal{A}$. So, to construct endomorphism algebras in the categorical context, we consider enriched categories from §3.11.

Take a monoidal category $(\mathcal{V}, \otimes^{\mathcal{V}}, \mathbb{1}^{\mathcal{V}})$ and a $\mathcal{V}$-category $\mathcal{A}$, as in §3.11.1. For $X \in \mathcal{A}$, the $\mathcal{V}$-**endomorphism algebra** $\text{End}_{\mathcal{V}}(X)$ of X consists of the following data:

(a) The object $\mathcal{A}(X, X)$ of $\mathcal{V}$;   with

(b) Multiplication morphism $\gamma_{X,X,X} : \mathcal{A}(X, X) \otimes^{\mathcal{V}} \mathcal{A}(X, X) \to \mathcal{A}(X, X)$;

(c) Unit morphism $\upsilon_X : \mathbb{1}^{\mathcal{V}} \to \mathcal{A}(X, X)$.

We leave it to the reader to then verify that $\text{End}_{\mathcal{V}}(X) \in \text{Alg}(\mathcal{V})$.

Example 4.9. (a) A (strict) left rigid category $\mathcal{C} := (\mathcal{C}, \otimes, \mathbb{1}, a, \ell, r, (-)^*)$ is enriched over itself, with Hom objects $\mathcal{C}(X, Y) := Y \otimes X^*$, for all $X, Y \in \mathcal{C}$ [Exercise 3.39]. Now we obtain that

$$\text{End}_{\mathcal{C}}(X) := (X \otimes X^*, \quad m := \text{id}_X \otimes \text{ev}_X^{\ell} \otimes \text{id}_{X^*}, \quad u := \text{coev}_X^{\ell}) \quad \in \text{Alg}(\mathcal{C}).$$

The associativity and unitality axioms follow from Figure 4.6 below.
(Note: we can replace $\mathcal{C}$ being left rigid, with just $X \in \mathcal{C}$ being left rigid here.)

(b) As a special case, recall that the category of finite-dimensional $\Bbbk$-vector spaces, FdVec, is (left) rigid [Example 3.34]. Exercise 4.11 asks us to show that, for $V \in \text{FdVec}$, the $\Bbbk$-algebra $\text{End}_{\Bbbk}(V)$ from §1.2.1 arises as follows:

$$\text{End}_{\text{FdVec}}(V) \cong \text{End}_{\Bbbk}(V) \quad \text{in Alg(FdVec)}.$$

We will also give the internal endomorphisms of §3.11.4 the structure of an algebra in a monoidal category later in §4.8.

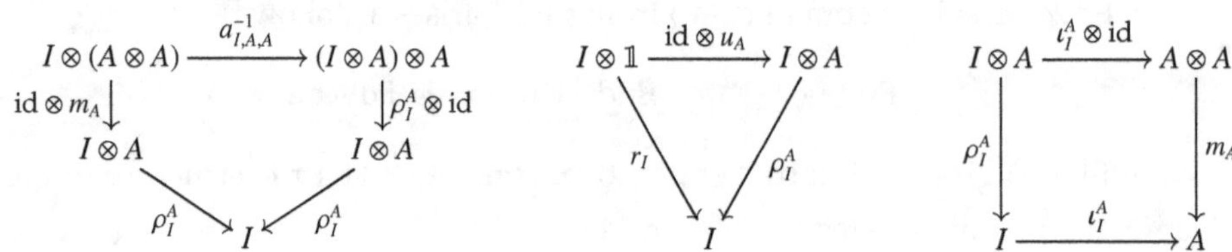

Figure 4.6: Associativity and unitality for $\mathrm{End}_{\mathcal{C}}(X)$, for $\mathcal{C}$ left rigid.

§4.2. Subalgebras and quotient algebras

Next, we study substructures and quotient structures for algebras in $\mathcal{C}$.

Standing hypothesis.   Assume that $\mathcal{C}$ is abelian monoidal as in §3.1.3.

---

§4.2.1. Subalgebras and ideals

A **subalgebra** of an algebra (A, m_A, u_A) in $\mathcal{C}$ is an algebra (B, m_B, u_B) in $\mathcal{C}$, such that B_{obj} is a subobject of A_{obj} via a mono $\iota_B^A : B \to A$, where ι_B^A is an algebra morphism. If unitality does not necessarily hold (that is, if u_B does not exist, or $\iota_B^A u_B \neq u_A$), then we refer to (B, ι_B^A, m_B) as a **nonunital subalgebra** of (A, m_A, u_A) in $\mathcal{C}$.

A **left ideal** of (A, m_A, u_A) consists of a subobject $(I, \iota_I^A : I \to A)$ equipped with a morphism $\lambda_I^A : A \otimes I \to I$ such that the following diagrams commute.

A **right ideal** of (A, m_A, u_A) consists of a subobject $(I, \iota_I^A : I \to A)$ equipped with a morphism $\rho_I^A : I \otimes A \to I$ such that the following diagrams commute.

A **(two-sided) ideal** of (A, m_A, u_A) is a tuple $(I, \iota_I^A, \lambda_I^A, \rho_I^A)$, such that $(I, \iota_I^A, \lambda_I^A)$ is a left ideal, (I, ι_I^A, ρ_I^A) is a right ideal of A, and the diagram below commutes.

Example 4.10. Take $(A, m_A, u_A) \in \mathsf{Alg}(\mathcal{C})$. We will see in Exercise 4.12 that:

(a) A_{obj} is an ideal of A, with $\lambda_A^A = \rho_A^A = m_A$ and $\iota_A^A = \mathrm{id}_A$;

(b) the zero object 0 is an ideal of A with $\lambda_0^A = {}_{A \otimes 0}\vec{0}$, and $\rho_0^A = {}_{0 \otimes A}\vec{0}$, and $\iota_0^A = \vec{0}_A$.

Ideals of (A, m_A, u_A) not of these types are referred to as **proper ideals** of A in $\mathcal{C}$.

In fact, ideals are nonunital subalgebras as we see below.

Proposition 4.11. *Let $(I, \iota_I^A, \lambda_I^A, \rho_I^A)$ be an ideal of an algebra (A, m_A, u_A) in $\mathcal{C}$.*

(a) *We have that $\lambda_I^A(\iota_I^A \otimes \mathrm{id}_I) = \rho_I^A(\mathrm{id}_I \otimes \iota_I^A)$ as morphisms $I \otimes I \to I$.*

(b) *Let m_I denote the morphism in part (a). Then (I, ι_I^A, m_I) is a nonunital subalgebra of (A, m_A, u_A) in $\mathcal{C}$.*

Proof. (a) Consider the following diagram.

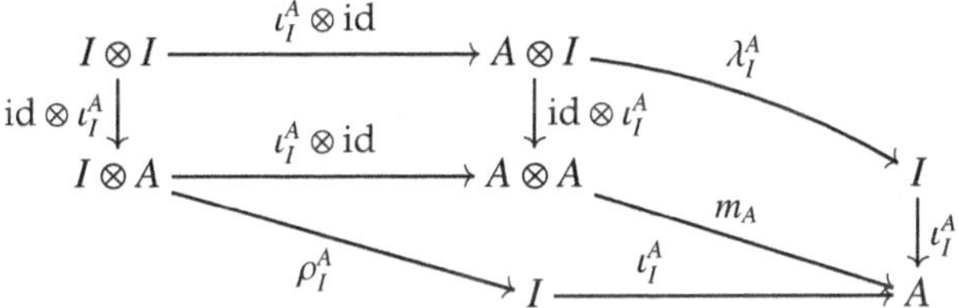

The left region commutes by the level exchange; the right region commutes as I is a left ideal; and the bottom region commutes as I is a right ideal. Since ι_I^A is monic, it is left cancellative. So, the commutative diagram above yields the result.

(b) The result holds by the commutative diagrams below.

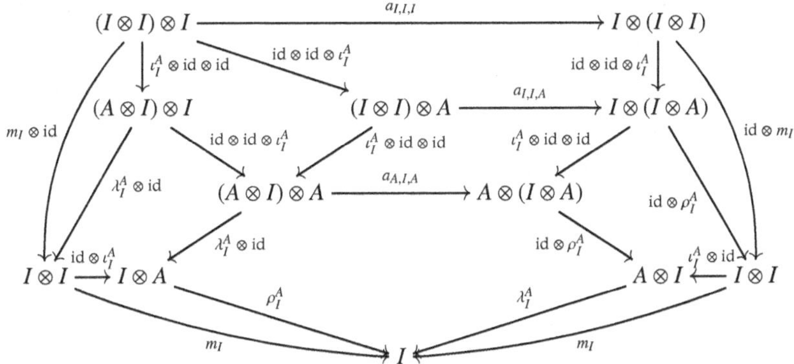

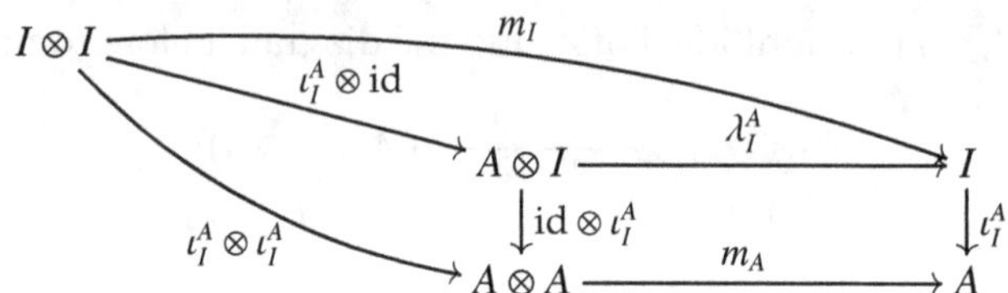

We leave it to the reader to think about why each sub-region commutes. □

In fact, if I is an ideal that 'contains' the unit of A, then I is isomorphic to A. We see this by the result below; the proof is reserved as Exercise 4.13.

Corollary 4.12. *Suppose that I is an ideal of an algebra A in C equipped with a morphism $u_I : \mathbb{1} \to I$ such that $u_A = \iota_I^A u_I$. Then, I is a unital subalgebra of A, with unit u_I, and I is isomorphic to A as algebras in C.* □

Next, we see that the kernel of an algebra morphism is an ideal. See §2.2.1vi.

Proposition 4.13. *If $\phi : (A, m, u) \to (A', m', u')$ is a morphism in $\mathsf{Alg}(C)$, then the kernel of $\phi_{\mathrm{obj}} : A_{\mathrm{obj}} \to A'_{\mathrm{obj}}$ forms an ideal of (A, m, u).*

Proof. We sketch the proof here, and leave it to the reader to complete the details in Exercise 4.14. Let I denote $\ker(\phi_{\mathrm{obj}})$ equipped with morphism $\alpha' : I \to A$ such that $\phi_{\mathrm{obj}} \alpha' = {}_I\vec{0}_{A'}$. Then, I is an ideal with

$$\iota_I^A := \alpha',$$

and with λ_I^A and ρ_I^A defined uniquely by the universal property of kernels below.

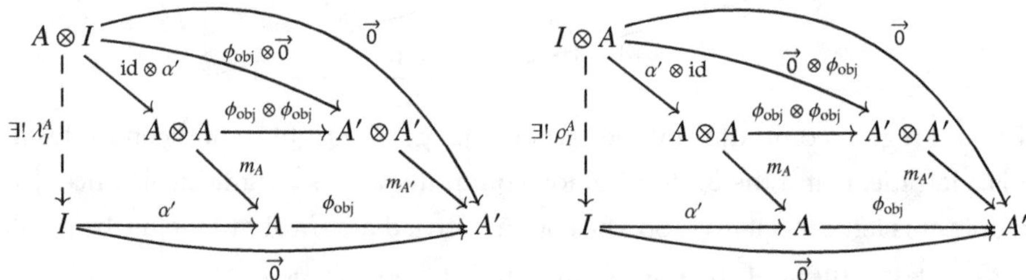

Here, we apply Lemma 3.3, along with the fact that the (pre-)composition of a zero morphism with any morphism is a zero morphism (see §2.2.1v).

Next, $\alpha' \lambda_I^A (\mathrm{id}_A \otimes \lambda_I^A) a_{A,A,I} = \alpha' \lambda_I^A (m_A \otimes \mathrm{id}_I)$ yields one of the compatibility conditions required for I to be an ideal. Namely, α' is monic, so it is left cancellative. Likewise, one can establish the other compatibility condition to conclude that I is an ideal of A. □

By taking $\phi : A \to A'$ to be the identity morphism or the zero morphism, respectively, we obtain that 0 and A are ideals of A; cf. Example 4.10. See also Exercise 4.15(i,ii) for more examples.

§4.2.2. Quotient algebras

Now we define quotient algebras in $\mathcal{C}$, which will require several careful commutative diagram arguments using the universal property of cokernels (see §2.2.1vi). A version of this material appeared in the work of Walton and Yadav [2023].

Standing hypothesis. Assume that $\mathcal{C}$ is abelian monoidal as in §3.1.3. Also assume that the endofunctors $(X \otimes -)$ and $(- \otimes X)$ on $\mathcal{C}$ are right exact, for each $X \in \mathcal{C}$ (see §2.8.1); see Remark 4.14, along with Figure 3.15.

Remark 4.14. We can alter the second hypothesis above via the statements below.

(a) The category $\mathcal{C}$ is, in addition, right (resp., left) rigid monoidal.

(b) The category $\mathcal{C}$ is, in addition, left (resp., right) closed monoidal.

(c) The endofunctor $(X \otimes -)$ (resp., $(- \otimes X)$) of $\mathcal{C}$ has a right adjoint, for each $X \in \mathcal{C}$.

(d) The endofunctor $(X \otimes -)$ (resp., $(- \otimes X)$) of $\mathcal{C}$ is right exact, for each $X \in \mathcal{C}$.

(e) For any morphism ϕ in $\mathcal{C}$, we have that $(X \otimes -)$ (resp., $(- \otimes X)$) preserves the cokernel of ϕ for each $X \in \mathcal{C}$, that is,

$$X \otimes \mathrm{coker}(\phi) \cong \mathrm{coker}(\mathrm{id}_X \otimes \phi) \qquad (\text{resp.,} \quad \mathrm{coker}(\phi) \otimes X \cong \mathrm{coker}(\phi \otimes \mathrm{id}_X)).$$

Indeed, the implications below hold due to previous results.

$$(\text{a}) \underset{\text{Prop. 3.33(a,d)}}{\overset{\text{Exer. 3.40}}{\rightleftarrows}} (\text{b}) \overset{\text{Def. §3.11.3}}{\leftrightarrows} (\text{c}) \overset{\text{Prop. 2.49(b)}}{\longrightarrow} (\text{d}) \overset{\text{Prop. 2.49(a)}}{\longleftrightarrow} (\text{e}).$$

We also have that (c) $\Leftarrow$ (d) when $\mathcal{C}$ is finite and linear [Corollary 2.62].

To proceed, we will need the constructions and preliminary results below. First, the right exactness condition above yields the useful lemma below.

Lemma 4.15. *Take morphisms $f_1 : X_1 \to Y_1$ and $f_2 : X_2 \to Y_2$ in $\mathcal{C}$, along with an object $Z \in \mathcal{C}$. Then, the following statements hold.*

(a) *If f_1 and f_2 are epic in $\mathcal{C}$, then so are the morphisms $f_1 \otimes \mathrm{id}_Z$ and $\mathrm{id}_Z \otimes f_2$.*

(b) *If f_1 and f_2 are epic in $\mathcal{C}$, then so is $f_1 \otimes f_2$.*

Proof. (a) Since f_1 and f_2 are epic, the cokernels of these morphisms are zero morphisms by Lemma 2.43(b). This implies that the cokernel of $f_1 \otimes \mathrm{id}_Z$ and $\mathrm{id}_Z \otimes f_2$ are also zero morphisms by Remark 4.14(d)$\Rightarrow$(e). Therefore, $f_1 \otimes \mathrm{id}_Z$ and $\mathrm{id}_Z \otimes f_2$ are epic by Lemma 2.43(b).

(b) This holds by part (a) and Exercise 2.1(a) since $f_1 \otimes f_2 = (f_1 \otimes \mathrm{id}_{Y_2})(\mathrm{id}_{X_1} \otimes f_2)$. $\square$

Next, recall the pushout of morphisms from §2.2.1iii. Let $f_1 : X_1 \to Y_1$ and $f_2 : X_2 \to Y_2$ be morphisms in $\mathcal{C}$. We define their **pushout product** to be the unique morphism $f_1 \sqcup f_2 := f_1 \sqcup_{X_1 \otimes X_2} f_2$ in the commutative diagram below.

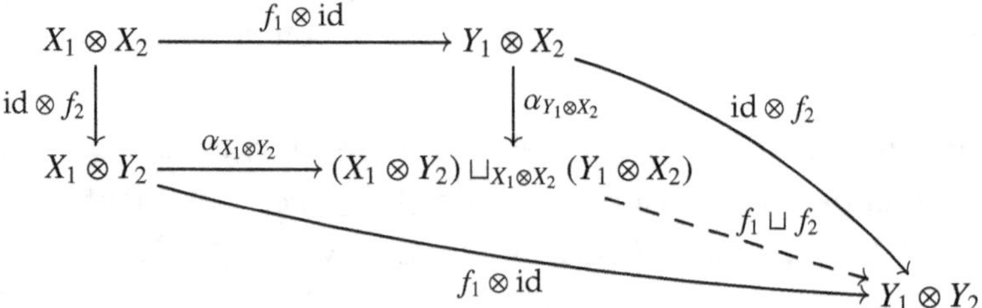

Consider the preliminary results below about the pushout product above.

Lemma 4.16. *Retain the notation above. Take a morphism* $g : Y_1 \otimes Y_2 \to Z$. *Then,* $g(f_1 \sqcup f_2) = \vec{0}$ *when* $g(f_1 \otimes \mathrm{id}_{Y_2}) = \vec{0}$ *and* $g(\mathrm{id}_{Y_1} \otimes f_2) = \vec{0}$.

Proof. By the definition of a pushout, we get the commutative diagram below.

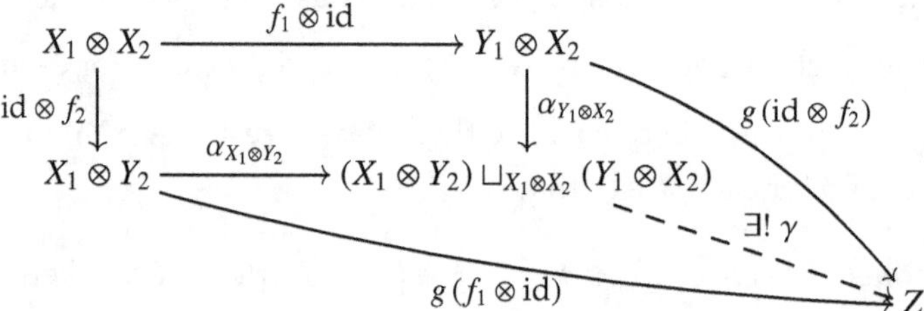

Now, $\gamma = g(f_1 \sqcup f_2)$ as this choice of γ makes the diagram commute and by γ being unique. Moreover, $g(f_1 \otimes \mathrm{id}_{Y_2}) = \vec{0}$ and $g(\mathrm{id}_{Y_1} \otimes f_2) = \vec{0}$ by the hypothesis. So, $\gamma = \vec{0}$ makes the diagram commute. By uniqueness, $g(f_1 \sqcup f_2) = \vec{0}$. $\qquad\square$

Lemma 4.17. *Retain the notation above. Then, the cokernel of the pushout product* $f_1 \sqcup f_2$ *of* f_1 *and* f_2 *in* $\mathcal{C}$ *is isomorphic to* $\mathrm{coker}(f_1) \otimes \mathrm{coker}(f_2)$.

Proof. Recall the cokernel diagrams below, for $i = 1, 2$:

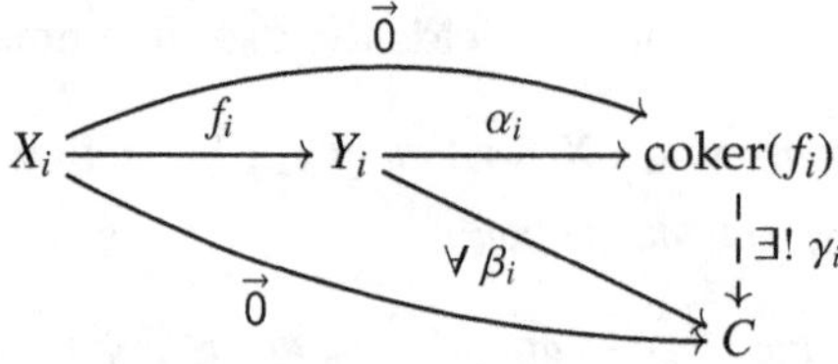

Next, consider the diagram below for the cokernel of $f_1 \sqcup f_2$, with accompanying morphism $\alpha_\sqcup : Y_1 \otimes Y_2 \to \mathrm{coker}(f_1 \sqcup f_2)$.

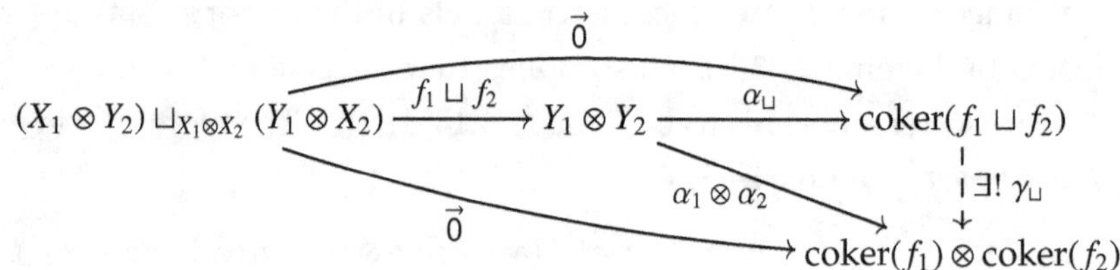

This diagram commutes since $(\alpha_1 \otimes \alpha_2)(f_1 \otimes \mathrm{id}) = \vec{0}$ and $(\alpha_1 \otimes \alpha_2)(\mathrm{id} \otimes f_2) = \vec{0}$ [Lemma 3.3], and thus, $(\alpha_1 \otimes \alpha_2)(f_1 \sqcup f_2) = \vec{0}$ [Lemma 4.16].

Now we construct a morphism $\gamma_\otimes : \mathrm{coker}(f_1) \otimes \mathrm{coker}(f_2) \to \mathrm{coker}(f_1 \sqcup f_2)$. Note that $\alpha_\sqcup (\mathrm{id}_{Y_1} \otimes f_2) = \alpha_\sqcup (f_1 \sqcup f_2)\alpha_{Y_1 \otimes X_2} = \vec{0}$. So, we get by the universal property of cokernels the morphisms $\widetilde{\alpha}$ and $\widetilde{\gamma}$ in the commutative diagram below.

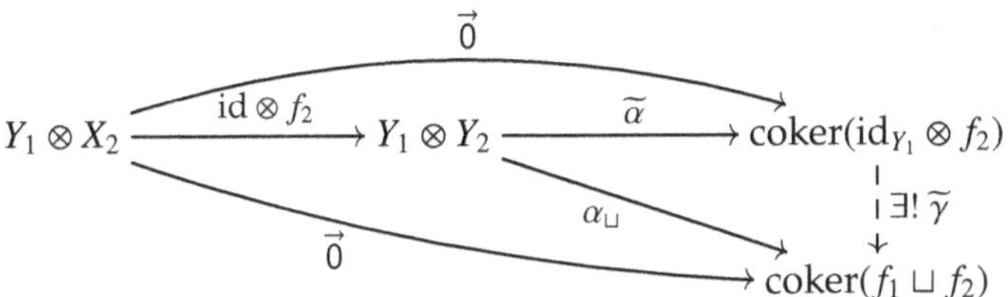

In fact, $\mathrm{coker}(\mathrm{id}_{Y_1} \otimes f_2) \cong Y_1 \otimes \mathrm{coker}(f_2)$ and $\widetilde{\alpha} = \mathrm{id}_{Y_1} \otimes \alpha_2$ by Remark 4.14(d)$\Rightarrow$(e). Then, we construct the morphism $\gamma_\otimes$ via the universal property of cokernels in the commutative diagram below.

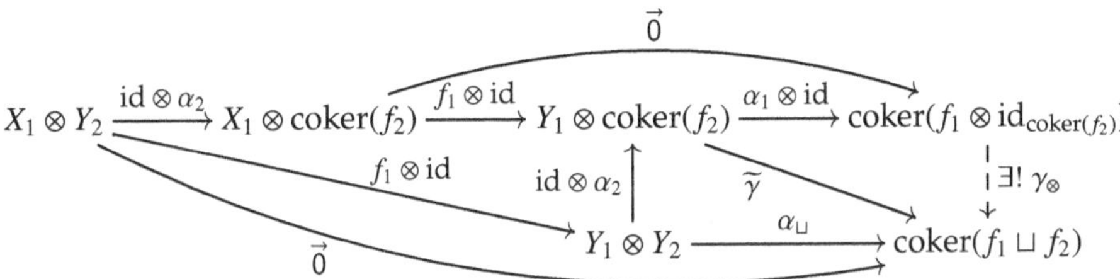

Namely, $\widetilde{\gamma}(f_1 \otimes \mathrm{id})(\mathrm{id} \otimes \alpha_2) = \alpha_\sqcup (f_1 \otimes \mathrm{id}) = \vec{0}$. Since α_2 is epic (and right-cancellative), so is $\mathrm{id} \otimes \alpha_2$ [Lemma 4.15(a)]. Therefore, $\widetilde{\gamma}(f_1 \otimes \mathrm{id}) = \vec{0}$, as required to construct the morphism $\gamma_\otimes$. Moreover, $\mathrm{coker}(f_1 \otimes \mathrm{id}_{\mathrm{coker}(f_2)}) \cong \mathrm{coker}(f_1) \otimes \mathrm{coker}(f_2)$ by Remark 4.14(d)$\Rightarrow$(e), and this is the domain of the morphism $\gamma_\otimes$.

Finally, the morphisms $\gamma_\sqcup$ and $\gamma_\otimes$ are mutually inverse, due to their uniqueness, and due to the commutative diagrams below.

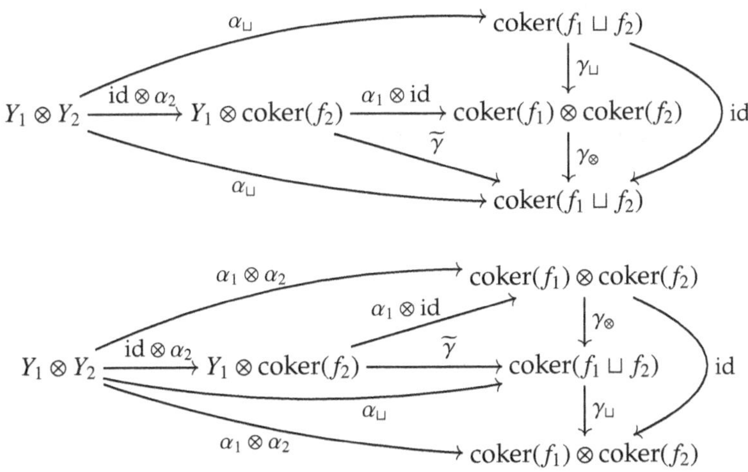

This concludes the proof of the lemma. $\qquad\qquad\square$

Now the final result of this section defines the **quotient algebra** of an algebra A in $\mathcal{C}$ by an ideal I of A in $\mathcal{C}$.

Proposition 4.18. *Take $(A, m_A, u_A) \in \mathsf{Alg}(\mathcal{C})$, with an ideal $(I, \iota_I^A, \lambda_I^A, \rho_I^A)$ of A. Then, the object $A/I := \mathrm{coker}(\iota_I^A)$ equipped with epi $\pi_I^A : A \to A/I$, forms an algebra:*

$$(A/I, \quad \overline{m} : A/I \otimes A/I \to A/I, \quad \overline{u} : \mathbb{1} \to A/I) \in \mathsf{Alg}(\mathcal{C}).$$

Here, the morphisms $\overline{m}$ and $\overline{u}$ are presented respectively in (4.20) and (4.21) below, which makes π_I^A a morphism of algebras in $\mathcal{C}$.

Proof. By Lemma 4.17, we have that

$$A/I \otimes A/I := \mathrm{coker}(\iota_I^A) \otimes \mathrm{coker}(\iota_I^A) \cong \mathrm{coker}(\iota_I^A \sqcup \iota_I^A). \tag{4.19}$$

We will define the multiplication morphism $\overline{m}$ via this iso by verifying that $\pi_I^A \, m \, (\iota_I^A \sqcup \iota_I^A) = \vec{0}$. Namely, the latter equation shows there exists a unique morphism $\overline{m}$ such that the diagram below commutes.

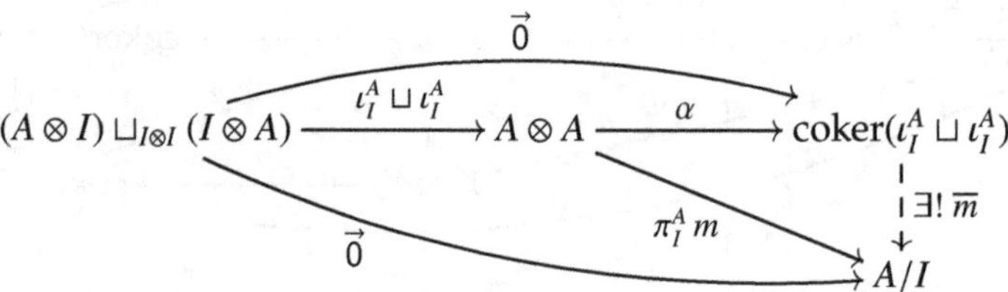

Indeed, we have that

$$\pi_I^A \, m \, (\iota_I^A \otimes \mathrm{id}_A) = \pi_I^A \, \iota_I^A \, \rho_I^A = \vec{0} \, \rho_I^A = \vec{0}$$

because I is a right ideal. Likewise, $\pi_I^A \, m \, (\mathrm{id}_A \otimes \iota_I^A) = \vec{0}$ because I is a left ideal. So, $\pi_I^A \, m \, (\iota_I^A \sqcup \iota_I^A) = \vec{0}$ by Lemma 4.16. Thus, the morphism $\overline{m}$ exists as in the commutative diagram above. By (4.19), we have that $\alpha = \pi_I^A \otimes \pi_I^A$. So,

$$\overline{m} \, (\pi_I^A \otimes \pi_I^A) = \pi_I^A \, m. \tag{4.20}$$

Next, $\overline{m}$ is associative partly due to the diagram below, which commutes by (4.20) and by the associativity of m.

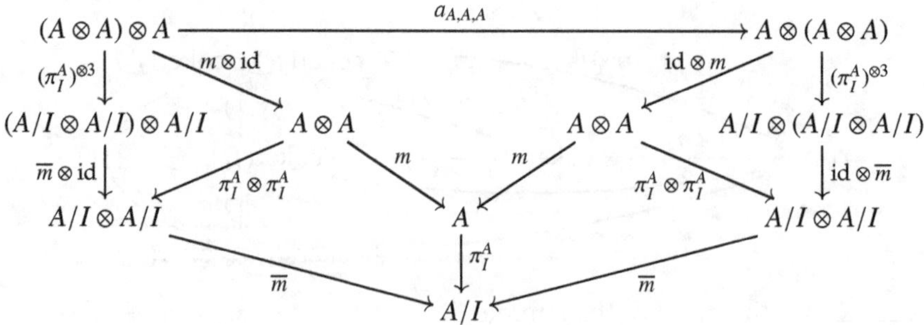

We then get the equation below by the naturality of the associativity constraint:

$$\overline{m}\,(\overline{m} \otimes \mathrm{id})\,(\pi_I^A \otimes \pi_I^A \otimes \pi_I^A) \;=\; \overline{m}\,(\mathrm{id} \otimes \overline{m})\,(\pi_I^A \otimes \pi_I^A \otimes \pi_I^A)\,a_{A,A,A}$$

$$=\; \overline{m}\,(\mathrm{id} \otimes \overline{m})\,a_{A/I,A/I,A/I}\,(\pi_I^A \otimes \pi_I^A \otimes \pi_I^A).$$

Since π_I^A is epic, $\pi_I^A \otimes \pi_I^A \otimes \pi_I^A$ is epic (and is right cancellative) by Lemma 4.15(b). Thus, $\overline{m}$ is associative by the equation above.

Lastly, $\overline{m}$ is unital by using the definition of the monoidal unit for A/I below.

$$\overline{u} := \pi_I^A u : \; \mathbb{1} \longrightarrow A/I. \tag{4.21}$$

Left unitality holds by the commutative diagram below and by $\mathrm{id} \otimes \pi_I^A$ being epic [Lemma 4.15(a)]. In particular, the middle left (resp., top left, top right, right, bottom) region commutes by the left unitality of A (resp., by level exchange, by the definition of $\overline{u}$, by (4.20), by the naturality of ℓ).

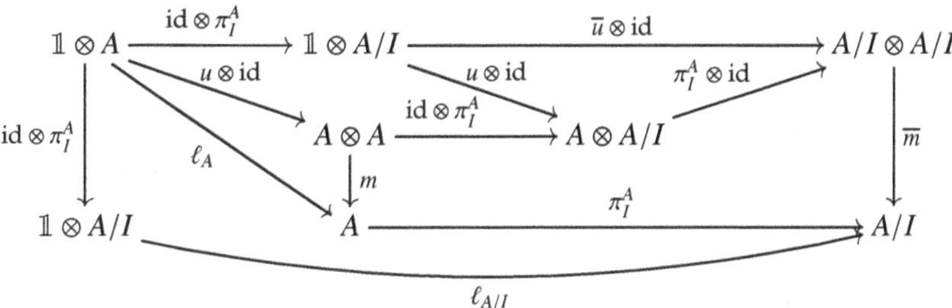

A similar computation yields the right unitality axiom for $\overline{u}$. $\qquad\square$

Continuing Examples 4.1 and 4.10, we have that an algebra A and the zero algebra are quotient algebras of A by the zero ideal and by A, respectively. See Exercise 4.15(ii,iii) and Exercise 4.44 (after reading §4.6) for more examples.

§4.3. Algebras via adjunction

We study here two constructions of algebras in monoidal categories that arise via an adjunction of functors (see §2.5): coinduced algebras and monads.

§4.3.1. Coinduced algebras via Doctrinal Adjunction

Recall from Proposition 4.3 that monoidal functors send algebras to algebras. Now to proceed, we consider a version of *Doctrinal Adjunction* due to Kelly [1974].

Theorem 4.22 (Doctrinal Adjunction, special case). *Let C and $\mathcal{D}$ be monoidal categories, and take an adjunction between the underlying categories,*

$$(F : C \to \mathcal{D}) \dashv (G : \mathcal{D} \to C),$$

with $\eta : \mathrm{Id}_C \Rightarrow GF$ (unit), $\varepsilon : FG \Rightarrow \mathrm{Id}_{\mathcal{D}}$ (counit). Assume that F is strong monoidal. Then, the following statements hold.

(a) *We have that G is monoidal, with $G^{(2)}_{Y,Y'}$ (for $Y, Y' \in \mathcal{D}$) and $G^{(0)}$ defined as follows:*

$$G^{(2)}_{Y,Y'} : G(Y) \otimes^{\mathcal{C}} G(Y') \xrightarrow{\eta_{G(Y) \otimes G(Y')}} GF\big(G(Y) \otimes^{\mathcal{C}} G(Y')\big) \xrightarrow{G\big(F^{(-2)}_{G(Y),G(Y')}\big)} G\big(FG(Y) \otimes^{\mathcal{D}} FG(Y')\big) \xrightarrow{G(\varepsilon_Y \otimes \varepsilon_{Y'})} G\big(Y \otimes^{\mathcal{D}} Y'\big),$$

$$G^{(0)} : \mathbb{1}^{\mathcal{C}} \xrightarrow{\eta_{\mathbb{1}}} GF\big(\mathbb{1}^{\mathcal{C}}\big) \xrightarrow{G\big(F^{(-0)}\big)} G\big(\mathbb{1}^{\mathcal{D}}\big).$$

(b) *We have that η and ε are monoidal, that is, for all $X, X' \in \mathcal{C}$ and $Y, Y' \in \mathcal{D}$:*

$$\eta_{X \otimes X'} \, \mathrm{Id}^{(2)}_{X,X'} \;=\; (GF)^{(2)}_{X,X'} \, (\eta_X \otimes \eta_{X'}), \qquad (GF)^{(0)} \;=\; \eta_{\mathbb{1}^{\mathcal{C}}} \, \mathrm{Id}^{(0)},$$

$$\varepsilon_{Y \otimes Y'} \, (FG)^{(2)}_{Y,Y'} \;=\; \mathrm{Id}^{(2)}_{Y,Y'} \, (\varepsilon_Y \otimes \varepsilon_{Y'}), \qquad \mathrm{Id}^{(0)} \;=\; \varepsilon_{\mathbb{1}^{\mathcal{D}}} \, (FG)^{(0)}.$$

(c) *For $A \in \mathsf{Alg}(\mathcal{C})$ and $B \in \mathsf{Alg}(\mathcal{D})$, the component morphisms $\eta_A : A \to GF(A)$ and $\varepsilon_B : FG(B) \to B$ are algebra morphisms in $\mathcal{C}$ and $\mathcal{D}$, respectively.*

(d) *We have that η_A is monic (resp., split-epic, an iso) when F is faithful (resp., full, fully faithful). Likewise, ε_B is epic (resp., split-monic, an iso) when G is faithful (resp., full, fully faithful).*

Proof. First, we sketch the proof of part (a), and leave the details to the reader. The monoidal associativity axiom holds by the commutative diagram in Figure 4.8 (in the strict case via Theorem 3.26). Moreover, the monoidal left unit axiom holds by the commutative diagram in Figure 4.7. We leave the monoidal right unit axiom to the reader as part of Exercise 4.16. Parts (b,c) comprise Exercise 4.17. Part (d) follows from Exercise 2.41(a,b). $\qquad \square$

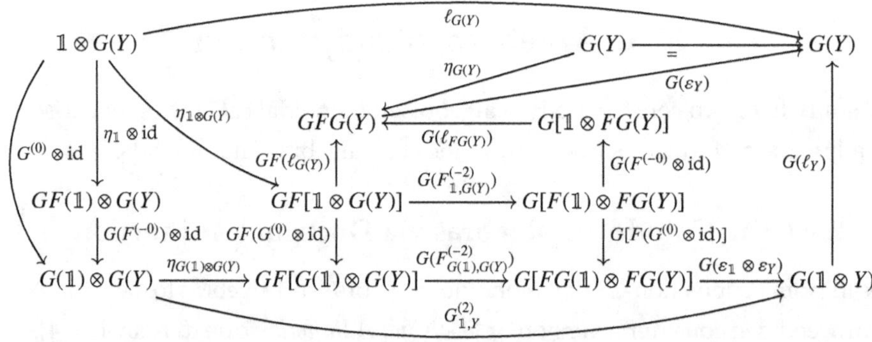

Figure 4.7: Doctrinal Adjunction: left unitality. The left and bottom regions commute by definition. The left internal and top regions (resp., bottom internal) commute by the naturality of η (resp., of $F^{(-2)}$). The right region commutes by the naturality of $F^{(-2)}$ and of ε, and a triangle identity. The triangle is a triangle identity. The middle region commutes by the left unit axiom of F.

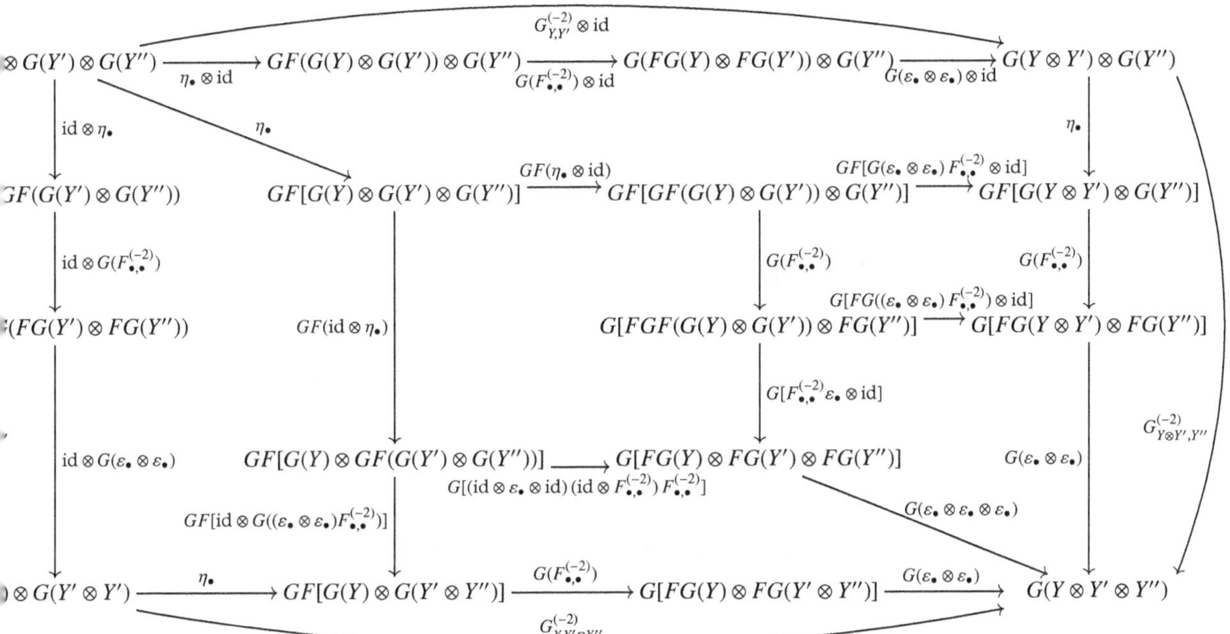

Figure 4.8: Doctrinal Adjunction: associativity. Bullets replace subscripts. The outer regions commute by definition. The left (resp., top, right, bottom-right, bottom) internal region commutes by the naturality of η (resp., of η, of $F^{(-2)}$, of ε, of ε and $F^{(-2)}$). The middle region commutes by the naturality of η and of $F^{(-2)}$, a triangle identity, and the associativity axiom of F.

Cooking up examples of algebras via Doctrinal Adjunction is Exercise 4.19. But our main example here involves restriction and coinduction functors from §1.4.4.

Example 4.23. Take a morphism $\phi : H \to G$ of finite groups. Then, we have the adjunction given below (cf., Exercise 2.45(b)):

$$(\mathrm{Res}^G_H : G\text{-Mod} \to H\text{-Mod}) \dashv (\mathrm{Coind}^G_H : H\text{-Mod} \to G\text{-Mod}).$$

Indeed, by the material in §1.4.4, we have that:

- $\mathrm{Res}^G_H\left(V, \, \triangleright^G_V : G \times V \to V\right) = \left(V, \, \triangleright^H_V := \triangleright^G_V (\phi \times \mathrm{id}_V) : H \times V \to V\right);$ and

- $\mathrm{Coind}^G_H({}_HW) = \mathrm{Hom}_{H\text{-Mod}}(\Bbbk G, W)$, where $(g \triangleright f)(g') := f(g'g)$ for $g, g' \in G$ and $f \in \mathrm{Hom}_{H\text{-Mod}}(\Bbbk G, W)$.

We also have that Res^G_H is strong monoidal [Exercise 4.8(c.i)]. Now Coind^G_H is monoidal by Doctrinal Adjunction. So by Proposition 4.3, we get that:

$$\mathrm{Coind}^G_H(A) := \mathrm{Hom}_{H\text{-Mod}}(\Bbbk G, A) \in \mathrm{Alg}(G\text{-Mod}), \quad \text{for } A \in \mathrm{Alg}(H\text{-Mod}).$$

We refer to $\mathrm{Coind}^G_H(A)$ as a **coinduced algebra** of A.

In particular, if $H = \langle e \rangle$, and A is the unit algebra $\Bbbk$ in $\mathrm{Alg}(\mathrm{Vec})$ [Example 4.1(a)], then $(\Bbbk G)^* := \mathrm{Hom}_{\mathrm{Vec}}(\Bbbk G, \Bbbk) \in \mathrm{Alg}(G\text{-Mod})$ (cf., Exercise 4.2). The details of this discussion are explored in Exercise 4.18.

§4.3.2. Monads via adjunction

We turn our attention to algebras in the (strict) monoidal category of endofunctors, especially those that we build with adjunctions. Consider the terminology below.

A **monad** on a category $\mathcal{A}$ is an algebra in the monoidal category $(\mathrm{End}(\mathcal{A}), \circ, \mathrm{Id}_{\mathcal{A}})$. Namely, a monad consists of the following data.

(a) An endofunctor $T : \mathcal{A} \to \mathcal{A}$.

(b) (**multiplication natural transformation**) A natural transformation $\mu : T \circ T \Rightarrow T$.

(c) (**unit natural transformation**) A natural transformation $\eta : \mathrm{Id}_{\mathcal{A}} \Rightarrow T$.

This data must satisfy the commutative diagrams below for associativity and unitality. Here, T^n denotes the n-fold composition of T with itself. Moreover, $\bullet \Rightarrow \bullet \Rightarrow \bullet$ represents the vertical composition of natural transformations.

$$
\begin{array}{ccc}
T^3 \xrightarrow{\mu * T} T^2 & \qquad T \xrightarrow{\eta * T} T^2 & \qquad T^2 \xleftarrow{T * \eta} T \\
\end{array}
$$

That is, for all $X \in \mathcal{A}$, the diagrams below commute.

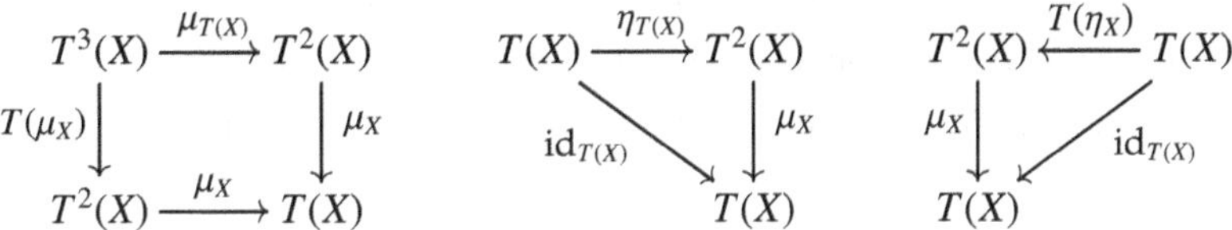

A **monad morphism** between two monads (T, μ, η) and (T', μ', η') on $\mathcal{A}$ is a natural transformation $\phi : T \Rightarrow T'$ such that the following diagrams commute.

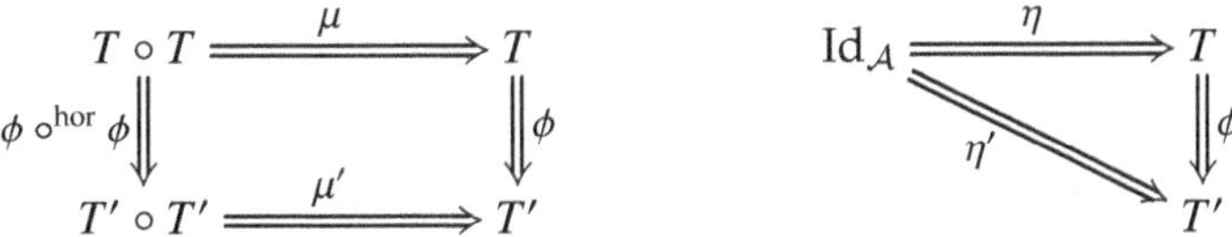

That is, for all $X \in \mathcal{A}$, the diagrams below commute.

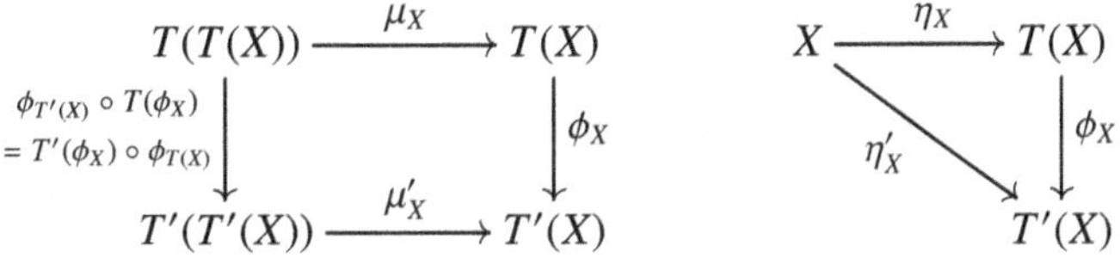

With the objects and morphisms defined above, we can consider the category:

$$\mathsf{Monad}(\mathcal{A}) := \mathsf{Alg}((\mathsf{End}(\mathcal{A}), \circ, \mathsf{Id}_{\mathcal{A}})).$$

Example 4.24. The **identity monad** on $\mathcal{A}$ is given as the endofunctor $T = \mathsf{Id}_{\mathcal{A}}$, with multiplication natural transformation $\mu = \{\mu_X := \mathrm{id}_X\}_{X \in \mathcal{A}}$, and unit natural transformation $\eta = \{\eta_X := \mathrm{id}_X\}_{X \in \mathcal{A}}$. We write this as $\mathsf{Id} \in \mathsf{Monad}(\mathcal{A})$.

Example 4.25. Given an algebra (A, m, u) in a strict monoidal category $(\mathcal{C}, \otimes, \mathbb{1})$, the endofunctor $A \otimes - : \mathcal{C} \to \mathcal{C}$ is a monad on $\mathcal{C}$ with operations:

$$\mu = \{\mu_X := m \otimes \mathrm{id}_X : A \otimes A \otimes X \to A \otimes X\}_{X \in \mathcal{C}},$$

$$\eta = \{\eta_X := u \otimes \mathrm{id}_X : X \to A \otimes X\}_{X \in \mathcal{C}}.$$

Namely, the associativity and unitality axioms of $(A \otimes -, \mu, \eta)$ follow from the associativity and unitality axioms of (A, m, u).

Conversely, given a monad $(A \otimes -, \mu, \eta)$ on $\mathcal{C}$, we obtain that $(A, \mu_{\mathbb{1}}, \eta_{\mathbb{1}})$ is an algebra in $\mathcal{C}$. Verifying the details is Exercise 4.20.

Other examples of monads are explored in Exercises 4.21–4.23. The next example however, is most important, as it gives an abundance of monads in the literature.

Example 4.26. For an adjunction $(F : \mathcal{A} \to \mathcal{B}) \dashv (G : \mathcal{B} \to \mathcal{A})$, we obtain that

$$(GF, \ G\varepsilon F, \ \eta) \in \mathsf{Monad}(\mathcal{A}).$$

Here, $G\varepsilon F := G * \varepsilon * F$. We also abuse the notation "η" using it as both the unit of the monad and the unit of the adjunction, which happen to be equal in this case. We refer to this construction as an **adjunction monad**. Moreover, Examples 4.24 and 4.25 are special cases of this construction. Checking the details is Exercise 4.24.

On the other hand, the endofunctor FG on $\mathcal{B}$, for F and G above, admits the structure of a *comonad* on $\mathcal{B}$. This will be discussed in a future volume.

We also note that a converse of the result in Example 4.26 will be presented in §4.4.3ii. Namely, we will see that every monad arises from an adjunction.

§4.4. Modules and bimodules in monoidal categories

Next, we discuss modules and bimodules over algebras in monoidal categories, building on the material in §1.3 for (bi)modules over $\Bbbk$-algebras.

Remark 4.27. Towards generalizing the material in §1.3.1, one may want to define a representation of an algebra A in a monoidal category $\mathcal{C}$ to be an object X in $\mathcal{C}$ equipped with an algebra morphism $A \to \mathrm{End}_{\mathcal{C}}(X)$ in $\mathcal{C}$. However, $\mathrm{End}_{\mathcal{C}}(X)$ is not necessarily an algebra (nor an object) in $\mathcal{C}$. To make progress, one can assume that $\mathcal{C}$ is enriched over itself, and use the $\mathcal{C}$-endomorphism algebra presented in §4.1.3. But we will skip this here, and encourage the reader to explore this if curious.

§4.4.1. Modules

i. Definitions

A **left module** over an algebra (A, m, u) in $\mathcal{C}$ consists of the following data.

(a) An object M in $\mathcal{C}$.

(b) (**left action morphism**) A morphism $\triangleright := \triangleright_M := \triangleright_M^A : A \otimes M \to M$ in $\mathcal{C}$.

This data must satisfy the commutative diagrams below.

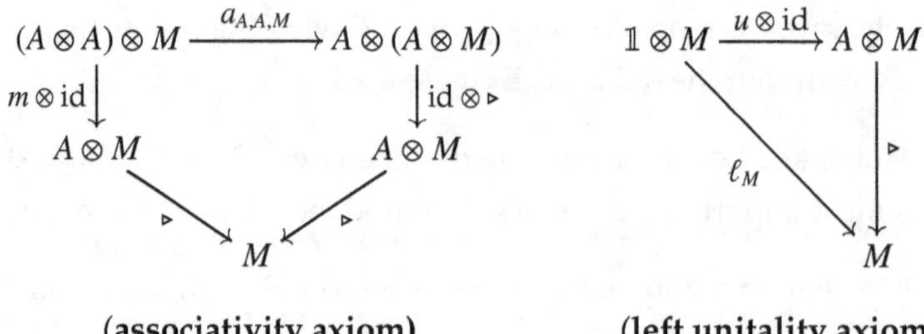

<table>
<tr><td style="text-align:center">(associativity axiom)</td><td style="text-align:center">(left unitality axiom)</td></tr>
</table>

Likewise, a **right module** over $A \in \mathsf{Alg}(C)$ consists of the following data.

(a) An object M in C.

(b) (**right action morphism**) A morphism $\triangleleft := \triangleleft_M := \triangleleft_M^A : M \otimes A \to M$ in C.

This data must satisfy the commutative diagrams below.

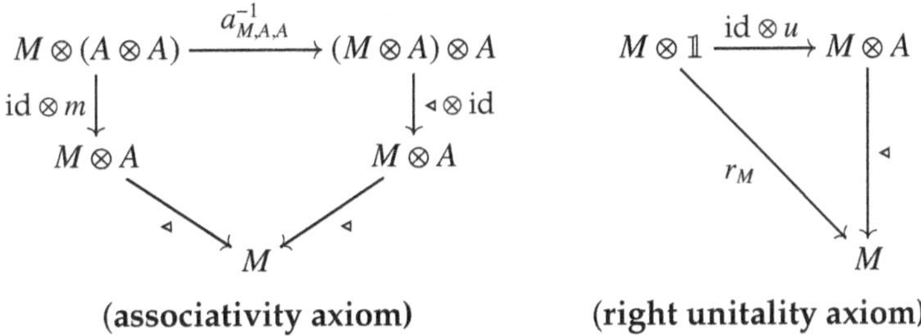

(**associativity axiom**)   (**right unitality axiom**)

The string diagrams for modules in strict C are given in Figure 4.9.

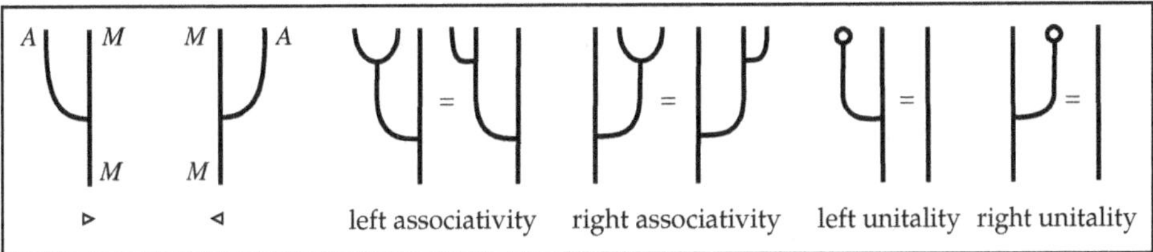

Figure 4.9: String diagrams for modules in (strict) monoidal categories.

Examples of (left) modules over algebras in monoidal categories include the following. Other examples are explored in Exercise 4.26.

Example 4.28. Let (A, m, u) be an algebra in C.

(a) The **regular left A-module** in C is the object $A := A_{\mathrm{reg}}$ with left action $\triangleright := m$. Indeed, the left module associativity and left module unitality axioms hold by the associativity and left unitality axioms for the algebra (A, m, u).

(b) It follows from definition that a left ideal $(I,\ \iota_I^A : I \to A,\ \lambda_I^A : A \otimes I \to I)$ of A is a left A-module in C with $M := I$ and $\triangleright := \lambda_I^A$.

(c) The **free left A-module on an object** X in C is the object $A \otimes X$ with left action $\triangleright := (m \otimes \mathrm{id}_X)\, a_{A,A,X}^{-1} : A \otimes (A \otimes X) \to A \otimes X$. See Exercise 4.25.

Given two left A-modules $(M, \triangleright)$ and $(M', \triangleright')$ in C, a morphism $\phi : M \to M'$ in C is a **left A-module morphism** if the following diagram commutes.

$$
\begin{array}{ccc}
A \otimes M & \xrightarrow{\ \triangleright\ } & M \\
{\scriptstyle \mathrm{id} \otimes \phi}\downarrow & & \downarrow{\scriptstyle \phi} \\
A \otimes M' & \xrightarrow{\ \triangleright'\ } & M'
\end{array}
$$

Now we can form the **category of left A-modules in C**:

- A-Mod(C): left A-modules in C and their morphisms in C.

We denote the collection of left A-module morphisms $(M, \triangleright) \to (M', \triangleright')$ by $\mathrm{Hom}_{A\text{-Mod}(C)}(M, M')$, which is a subcollection of $\mathrm{Hom}_C(M_{\mathrm{obj}}, M'_{\mathrm{obj}})$.

A **monic** (resp., **epic, iso-**) **morphism of left A-modules** in C is a left A-module morphism that is a mono (resp., an epi, an iso) in C.

Example 4.29. Take the unit algebra $(\mathbb{1}, \ell_{\mathbb{1}}, \mathrm{id}_{\mathbb{1}})$ in C [Example 4.1(a)]. We identify $\mathbb{1}$-Mod(C) with C. Namely, if $(M, \triangleright_M) \in \mathbb{1}$-Mod$(C)$, then the left unitality axiom implies that $\triangleright_M = \triangleright_M (\mathrm{id}_{\mathbb{1}} \otimes \mathrm{id}_M) = \triangleright_M (u_{\mathbb{1}} \otimes \mathrm{id}_M) = \ell_M$.

Likewise, the discussion above can be adapted for right A-modules in C, and we can form the **category of right A-modules in C**:

- Mod-$A(C)$: right A-modules in C and their morphisms in C.

ii. Substructures and quotient structures

Standing hypothesis. Assume that C is abelian monoidal as in §3.1.3.

Take a left A-module $(M, \triangleright_M)$ in C. A **left A-submodule** of $(M, \triangleright_M)$ consists of:

(a) A subobject $(N, \iota_N^M : N \to M)$ of M in C,

(b) A left action morphism, $\triangleright_N : A \otimes N \to N$, in C

such that the following conditions hold:

- $(N, \triangleright_N)$ is a left A-module in C; and
- ι_N^M is a morphism of left A-modules in C.

Given a left A-submodule $(N, \iota_N^M, \triangleright_N)$ of $(M, \triangleright_M)$ in C, a **left A-quotient module** of M by N consists of:

(a) An object $M/N := \mathrm{coker}(\iota_N^M)$, equipped with morphism $\pi_N^M : M \to M/N$ in C,

(b) A left action morphism $\triangleright_{M/N} : A \otimes M/N \to M/N$ in C,

such that the following conditions hold:

- $(M/N, \triangleright_{M/N})$ is a left A-module in C; and
- π_N^M is a morphism of left A-modules in C.

Note that if the endofunctor $A \otimes -$ on C is right exact, then such a left action morphism $\triangleright_{M/N} : A \otimes M/N \to M/N$ exists. See Exercise 4.27 and Remark 4.14.

Likewise, we can define a **right A-submodule** and a **right A-quotient module** of a right A-module in C.

iii. External action

So far, we have discussed categories of modules *within* C (i.e., internal actions), but these categories of modules also arise as modules *over* C (i.e., external action).

Proposition 4.30. *Let A be an algebra in C. Then, A-Mod(C) (resp., Mod-$A(C)$) is a right (resp., left) C-module category as in §3.3.1. Here,*

$$\vartriangleleft_A : A\text{-Mod}(C) \times C \to A\text{-Mod}(C), \quad ((M,\vartriangleright),X) \mapsto (M \otimes X, \; \blacktriangleright := (\vartriangleright \otimes \operatorname{id}_X)\, a^{-1}_{A,M,X}),$$

$$\vartriangleright_A : C \times \text{Mod-}A(C) \to \text{Mod-}A(C), \quad (X,(M,\vartriangleleft)) \mapsto (X \otimes M, \; \blacktriangleleft := (\operatorname{id}_X \otimes \vartriangleleft)\, a_{X,M,A}). \qquad \square$$

We leave the proof of this result to Exercise 4.28. We also have that the converse of this result holds under certain conditions. Namely, there are conditions when a C-module category is equivalent to a category of modules over an algebra A in C. Here, A is an *internal End algebra*, and we will discuss this in detail later in §4.8.

iv. Properties

Next, we discuss some properties of A-Mod(C) and Mod-$A(C)$ that are induced by properties of C and A. More properties of A-Mod(C) will be discussed later in §4.9.

Proposition 4.31. *Take $A \in \text{Alg}(C)$. Then, the following statements hold.*

(a) *A-Mod(C) (resp., Mod-$A(C)$) is preadditive if C is preadditive and $A \otimes - : C \to C$ (resp., $- \otimes A : C \to C$) is additive.*

(b) *A-Mod(C) (resp., Mod-$A(C)$) is linear if C is linear and the functor $A \otimes - : C \to C$ (resp., $- \otimes A : C \to C$) is linear.*

(c) *A-Mod(C) (resp., Mod-$A(C)$) is additive if C is additive and the functor $A \otimes - : C \to C$ (resp., $- \otimes A : C \to C$) is additive.*

(d) *A-Mod(C) (resp., Mod-$A(C)$) is abelian if C is abelian and the functor $A \otimes - : C \to C$ (resp., $- \otimes A : C \to C$) is additive and exact.*

Proof. We sketch the proof for left modules, and leave the rest of the details as Exercise 4.29. To start, we leave the proof of parts (a,b) as an exercise.

For part (c), assume that C is additive. Part (a) implies that A-Mod(C) is preadditive. Next, take the zero object 0 of C. Check that $(0, {}_{A \otimes 0}\vec{0}) \in A\text{-Mod}(C)$; this serves as the zero object of A-Mod(C). Moreover, for modules $(M_1, \vartriangleright_1), (M_2, \vartriangleright_2) \in A\text{-Mod}(C)$, their binary biproduct is defined by

$$(M_1, \vartriangleright_1) \,\square\, (M_2, \vartriangleright_2) := (M_1 \,\square\, M_2, \; \blacktriangleright := (\vartriangleright_1 \,\square\, \vartriangleright_2)\psi),$$

where $\psi : A \otimes (M_1 \,\square\, M_2) \xrightarrow{\sim} (A \otimes M_1) \,\square\, (A \otimes M_2)$ exists by Lemma 3.4. Thus, A-Mod(C) is additive.

For part (d), recall that C is abelian and $A \otimes - : C \to C$ is exact. Part (c) implies that A-Mod(C) is additive. Take a morphism $\phi : (M, \triangleright) \to (M', \triangleright') \in A$-Mod($C$). Since C is abelian, the underlying morphism $\phi_{\text{obj}} : M \to M'$ in C has a kernel and cokernel in C. Using the right exactness of $(A \otimes -)$, we equip coker(ϕ_{obj}) with an action map $\triangleright$ that makes it a left A-module in C. (See Lemma 4.15 and other results in §4.2.2 for similar arguments.) Here, (coker(ϕ_{obj}), $\triangleright$) is the cokernel of ϕ in A-Mod(C). Likewise, $(A \otimes -)$ being left exact implies that $A \otimes \ker(\phi_{\text{obj}}) = \ker(\text{id}_A \otimes \phi)$. This yields an action morphism $A \otimes \ker(\phi_{\text{obj}}) \to \ker(\phi_{\text{obj}})$ by the universal property of kernels, which then defines the kernel of ϕ in A-Mod(C). Thus, each morphism in A-Mod(C) has a kernel and a cokernel in A-Mod(C). We leave to the reader to show that every mono (resp., epi) in A-Mod(C) is a kernel (resp., cokernel) of a morphism in A-Mod(C). Hence, A-Mod(C) is abelian. $\qquad\square$

Note that for part (d), if C is assumed to be abelian and rigid, then the functors $(A \otimes -)$ and $(- \otimes A)$ are exact for free by Proposition 3.68.

§4.4.2. Bimodules

i. Definitions

Take algebras (A, m_A, u_A), (B_1, m_1, u_1), (B_2, m_2, u_2) in C. A (B_1, B_2)-**bimodule** in C consists of the following data.

(a) An object M in C.

(b) (**left action morphism**) A morphism $\triangleright : B_1 \otimes M \to M$ in C.

(c) (**right action morphism**) A morphism $\triangleleft : M \otimes B_2 \to M$ in C.

This data must satisfy the conditions below.

- $(M, \triangleright) \in B_1$-Mod($C$).

- $(M, \triangleleft) \in$ Mod-B_2(C).

- (**middle associativity axiom**) $\triangleleft (\triangleright \otimes \text{id}_{B_2}) = \triangleright (\text{id}_{B_1} \otimes \triangleleft) a_{B_1, M, B_2}$, depicted in Figure 4.10 in the strict case.

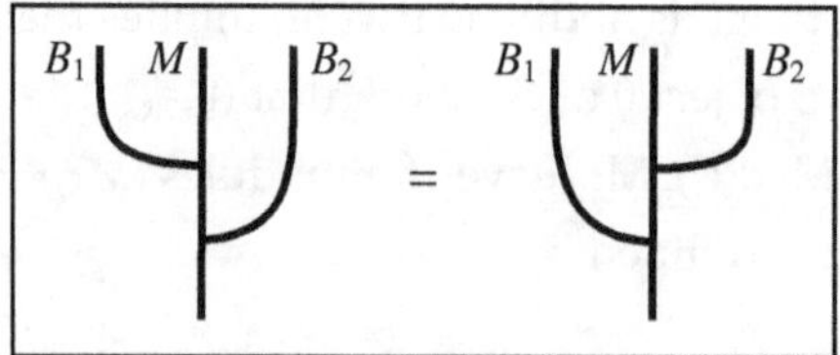

Figure 4.10: Middle associativity constraint for bimodules in C.

We refer to an (A, A)-bimodule in C as an A-**bimodule** in C.

Example 4.32. The following are examples of bimodules in monoidal categories.

(a) The **regular A-bimodule** in $\mathcal{C}$ is $A_{\mathrm{reg}} = (A, \, \rhd := m_A, \, \lhd := m_A)$.

(b) An ideal $(I, \iota_I^A : I \to A, \, \lambda_I^A : A \otimes I \to I, \, \rho_I^A : I \otimes A \to I)$ of A is an A-bimodule in $\mathcal{C}$ with $M := I$ and $\rhd := \lambda_I^A$ and $\lhd := \rho_I^A$.

(c) The **free (B_1, B_2)-bimodule on an object** X in $\mathcal{C}$ is $(B_1 \otimes X) \otimes B_2$ with actions:

$$\rhd := (m_1 \otimes \mathrm{id}_{X \otimes B_2})(a^{-1}_{B_1,B_1,X} \otimes \mathrm{id}_{B_2}) a^{-1}_{B_1, B_1 \otimes X, B_2} : B_1 \otimes ((B_1 \otimes X) \otimes B_2) \to (B_1 \otimes X) \otimes B_2,$$

$$\lhd := (\mathrm{id}_{B_1 \otimes X} \otimes m_2) a_{B_1 \otimes X, B_2, B_2} : ((B_1 \otimes X) \otimes B_2) \otimes B_2 \to (B_1 \otimes X) \otimes B_2.$$

Given (B_1, B_2)-bimodules $(M, \rhd, \lhd)$ and $(M', \rhd', \lhd')$ in $\mathcal{C}$, a morphism $\phi : M \to M'$ in $\mathcal{C}$ is a (B_1, B_2)-**bimodule morphism** if it is in B_1-$\mathrm{Mod}(\mathcal{C})$ and in Mod-$B_2(\mathcal{C})$.

Now we form the **category of** (B_1, B_2)-**bimodules in** $\mathcal{C}$:

- (B_1, B_2)-$\mathrm{Bimod}(\mathcal{C})$: (B_1, B_2)-bimodules in $\mathcal{C}$ and their morphisms in $\mathcal{C}$.

The collection of (B_1, B_2)-bimodule morphisms $(M, \rhd, \lhd) \to (M', \rhd', \lhd')$ is denoted by $\mathrm{Hom}_{(B_1,B_2)\text{-}\mathrm{Bimod}(\mathcal{C})}(M, M')$, which is a subcollection of $\mathrm{Hom}_{\mathcal{C}}(M_{\mathrm{obj}}, M'_{\mathrm{obj}})$.

A (B_1, B_2)-bimodule morphism is **monic** (resp., **epic**, an **iso**) it if is a mono (resp., an epi, an iso) in $\mathcal{C}$.

ii. Substructures and quotient structures

Standing hypothesis. Assume that $\mathcal{C}$ is abelian monoidal as in §3.1.3.

Take a (B_1, B_2)-bimodule $(M, \rhd_M, \lhd_M)$ in $\mathcal{C}$. A (B_1, B_2)-**subbimodule** of $(M, \rhd_M, \lhd_M)$ is a subobject $(N, \iota_N^M : N \to M)$ in $\mathcal{C}$ equipped with morphisms $\rhd_N$ and $\lhd_N$ in $\mathcal{C}$ such that $(N, \rhd_N) \in B_1$-$\mathrm{Mod}(\mathcal{C})$, and $(N, \lhd_N) \in \mathrm{Mod}$-$B_2(\mathcal{C})$, and $\iota_N^M \in (B_1, B_2)$-$\mathrm{Bimod}(\mathcal{C})$.

Given a (B_1, B_2)-subbimodule $(N, \iota_N^M, \rhd_N, \lhd_N)$ of $(M, \rhd_M, \lhd_M)$ in $\mathcal{C}$, we have that a (B_1, B_2)-**quotient bimodule of** M **by** N is an object $M/N := \mathrm{coker}(\iota_N^M)$, equipped with a morphism $\pi_N^M : M \to M/N$ in $\mathcal{C}$, and action morphisms $\rhd_{M/N}$ and $\lhd_{M/N}$ in $\mathcal{C}$, such that $\pi_N^M \in (B_1, B_2)$-$\mathrm{Bimod}(\mathcal{C})$.

iii. On external action

In contrast to Proposition 4.30 for external actions on categories on modules, (B_1, B_2)-$\mathrm{Bimod}(\mathcal{C})$ is not a $\mathcal{C}$-bimodule category (as in §3.3.3) in a canonical way. In fact, we will see later in §4.7.1 that (B_1, B_2)-$\mathrm{Bimod}(\mathcal{C})$ is equivalent to the category $\mathrm{Rex}_{\mathrm{Mod}\text{-}\mathcal{C}}(B_2\text{-}\mathrm{Mod}(\mathcal{C}), B_1\text{-}\mathrm{Mod}(\mathcal{C}))$ of right exact $\mathcal{C}$-module functors between B_2-$\mathrm{Mod}(\mathcal{C})$ and B_1-$\mathrm{Mod}(\mathcal{C})$. So, one natural way of making (B_1, B_2)-$\mathrm{Bimod}(\mathcal{C})$ is a bimodule category is to use Example 3.22, with $B_1 = B_2 =: A$, to get that A-$\mathrm{Bimod}(\mathcal{C})$ is a bimodule category over itself. We will also see in §4.5 that A-$\mathrm{Bimod}(\mathcal{C})$ admits the structure of a monoidal category.

iv. Properties

We now present some properties of A-Bimod(C) that are induced by properties of C and A below (cf. Proposition 4.31). We leave the proof to the reader.

Proposition 4.33. *Take $B_1, B_2 \in$ Alg(C). Then, the following statements hold.*

(a) *(B_1, B_2)-Bimod(C) is preadditive if C is preadditive and $(B_1 \otimes -), (- \otimes B_2) : C \to C$ are additive.*

(b) *(B_1, B_2)-Bimod(C) is linear if C is linear and $(B_1 \otimes -), (- \otimes B_2) : C \to C$ are linear.*

(c) *(B_1, B_2)-Bimod(C) is additive if C is additive and $(B_1 \otimes -), (- \otimes B_2) : C \to C$ are additive.*

(d) *(B_1, B_2)-Bimod(C) is abelian if C is abelian and if $(B_1 \otimes -), (- \otimes B_2) : C \to C$ are additive and exact.* □

Note that if C is assumed to be abelian and rigid in part (d), then the functors $(B_1 \otimes -)$ and $(- \otimes B_2)$ are exact for free by Proposition 3.68. Other features of A-Bimod(C) will be studied in §4.10.1.

§4.4.3. Eilenberg–Moore categories

Next, we turn our attention to modules over the monads introduced in §4.3.2. Again, we work in the strict monoidal case here. To begin, take a monad T on $\mathcal{A}$, that is, an algebra $T := (T, \mu, \eta)$ in the monoidal category $(\mathrm{End}(\mathcal{A}), \circ, \mathrm{Id}_{\mathcal{A}})$.

i. Modules over monads

First, we consider the category of left T-modules in End($\mathcal{A}$), with objects and morphisms defined as follows.

A **left T-module** is a pair (V, λ), where $V : \mathcal{A} \to \mathcal{A}$ is an endofunctor and $\lambda : TV \Rightarrow V$ is an **action natural transformation**, satisfying the commutative diagrams below.

$$
\begin{array}{ccc}
T^2 V & \xrightarrow{\ \mu * V\ } & TV \\
{\scriptstyle T * \lambda}\big\downarrow & & \big\downarrow{\scriptstyle \lambda} \\
TV & \xrightarrow{\ \lambda\ } & V
\end{array}
\qquad\qquad
\begin{array}{ccc}
V & \xrightarrow{\ \eta * V\ } & TV \\
 & {\scriptstyle \mathrm{Id}_V}\searrow & \big\downarrow{\scriptstyle \lambda} \\
 & & V
\end{array}
$$

Here, $\bullet \Rightarrow \bullet \Rightarrow \bullet$ represents the vertical composition of natural transformations.

That is, for all $X \in \mathcal{A}$, the diagrams below commute.

$$
\begin{array}{ccc}
T^2 V(X) & \xrightarrow{\mu_{V(X)}} & TV(X) \\
{\scriptstyle T(\lambda_X)}\big\downarrow & & \big\downarrow{\scriptstyle \lambda_X} \\
TV(X) & \xrightarrow{\lambda_X} & V(X)
\end{array}
\qquad
\begin{array}{ccc}
V(X) & \xrightarrow{\eta_{V(X)}} & TV(X) \\
& {\scriptstyle \mathrm{id}_{V(X)}}\searrow & \big\downarrow{\scriptstyle \lambda_X} \\
& & V(X)
\end{array}
$$

A **T-module morphism** between left T-modules (V, λ) and (V', λ') is a natural transformation $\phi : V \Rightarrow V'$ such that $\phi \circ^{\mathrm{ver}} \lambda = \lambda' \circ^{\mathrm{ver}} (\mathrm{Id}_T \circ^{\mathrm{hor}} \phi)$. That is, for each $X \in \mathcal{A}$, the diagram below commutes.

$$
\begin{array}{ccc}
T(V(X)) & \xrightarrow{\lambda_X} & V(X) \\
{\scriptstyle T(\phi_X)}\big\downarrow & & \big\downarrow{\scriptstyle \phi_X} \\
T(V'(X)) & \xrightarrow{\lambda'_X} & V'(X)
\end{array}
$$

But for applications and for a richer theory, it is very common to use a subcategory of $T\text{-Mod}(\mathrm{End}(\mathcal{A}))$ consisting of endofunctors V of $\mathcal{A}$ that sends each object $X \in \mathcal{A}$ to a fixed object $Y \in \mathcal{A}$. This is studied in the next part.

ii. Eilenberg-Moore (EM-)categories

An **Eilenberg-Moore (EM-)object of** T consists of the following data:

(a) An object Y in $\mathcal{A}$,

(b) An **action morphism** $\xi := \xi_Y : T(Y) \to Y$ in $\mathcal{A}$,

such that the following diagrams commute.

$$
\begin{array}{ccc}
T^2(Y) & \xrightarrow{\mu_Y} & T(Y) \\
{\scriptstyle T(\xi)}\big\downarrow & & \big\downarrow{\scriptstyle \xi} \\
T(Y) & \xrightarrow{\xi} & Y
\end{array}
\qquad
\begin{array}{ccc}
Y & \xrightarrow{\eta_Y} & T(Y) \\
& {\scriptstyle \mathrm{id}_Y}\searrow & \big\downarrow{\scriptstyle \xi} \\
& & Y
\end{array}
$$

EM-objects of T are also known as T-**algebras**.

A **morphism between EM-objects** (Y, ξ) and (Y', ξ') is a morphism $\phi : Y \to Y'$ in $\mathcal{A}$, such that the following diagram commutes.

$$
\begin{array}{ccc}
T(Y) & \xrightarrow{\xi} & Y \\
{\scriptstyle T(\phi)}\big\downarrow & & \big\downarrow{\scriptstyle \phi} \\
T(Y') & \xrightarrow{\xi'} & Y'
\end{array}
$$

Now we can form the **Eilenberg-Moore (EM-)category of a monad** T **on** $\mathcal{A}$:

- $\mathcal{A}^T$: Eilenberg-Moore objects of T on $\mathcal{A}$ and their morphisms.

Continuing Examples 4.24 and 4.25 of monads in §4.3.2, we have the following examples of EM-categories. Verifying these examples is Exercise 4.30.

Example 4.34. Given $\mathrm{Id} \in \mathrm{Monad}(\mathcal{A})$ from Example 4.24, we obtain that $\mathcal{A}^{\mathrm{Id}} \cong \mathcal{A}$.

Example 4.35. Given $(A \otimes -, \mu, \eta) \in \mathrm{Monad}(\mathcal{C})$ from Example 4.25, we obtain that

$$\mathcal{C}^{(A\otimes-)} \cong A\text{-}\mathrm{Mod}(\mathcal{C}).$$

Remark 4.36. With Example 4.35, questions about algebras A and their modules in monoidal categories can be translated to questions about monads $(A \otimes -)$ and their EM-categories. For instance, the study of properties of A-$\mathrm{Mod}(\mathcal{C})$ inherited from both $\mathcal{C}$ and the algebra A in $\mathcal{C}$ (e.g., Proposition 4.31) is a study of the properties of $\mathcal{C}^{(A\otimes-)}$ inherited from both $\mathcal{C}$ and the monad $(A \otimes -)$ on $\mathcal{C}$.

See Exercises 4.31 and 4.32 to explore more examples of EM-categories.

Next, as promised in §4.3.2, we will show that not only do adjunctions produce monads, but we also get that every monad arises from an adjunction. This will be achieved by generalizing the Free-Forget adjunction,

$$(\mathrm{Free} : \mathcal{C} \to A\text{-}\mathrm{Mod}(\mathcal{C})) \dashv (\mathrm{Forg} : A\text{-}\mathrm{Mod}(\mathcal{C}) \to \mathcal{C}),$$

introduced in Exercise 4.25, in view of Example 4.35 above.

Theorem 4.37. *Let (T, μ, η) be a monad on $\mathcal{A}$. Then, the following statements hold.*

(a) *There exists a forgetful functor:*

$$\mathrm{Forg}^T : \mathcal{A}^T \to \mathcal{A}, \quad (Y, \xi) \mapsto Y.$$

(b) *There exists a free functor:*

$$\mathrm{Free}^T : \mathcal{A} \to \mathcal{A}^T, \quad Y \mapsto (T(Y), \mu_Y).$$

Here, $(T(Y), \mu_Y)$ is known as the **free EM-object of T on Y.**

(c) *The functors above form an adjunction: $\mathrm{Free}^T \dashv \mathrm{Forg}^T$.*

(d) *We recover the monad T as the monad $\mathrm{Forg}^T \circ \mathrm{Free}^T$ on $\mathcal{A}$ from Example 4.26.*

Proof. We leave the proof of parts (a), (b), and (d) to the reader as Exercise 4.33.

For part (c), we need to build a unit η^T and counit ε^T of the desired adjunction. First, define the components of η^T to be the morphisms in $\mathcal{A}$ below:

$$\eta_Y^T := \eta_Y : Y \to T(Y) = \mathrm{Forg}^T(\mathrm{Free}^T(Y)),$$

for $Y \in \mathcal{A}$. Next, define the components of ε^T to be the morphisms in $\mathcal{A}^T$ below:

$$\varepsilon_{(Y,\xi)}^T : \mathrm{Free}^T(\mathrm{Forg}^T(Y, \xi)) = (T(Y), \mu_Y) \to (Y, \xi),$$

a morphism in $\mathcal{A}^T$ given by $\xi : T(Y) \to Y$, since $\xi \circ \mu_Y = \xi \circ T(\xi)$. We piece together these components to build the unit and counit of $\mathrm{Free}^T \dashv \mathrm{Forg}^T$:

$$\eta^T : \mathrm{Id}_{\mathcal{A}} \Rightarrow \mathrm{Forg}^T \circ \mathrm{Free}^T, \qquad \varepsilon^T : \mathrm{Free}^T \circ \mathrm{Forg}^T \Rightarrow \mathrm{Id}_{\mathcal{A}^T}.$$

Lastly, we need to show that the adjunction triangle identities hold; we will verify one identity and leave the other to the reader. For $Y \in \mathcal{A}$, we have that

$$[\varepsilon^T_{\mathrm{Free}^T(Y)} \circ \mathrm{Free}^T(\eta^T_Y)](T(Y), \mu_Y) = [\varepsilon^T_{(T(Y), \mu_Y)} \circ \mathrm{Free}^T(Y \to T(Y))](T(Y), \mu_Y)$$

$$= \varepsilon^T_{(T(Y), \mu_Y)}(T^2(Y), \mu_{T(Y)}) = [\mathrm{Id}_{(T(Y), \mu_Y)}](T(Y), \mu_Y).$$

The remainder of the proof of part (c) is part of Exercise 4.33. $\qquad\square$

iii. Beck's Monadicity Theorem

Recall that an adjunction $(F : \mathcal{A} \to \mathcal{B}) \dashv (G : \mathcal{B} \to \mathcal{A})$ yields an adjunction monad GF [Example 4.26]. Further, by Theorem 4.37, we get an adjunction

$$(\mathrm{Free}^{GF} : \mathcal{A} \to \mathcal{A}^{GF}) \dashv (\mathrm{Forg}^{GF} : \mathcal{A}^{GF} \to \mathcal{A}),$$

where $GF = \mathrm{Forg}^{GF} \circ \mathrm{Free}^{GF}$. To contrast these two adjunctions involving $\mathcal{A}$, one uses the **comparison functor**, defined as follows:

$$K^{GF} : \mathcal{B} \to \mathcal{A}^{GF}, \quad Y \mapsto \left(G(Y), G(\varepsilon^{F \dashv G}_Y) : GFG(Y) \to G(Y)\right).$$

We call $F \dashv G$ **monadic** if K^{GF} is an equivalence of categories. For instance, $\left((A \otimes -) : \mathcal{C} \to A\text{-Mod}(\mathcal{C})\right) \dashv \left(\mathrm{Forg} : A\text{-Mod}(\mathcal{C}) \to \mathcal{C}\right)$ is monadic due to Example 4.35.

Precise conditions to get that a comparison functor is an equivalence (or in some references, an isomorphism) of categories is provided by the theorem below. Towards this, we consider certain coequalizers in categories. For some parallel morphisms $f, g : X \to Y$ in a category, $(\mathrm{coeq}(f, g), \ \alpha : Y \to \mathrm{coeq}(f, g))$ is said to be **split** if there exist morphisms $t : Y \to X$ and $s : \mathrm{coeq}(f, g) \to Y$ such that $\alpha s = \mathrm{id}_{\mathrm{coeq}(f,g)}$, and $ft = \mathrm{id}_Y$, and $gt = s\alpha$.

Theorem 4.38 (Beck's Monadicity Theorem). *An adjunction*

$$(F : \mathcal{A} \to \mathcal{B}) \dashv (G : \mathcal{B} \to \mathcal{A})$$

is monadic if and only if the following conditions hold:

(a) *G reflects isos; and*

(b) *Given morphisms $f, g : X \to Y$ in $\mathcal{B}$ such that $G(f), G(g) : G(X) \to G(Y)$ has a split coequalizer in $\mathcal{A}$, there exists a coequalizer $\mathrm{coeq}(f, g)$ in $\mathcal{B}$, and we obtain that $G(\mathrm{coeq}(f, g)) \cong \mathrm{coeq}(G(f), G(g))$ in $\mathcal{A}$.* $\qquad\square$

We refer to Section VI.7 of MacLane [1971] and Section 6.6 of Richter [2020] for the proof and further details.

Chapter 4. Algebras in monoidal categories

iv. Morphisms of monads versus functors between EM-categories

Next, we will see that a morphism of monads yields a functor between EM-categories, similar to how a morphism of $\Bbbk$-algebras yields a functor between categories of modules over $\Bbbk$ (see the restriction functor from §2.3.2 and §1.4.4). Conversely, we will also see that certain functors between EM-categories come from morphisms of monads. These correspondences are contravariant.

Theorem 4.39. *Let (T, μ, η) and (T', μ', η') be monads on $\mathcal{A}$. Also, use F (resp., U) to denote the free (resp., forgetful) functors constructed in Theorem 4.37. Then, the following statements hold.*

(a) *A monad morphism $\phi : T \Rightarrow T'$ in $\mathcal{A}$ yields a functor:*

$$\phi^\# : \mathcal{A}^{T'} \to \mathcal{A}^T, \quad (Y, \xi' : T'(Y) \to Y) \mapsto (Y, \xi' \phi_Y : T(Y) \to Y)$$

Moreover, $U^{T'} = U^T \phi^\#$ as functors from $\mathcal{A}^{T'}$ to $\mathcal{A}$.

(b) *Suppose that $\psi : \mathcal{A}^{T'} \to \mathcal{A}^T$ is a functor, such that the diagram below commutes.*

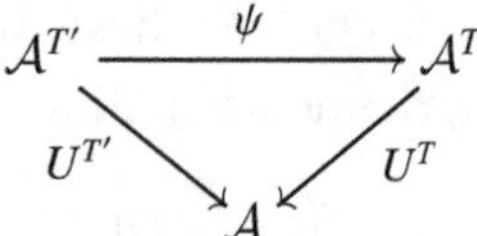

Then, via Theorem 4.37(d), ψ yields a morphism of monads,

$$\psi_\# : T = U^T F^T \xrightarrow{\; U^T F^T \eta^{T'} \;} \begin{matrix} U^T F^T U^{T'} F^{T'} \\ \| \text{ hyp.} \\ U^T F^T U^T \psi F^{T'} \end{matrix} \xrightarrow{\; U^T \varepsilon^T \psi F^{T'} \;} \begin{matrix} U^T \psi F^{T'} \\ \| \text{ hyp.} \\ U^{T'} F^{T'} = T'. \end{matrix}$$

Here, we suppress the notation $*$ for whiskering.

(c) *We obtain that $(\phi^\#)_\# = \phi$ as morphisms of monads on $\mathcal{A}$, and $(\psi_\#)^\# = \psi$ as functors between EM-categories over $\mathcal{A}$.*

Proof. We will provide some details of the proof of parts (a) and (b) below, and leave the rest of the details and part (c) to the reader as Exercise 4.34.

For part (a), note that $(Y, \xi := \xi' \phi_Y) \in \mathcal{A}^T$ via the commutative diagrams below.

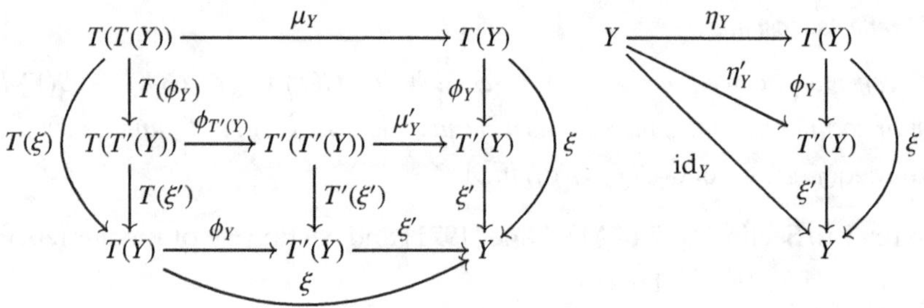

Indeed, for the left diagram: the top region commutes due to ϕ being a monad morphism; the left region commutes due to the naturality of ϕ; and the right region commutes due to $(Y, \xi') \in \mathcal{A}^{T'}$. Moreover, for the right diagram: the top region commutes since ϕ is a monad morphism; and the bottom region commutes since $(Y, \xi') \in \mathcal{A}^{T'}$. We leave the rest of the proof of part (a) to the reader.

For part (b), we have that $\psi_\#$ is a monad morphism, that is, $\psi_\# \circ^{\mathrm{ver}} \eta = \eta'$ and $\psi_\# \circ^{\mathrm{ver}} \mu = \mu' \circ^{\mathrm{ver}} (\psi_\# \circ^{\mathrm{hor}} \psi_\#)$ via the diagrams below.

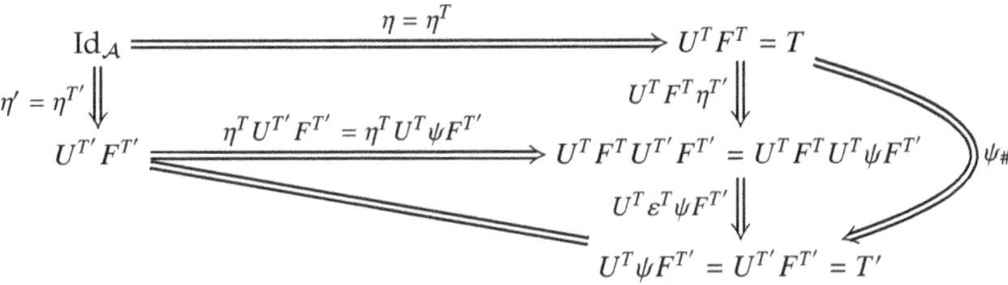

Here, the top region commutes by the naturality of η^T, and the bottom triangle commutes by an adjunction triangle axiom.

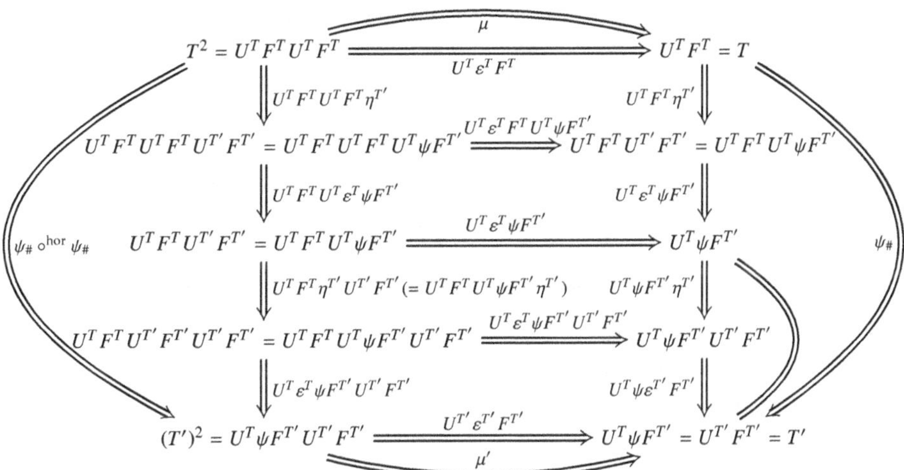

The top two regions commute by level exchange, and the right triangle commutes by an adjunction triangle axiom. The parenthetical substitution can be made by precomposition with the previous transformations. $\qquad\square$

As a direct consequence of the result above, along with Example 4.35 and Exercise 4.25, we obtain the result below.

Corollary 4.40. *Take* $(A, m, u), (A', m', u') \in \mathsf{Alg}(\mathcal{C})$. *Then, the following statements hold.*

(a) *An algebra morphism* $\phi : A \to A'$ *in* $\mathcal{C}$ *yields a functor:*

$$\phi^\# : A'\text{-}\mathsf{Mod}(\mathcal{C}) \to A\text{-}\mathsf{Mod}(\mathcal{C}), \quad (M, \rhd') \mapsto (M, \rhd'(\phi \otimes \mathrm{id}_M)).$$

Moreover, $\mathsf{Forg}' = \mathsf{Forg} \circ \phi^\#$ *as functors from* $A'\text{-}\mathsf{Mod}(\mathcal{C})$ *to* $\mathcal{C}$.

(b) *Suppose that $\psi : A'\text{-Mod}(\mathcal{C}) \to A\text{-Mod}(\mathcal{C})$ is a functor such that $\mathrm{Forg}' = \mathrm{Forg} \circ \psi$. That is, for $(M, \triangleright') \in A'\text{-Mod}(\mathcal{C})$, with $\psi(M, \triangleright'_M) =: (\psi(M), \psi(\triangleright'_M))$ in $A\text{-Mod}(\mathcal{C})$, we obtain that $\psi(M) \cong M$. Then, ψ yields an algebra morphism in C:*

$$\psi_\# : A \xrightarrow{\ \mathrm{id}\,\otimes\, u'\ } A \otimes A' \xrightarrow{\ \psi(\triangleright'_{A'}) = \psi(m')\ } A'.$$

(c) *We obtain that $(\phi^\#)_\# = \phi$ as algebra morphisms in C, and $(\psi_\#)^\# = \psi$ as functors between categories of modules in C.* $\qquad\square$

Example 4.41. Take $A = \mathbb{1}$ be the unit algebra in $\mathcal{C}$. So, $m_A = \ell_\mathbb{1}$ and $u_A = \mathrm{id}_\mathbb{1}$. (Even though $\mathcal{C}$ is strict in this section, we use the morphisms ℓ and r as placeholders here.) For Corollary 4.40(a), take $\phi = u' : \mathbb{1} \to A'$, which is in $\mathsf{Alg}(\mathcal{C})$. Then,

$$(u')^\# : A'\text{-Mod}(\mathcal{C}) \to \mathbb{1}\text{-Mod}(\mathcal{C}) \overset{\mathrm{Ex.4.29}}{\cong} C$$

is the forgetful functor. Indeed, we have that

$$(u')^\#(M, \triangleright') = (M, \ \triangleright'\,(u' \otimes \mathrm{id}_M) : \mathbb{1} \otimes M \to M) \ \in \mathbb{1}\text{-Mod}(\mathcal{C})$$

corresponds to $M \in C$, since $\triangleright'\,(u' \otimes \mathrm{id}_M) = \ell_M$ by the left unit axiom.

For Corollary 4.40(b), let $\psi : A'\text{-Mod}(\mathcal{C}) \to \mathbb{1}\text{-Mod}(\mathcal{C}) \cong C$ be the forgetful functor. Then, $\psi(m') = \ell_{A'}$. Therefore, $\psi_\# = \ell_{A'}\,(\mathrm{id}_\mathbb{1} \otimes u')\,r_\mathbb{1}^{-1} = u'\,\ell_\mathbb{1}\,r_\mathbb{1}^{-1} = u'$; see Example 4.29.

Modifying the example above for group algebras over a field is Exercise 4.35.

The discussion in this section, along with §4.3.2, is only the beginning of the theory of monads and their modules, especially their Eilenberg-Moore categories. We highly recommend the references presented in §4.13, along with the textbook by Böhm [2018] and Chapter VI of MacLane [1971], for further exploration.

§4.5. Algebraic operations in monoidal categories

Analogous to the material in Section 1.4, we perform various operations on algebras and (bi)modules in monoidal categories to create new ones from old ones. In some cases, algebras or (bi)modules form monoidal categories under such operations. For ease, we impose the following hypothesis.

Standing hypothesis. Assume that $\mathcal{C}$ is abelian monoidal as in §3.1.3.

---

§4.5.1. Operations on algebras in monoidal categories

Fix algebras (A_1, m_1, u_1), (A_2, m_2, u_2), (A, m_A, u_A), and (B, m_B, u_B) in C.

i. Biproduct of algebras in $\mathcal{C}$

Recall §§2.2.1ii, 2.2.2ii on how to form (bi)products of objects. Here, we define a biproduct of algebras in $\mathcal{C}$,

$$(A_1, m_1, u_1) \,\square\, (A_2, m_2, u_2) := (A_1 \,\square\, A_2,\ m_\square,\ u_\square) \ \in \mathsf{Alg}(\mathcal{C}),$$

which will generalize Proposition 1.15 in §1.4.1 for $\mathcal{C} = \mathsf{Vec}$.

To proceed, the multiplication morphism $m_\square$ above is defined by the composition of morphisms below.

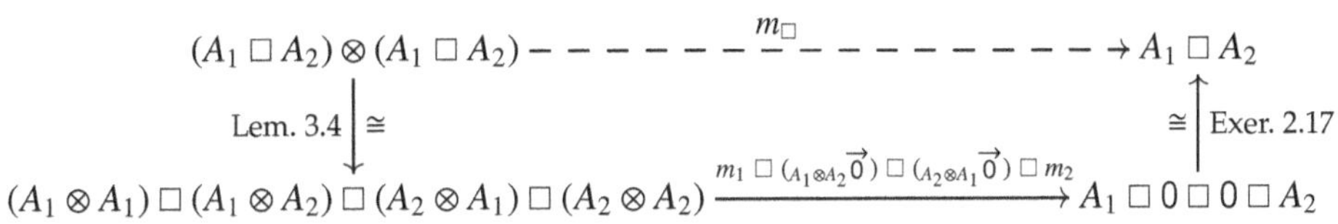

$$(A_1 \,\square\, A_2) \otimes (A_1 \,\square\, A_2) \xdashrightarrow{\quad\quad\quad m_\square \quad\quad\quad} A_1 \,\square\, A_2$$
$$\text{Lem. 3.4} \downarrow \cong \qquad\qquad\qquad\qquad \cong \uparrow \text{Exer. 2.17}$$
$$(A_1 \otimes A_1) \,\square\, (A_1 \otimes A_2) \,\square\, (A_2 \otimes A_1) \,\square\, (A_2 \otimes A_2) \xrightarrow{m_1 \,\square\, (_{A_1 \otimes A_2}\vec{0}) \,\square\, (_{A_2 \otimes A_1}\vec{0}) \,\square\, m_2} A_1 \,\square\, 0 \,\square\, 0 \,\square\, A_2$$

Moreover, the unit morphism $u_\square$ is defined by the universal property of products as shown below.

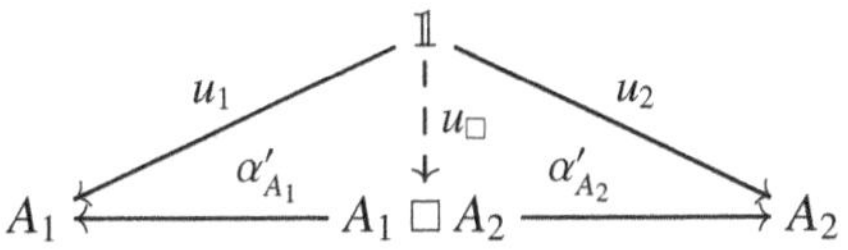

$$
\begin{array}{ccccc}
 & & \mathbb{1} & & \\
 & {}^{u_1}\!\!\nearrow & \downarrow{\scriptstyle u_\square} & \nwarrow^{u_2} & \\
A_1 & \xleftarrow{\ \alpha'_{A_1}\ } & A_1 \,\square\, A_2 & \xrightarrow{\ \alpha'_{A_2}\ } & A_2
\end{array}
$$

With this, we achieve that $(A_1 \,\square\, A_2,\ m_\square,\ u_\square) \in \mathsf{Alg}(\mathcal{C})$, and we leave the details to the reader as Exercise 4.36. Note that this can be extended to define the biproduct of finitely many algebras in $\mathcal{C}$.

We say that an algebra A in $\mathcal{C}$ is **indecomposable** if it is not isomorphic to a biproduct of nonzero algebras in $\mathcal{C}$ (see Example 4.1(b)).

Now with the biproduct, we can endow the category $\mathsf{Alg}(\mathcal{C})$ of algebras in $\mathcal{C}$ with a monoidal structure as follows.

- $\mathsf{Alg}(\mathcal{C})_\square$: objects are algebras in $\mathcal{C}$, with monoidal product $\square$, and with unit object being the zero algebra in $\mathcal{C}$ from Example 4.1(b).

ii. On products and Homs of algebras in $\mathcal{C}$

We discuss generalizations of the tensor product, free products, and Homs of $\Bbbk$-algebras in §§1.4.2v,vi and §1.4.3iii to the monoidal setting.

To define the monoidal product of algebras, $A \otimes B$, in $\mathcal{C}$, one requires an iso $c : A \otimes B \xrightarrow{\sim} B \otimes A$ of objects in $\mathcal{C}$ that is analogous to the flip map in Proposition 1.19. Such a map c can be obtained via a *braiding*, which is discussed in a future volume. On the other hand, one can define a free product of algebras, $A \circledast B$, in $\mathcal{C}$ exactly as in §1.4.2vi.

Lastly, as discussed in §1.4.3iii, there are issues with putting an algebra structure on the collection of morphisms, $\mathrm{Hom}_{\mathcal{C}}(A, B)$, even if $\mathcal{C}$ is enriched over itself. This can be remedied if B is a *coalgebra* in $\mathcal{C}$, which is discussed in a future volume.

§4.5.2. Operations on (bi)modules in monoidal categories

Fix $(A, m_A, u_A), (B_1, m_1, u_1), (B_2, m_2, u_2) \in \mathrm{Alg}(\mathcal{C})$. Next, we discuss operations to form new (bi)modules over algebras in monoidal categories from old ones.

i. Biproduct of (bi)modules in $\mathcal{C}$

Take two left A-modules $(M_1, \triangleright_1)$ and $(M_2, \triangleright_2)$ in $\mathcal{C}$. Then, the biproduct of underlying objects $M_1 \,\square\, M_2$ admits the structure of a left A-module in $\mathcal{C}$, which generalizes Proposition 1.14 in §1.4.1 (when $\mathcal{C} = \mathrm{Vec}$). The action morphism is defined below:

$$\triangleright : A \otimes (M_1 \,\square\, M_2) \xrightarrow{\;\cong\; \mathrm{Lem.3.4}\;} (A \otimes M_1) \,\square\, (A \otimes M_2) \xrightarrow{\;\triangleright_1 \,\square\, \triangleright_2\;} M_1 \,\square\, M_2.$$

We leave it to the reader to verify that $(M_1 \,\square\, M_2, \triangleright)$ is indeed in $A\text{-Mod}(\mathcal{C})$. This can also be extended to form the biproduct of finitely many left A-modules in $\mathcal{C}$.

We say that a left A-module in $\mathcal{C}$ is **indecomposable** if it is not isomorphic to a biproduct of nonzero left A-modules in $\mathcal{C}$.

Now, we can endow the category $A\text{-Mod}(\mathcal{C})$ with a monoidal structure as follows.

- $A\text{-Mod}(\mathcal{C})_{\square}$: objects are left A-modules in $\mathcal{C}$, with monoidal product $\square$, and with unit object being the zero A-module in $\mathcal{C}$ given in Examples 4.10(b) and 4.28(b).

Moreover, the discussion above can be translated to forming a biproduct of finitely many right A-modules in $\mathcal{C}$, and further, a biproduct of finitely many (B_1, B_2)-bimodules in $\mathcal{C}$. We can also form the monoidal categories below.

- $\mathrm{Mod}\text{-}A(\mathcal{C})_{\square}$
- $(B_1, B_2)\text{-Bimod}(\mathcal{C})_{\square}$

ii. Monoidal product of (bi)modules in $\mathcal{C}$

Towards generalizing §1.4.2i (where $\mathcal{C} = \mathrm{Vec}$), take $(M, \triangleright_M) \in B_1\text{-Mod}(\mathcal{C})$ and $(N, \triangleleft_N) \in \mathrm{Mod}\text{-}B_2(\mathcal{C})$. Then, the object $M \otimes N$ is a (B_1, B_2)-bimodule in $\mathcal{C}$ with:

$$\triangleright := (\triangleright_M \otimes \mathrm{id}_N)\, a^{-1}_{B_1, M, N} : B_1 \otimes (M \otimes N) \to M \otimes N,$$

$$\triangleleft := (\mathrm{id}_M \otimes \triangleleft_N)\, a_{M, N, B_2} : (M \otimes N) \otimes B_2 \to M \otimes N.$$

We leave it to the reader to verify that indeed $(M \otimes N, \triangleright, \triangleleft) \in (B_1, B_2)\text{-Bimod}(\mathcal{C})$.

iii. Monoidal product $\otimes_A$ of (bi)modules in C

To generalize §1.4.2iii (where $C = \mathsf{Vec}$), take $(M, \triangleleft) \in \mathsf{Mod}\text{-}A(C)$, $(N, \triangleright) \in A\text{-}\mathsf{Mod}(C)$. Then, we can form the object $M \otimes_A N$ in C as a coequalizer as follows (see §2.2.1iv):

$$(M \otimes A) \otimes N \underset{g := (\mathrm{id} \otimes \triangleright_N)\, a_{M,A,N}}{\overset{f := \triangleleft_M \otimes \mathrm{id}}{\rightrightarrows}} M \otimes N \xrightarrow{\;\;\alpha\;\;} \mathrm{coeq}(f,g) =: M \otimes_A N.$$

We refer to $M \otimes_A N$ as the **tensor product of M and N over A in C.**

Proposition 4.42. *Suppose that $(B_1 \otimes -)$ and $(- \otimes B_2)$ are right exact endofunctors of C. If $(M, \triangleright_M, \triangleleft_M)$ is a (B_1, A)-bimodule in C and $(N, \triangleright_N, \triangleleft_N)$ is a (A, B_2)-bimodule in C, then $M \otimes_A N$ is a (B_1, B_2)-bimodule in C.*

Proof. We will sketch why $M \otimes_A N$ is a left B_1-module in C, and leave the proof that $M \otimes_A N$ is a right B_2-module in C, and further, a (B_1, B_2)-bimodule in C, to the reader as Exercise 4.37. With the notation above, note that:

$$B_1 \otimes (M \otimes_A N) \;=\; B_1 \otimes \mathrm{coeq}(f,g) \;\cong\; \mathrm{coeq}(\mathrm{id}_{B_1} \otimes f,\ \mathrm{id}_{B_1} \otimes g).$$

This holds by Proposition 2.49(a) and Remark 4.14(d)$\Rightarrow$(e), and since coequalizers are cokernels [Exercise 2.13(d)]. Now the left B_1-action on $M \otimes_A N$ in C,

$$\blacktriangleright : B_1 \otimes (M \otimes_A N) \to M \otimes_A N,$$

is derived via both the iso above and the universal property of coequalizers, as depicted in the commutative diagram below.

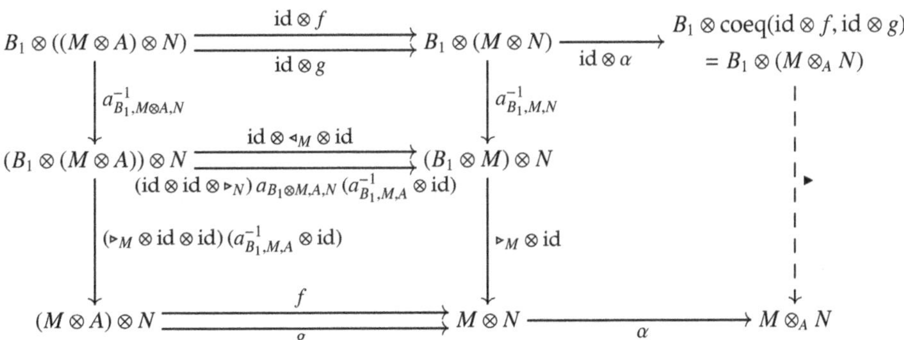

We defer the argument for the associativity and unitality of $\blacktriangleright$ to Exercise 4.37. $\square$

Example 4.43. Take right exact endofunctors $(B_1 \otimes -)$, $(- \otimes B_2)$ of C, and take $A = \mathbb{1}$ as in Example 4.1(a). Then, for $M \in (B_1, \mathbb{1})\text{-}\mathsf{Bimod}(C)$ and $N \in (\mathbb{1}, B_2)\text{-}\mathsf{Bimod}(C)$:

$$M \otimes_{\mathbb{1}} N \;\cong\; M \otimes N$$

as (B_1, B_2)-bimodules in C. Here, $M \otimes N \in (B_1, B_2)\text{-}\mathsf{Bimod}(C)$ as in §4.5.2ii. Verifying this is Exercise 4.38.

We also have the following identities for A-bimodules in C.

Lemma 4.44. *For A-bimodules M, N, P in C, we have the isos below in A-Bimod(C).*

(a) $(M \otimes_A N) \otimes_A P \cong M \otimes_A (N \otimes_A P)$.

(b) $M \otimes_A A_{\text{reg}} \cong M \cong A_{\text{reg}} \otimes_A M$.

Here, A_{reg} is the regular A-bimodule A with actions m_A from Example 4.32(a).

Proof. We will verify that $A \otimes_A M \cong M$ for part (b), and leave the rest of the proof as Exercise 4.39. To proceed, consider the morphism below:

$$\phi : M \xrightarrow{\ell_M^{-1}} \mathbb{1} \otimes M \xrightarrow{u_A \otimes \text{id}} A \otimes M \xrightarrow{\alpha} A \otimes_A M.$$

On the other hand, we can build a morphism $\psi : A \otimes_A M \to M$ from the universal property of coequalizers. Consider the diagram below

$$(A \otimes A) \otimes M \underset{(\text{id} \otimes \triangleright_M)\, a_{A,A,M}}{\overset{\triangleleft_A \otimes \text{id}}{\rightrightarrows}} A \otimes M \xrightarrow{\alpha} A \otimes_A M$$

Indeed, $\triangleleft_A = m_A$ and $\triangleright_M (m_A \otimes \text{id}_M) = \triangleright_M (\text{id}_A \otimes \triangleright_M) a_{A,A,M}$ by the left module associativity axiom. Next, we have that $\psi \phi = \triangleright_M (u_A \otimes \text{id}_M) \ell_M^{-1} = \text{id}_M$ by definition and by the left module unitality axiom. We also have that:

$$\begin{aligned}
\phi \psi \alpha \; &= \; \alpha (u_A \otimes \text{id}_M) \, \ell_M^{-1} \, \triangleright_M \\
&= \; \alpha (u_A \otimes \text{id}_M) (\text{id}_{\mathbb{1}} \otimes \triangleright_M) \, \ell_{A \otimes M}^{-1} \\
&= \; \alpha (\text{id}_A \otimes \triangleright_M) (u_A \otimes \text{id}_A \otimes \text{id}_M) \, \ell_{A \otimes M}^{-1} \\
&= \; \alpha (m_A \otimes \text{id}_M) \, a_{A,A,M}^{-1} (u_A \otimes \text{id}_A \otimes \text{id}_M) \, \ell_{A \otimes M}^{-1} \\
&= \; \alpha (m_A \otimes \text{id}_M) (u_A \otimes \text{id}_A \otimes \text{id}_M) \, a_{\mathbb{1},A,M}^{-1} \, \ell_{A \otimes M}^{-1} \; = \; \alpha.
\end{aligned}$$

These equalities hold, respectively, by definition, by the naturality of ℓ, by level exchange, by the coequalizer condition, by the naturality of a, and by the unitality of A with Exercise 3.1(a). Since α is epic [Exercise 2.13(a)], it is right-cancellative. Hence, $\phi \psi = \text{id}_{A \otimes_A M}$, and we have established that $A \otimes_A M \cong M$ in C. $\qquad \square$

So, with the result above, we can make A-Bimod(C) a monoidal category:

- A-Bimod(C): objects are A-bimodules in C, with monoidal product $\otimes_A$, and with unit object being the regular A-bimodule A_{reg} in C from Example 4.32(a).

iv. Homs and duals of (bi)modules in C

Remark 4.45. Towards generalizing the material in §1.4.3i, one would like $\text{Hom}_{A\text{-Mod}(C)}(M, N) \in A\text{-Mod}(C)$, for each pair of left A-modules M, N in C. However,

$\mathrm{Hom}_{A\text{-}\mathrm{Mod}(\mathcal{C})}(M, N)$ is not necessarily an object in $A\text{-}\mathrm{Mod}(\mathcal{C})$ (nor an object in $\mathcal{C}$). Instead, one may want to use enrichment theory (see §3.11) for such an operation. See Exercise 4.40 for an open-ended exploration.

On the other hand, duality (in the sense of §3.6.1) is a (bi)module operation. Namely, when the underlying object of a left (resp., right) A-module in $\mathcal{C}$ is rigid, then its dual is a right (resp., left) A-module. We see this in the following result; the proof is reserved as Exercise 4.41.

Proposition 4.46. *Take* $(M, \triangleright) \in A\text{-}\mathrm{Mod}(\mathcal{C})$ *and* $(N, \triangleleft) \in \mathrm{Mod}\text{-}A(\mathcal{C})$. *Suppose that* M *(resp., N) is a left (resp., right) rigid object in $\mathcal{C}$. Then, $M^* \in \mathrm{Mod}\text{-}A(\mathcal{C})$ and $^*N \in A\text{-}\mathrm{Mod}(\mathcal{C})$ with actions pictured in Figure 4.11 in the strict case.* $\qquad\square$

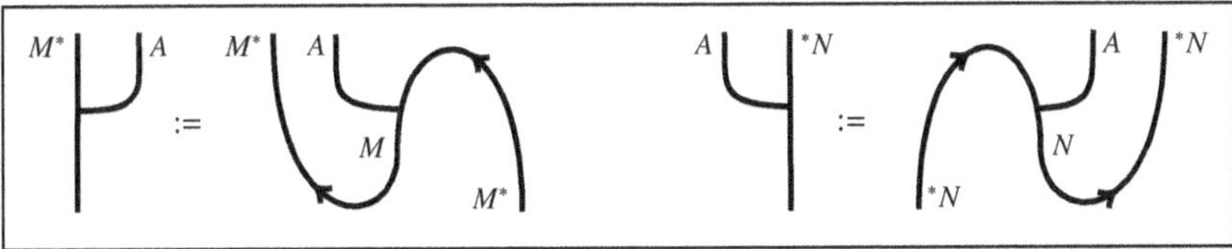

Figure 4.11: Duals of modules in monoidal categories.

§4.6. Graded algebras in the monoidal setting

The goal of this section is to generalize the graded algebras from §1.2.7i and their modules to the monoidal setting. We proceed in the strict case for ease via the Strictification Theorem [Theorem 3.26]. A version of this material appeared in the work of Liu Lopez and Walton [2023]. Here, we study gradings over the additive monoid of natural numbers $\mathbb{N}$, but this discussion can be generalized to gradings over any monoid N or group G.

§4.6.1. Graded objects

The **category of $\mathbb{N}$-graded objects in** $\mathcal{C}$ is the product category $\mathcal{C}^{\times \mathbb{N}}$ (see §2.1.3):

(a) Objects are $\mathbb{N}$-tuples $X := (X_i)_{i \in \mathbb{N}}$ of objects in $\mathcal{C}$, and

(b) Morphisms ϕ from X to Y are $\mathbb{N}$-tuples $(\phi : X_i \to Y_i)_{i \in \mathbb{N}}$ of morphisms in $\mathcal{C}$.

Note that for the monoidal category $(\underline{\mathbb{N}}, +, 0)$ from §3.1.2i, we have that

$$\mathcal{C}^{\times \mathbb{N}} \cong \mathrm{Fun}(\underline{\mathbb{N}}, \mathcal{C})$$

as categories. Verifying this, and the result below, is Exercise 4.42.

Proposition 4.47. *Several features of C are inherited by $C^{\times \mathbb{N}}$ as follows.*

(a) *If C is additive (or is abelian), then so is $C^{\times \mathbb{N}}$.*

(b) *If C is additive with biproduct $\square$ and zero object 0, then $C^{\times \mathbb{N}}$ is monoidal where:*

$$X \bar{\otimes} Y := \left(\square_{p=0}^{i} X_p \otimes Y_{i-p} \right)_{i \in \mathbb{N}}, \quad \bar{\mathbb{1}} := (\mathbb{1}_i)_{i \in \mathbb{N}}, \quad with \ \mathbb{1}_0 := \mathbb{1}^{C}, \ \mathbb{1}_{i \neq 0} := 0. \quad \square$$

The monoidal structure on $C^{\times \mathbb{N}}$ above is called the **Cauchy monoidal structure**. But as mentioned in Definition 2.1 of Aguiar and Mahajan [2010], there are other choices for a monoidal structure on $C^{\times \mathbb{N}}$ as given below.

- **Component-wise** (or **Hadamard**) **monoidal structure:**

$$X \otimes^c Y := (X_i \otimes Y_i)_{i \in \mathbb{N}} \qquad \text{and} \qquad \mathbb{1}^c := (\mathbb{1}^{C})_{i \in \mathbb{N}}.$$

- **Substitution monoidal structure** (for C additive, having infinite biproducts $\square$):

$$X \otimes^s Y := \left(\square_{p \geq 0} X_p \otimes \left(\square_{q_1 + \cdots + q_p = i} Y_{q_1} \otimes \cdots \otimes Y_{q_p} \right) \right)_{i \in \mathbb{N}}$$

$$\mathbb{1}^s := (\mathbb{1}_i)_{i \in \mathbb{N}}, \quad with \ \mathbb{1}_1 := \mathbb{1}^{C}, \ \mathbb{1}_{i \neq 1} := 0.$$

We will stick with the Cauchy monoidal structure on $C^{\times \mathbb{N}}$, but we encourage the reader to explore the material here via the other monoidal structures above.

For the rest of the section, we impose the following conditions.

Standing hypothesis. Assume that C is abelian monoidal as in §3.1.3.

---

§4.6.2. Graded algebras

Recall Proposition 4.47. The algebras of interest here are in the category below:

$$\mathsf{Alg}(C^{\times \mathbb{N}}) := \mathsf{Alg}(C^{\times \mathbb{N}}, \bar{\otimes}, \bar{\mathbb{1}}).$$

But for computations, it is easier to consider the following category.

Let $\mathbb{N}\text{-}\mathsf{GrAlg}(C)$ be the category whose objects consist of the data:

(a) a collection of objects $\{A_i\}_{i \in \mathbb{N}}$ in C,

(b) a collection of **multiplication morphisms** $\{m_{i,j} : A_i \otimes A_j \to A_{i+j}\}_{i,j \in \mathbb{N}}$ in C,

(c) a **unit morphism** $u_0 : \mathbb{1}^{C} \to A_0$ in C,

satisfying the associativity and unitality constraints,

- $m_{i+j,k} \, (m_{i,j} \otimes \mathrm{id}_{A_k}) \; = \; m_{i,j+k} \, (\mathrm{id}_{A_i} \otimes m_{j,k})$, for all $i, j, k \in \mathbb{N}$,
- $m_{0,i} \, (u_0 \otimes \mathrm{id}_{A_i}) \; = \; \mathrm{id}_{A_i} \; = \; m_{i,0} \, (\mathrm{id}_{A_i} \otimes u_0)$, for all $i \in \mathbb{N}$.

Morphisms $(\{A_i\}_i, \{m_{i,j}\}_{i,j}, u_0) \to (\{A_i'\}_i, \{m_{i,j}'\}_{i,j}, u_0')$ are defined to be tuples of morphisms in $\mathcal{C}$, namely $(\phi_i : A_i \to A_i')_{i\in\mathbb{N}}$, where $m_{i,j}'(\phi_i \otimes \phi_j) = \phi_{i+j} m_{i,j}$ for all $i, j \in \mathbb{N}$, and where $\phi_0 u_0 = u_0'$.

Proposition 4.48. *We have an isomorphism of categories:*

$$\mathsf{Alg}(\mathcal{C}^{\times \mathbb{N}}) \cong \mathbb{N}\text{-}\mathsf{GrAlg}(\mathcal{C}).$$

With this, we will interchange objects and morphisms $\mathsf{Alg}(\mathcal{C}^{\times \mathbb{N}})$ and $\mathbb{N}\text{-}\mathsf{GrAlg}(\mathcal{C})$ without mention. Moreover, we refer to these categories simultaneously as the **category of $\mathbb{N}$-graded algebras in $\mathcal{C}$**.

Proof of Proposition 4.48. We will sketch the proof and leave the details as Exercise 4.43. First, suppose that (A, m, u) is an object of $\mathsf{Alg}(\mathcal{C}^{\times \mathbb{N}})$. After taking degrees, the data of the multiplication map $m : A \bar{\otimes} A \to A$ amounts to a collection of morphisms: $\{m_k : \square_{p=0}^{k} A_p \otimes A_{k-p} \to A_k\}_{k\in\mathbb{N}}$ in $\mathcal{C}$. Also, consider the canonical morphisms from the biproduct construction: $\{\alpha_{l,q} : A_l \otimes A_{q-l} \to \square_{p=0}^{q} A_p \otimes A_{q-p}\}_{q\in\mathbb{N},\, l=0,\dots,q}$. Now the collection of multiplication morphisms in $\mathbb{N}\text{-}\mathsf{GrAlg}(\mathcal{C})$ is given as follows:

$$\{m_{i,j} := m_{i+j}\, \alpha_{i,i+j} : A_i \otimes A_j \to A_{i+j}\}_{i,j\in\mathbb{N}}.$$

Moreover, the unit morphism is defined by the degree 0 part of $u : \bar{\mathbb{1}} \to A$, namely by $u_0 : \mathbb{1}^{\mathcal{C}} \to A_0$.

Conversely, given the data $\{\{A_i\}_{i\in\mathbb{N}}, \{m_{i,j}\}_{i,j\in\mathbb{N}}, u_0\}$, take A to be $(A_i)_{i\in\mathbb{N}}$. Also, let $m : A \bar{\otimes} A \to A$ be the tuple of morphisms $(m_i : \square_{p=0}^{i} A_p \otimes A_{i-p} \to A_i)_{i\in\mathbb{N}}$, where m_i is the biproduct $\square_{p=0}^{i} m_{p,i-p}$ of certain given multiplication morphisms. Lastly, let $u : \bar{\mathbb{1}} \to A$ be the tuple of morphisms $(u_i : \mathbb{1}_i \to A_i)_{i\in\mathbb{N}}$, where u_0 is the given morphism $\mathbb{1}^{\mathcal{C}} \to A_0$ above, and $u_i := \vec{0}_{\mathbb{1}_i, A_i}$ for $i \neq 0$.

These correspondences yield a bijection between objects of $\mathsf{Alg}(\mathcal{C}^{\times \mathbb{N}})$ and of $\mathbb{N}\text{-}\mathsf{GrAlg}(\mathcal{C})$, which also extends to morphisms. $\square$

Example 4.49. We can generalize the tensor algebras over $\Bbbk$ from §1.2.2 as follows. Take an object $X \in \mathcal{C}$, and consider $X^{\otimes 0} := \mathbb{1}^{\mathcal{C}}$. Then, the $\mathcal{C}$**-tensor algebra** $T(X)$ **on** X is the $\mathbb{N}$-graded algebra in $\mathcal{C}$ defined as $T(X) = (X^{\otimes i})_{i\in\mathbb{N}}$ with

$$\{m_{i,j}^{T(X)} := \mathrm{id}_{X^{\otimes(i+j)}} : X^{\otimes i} \otimes X^{\otimes j} \to X^{\otimes(i+j)}\}_{i,j\in\mathbb{N}}, \qquad u_0^{T(X)} := \mathrm{id}_{\mathbb{1}} : \mathbb{1}^{\mathcal{C}} \to X^{\otimes 0}.$$

See Exercise 4.44 for practice. One can also generalize the path algebras over $\Bbbk$ from §1.2.5 (especially via Exercise 1.21). See the open-ended Exercise 4.45.

Remark 4.50. One may want to form $\mathcal{C}$-*symmetric algebras* $S(X)$, or form $\mathcal{C}$-*exterior algebras* $\Lambda(X)$, generalizing the versions of such $\Bbbk$-algebras in §§1.2.3, 1.2.4. But one needs suitable isos $X \otimes Y \xrightarrow{\sim} Y \otimes X$, for $X, Y \in \mathcal{C}$ to do so; this is achieved with *braidings* studied in a future volume.

Remark 4.51. The notions above for graded algebras can be generalized to examine filtered algebras in monoidal categories, achieved by working in $\mathsf{Fun}(\mathbb{N}_{\leq}, \mathcal{C})$ in place of $\mathsf{Fun}(\mathbb{N}, \mathcal{C})$, for the monoidal category $\mathbb{N}_{\leq}$ given in §3.1.2i. One can form associated graded algebras in monoidal categories as well. See, e.g., the work of Walton and Yadav [2023] for details, and compare to §§1.2.7ii,iii.

§4.6.3. Graded modules

The modules of interest here are objects in the category below:

$$A\text{-}\mathsf{Mod}(\mathcal{C}^{\times \mathbb{N}}), \quad \text{for } A \in \mathsf{Alg}(\mathcal{C}^{\times \mathbb{N}}).$$

But for computations, it is easier to consider instead the following category.

Take $A := (\{A_i\}_i,\ \{m_{i,j}\}_{i,j},\ u_0) \in \mathbb{N}\text{-}\mathsf{GrAlg}(\mathcal{C})$, corresponding to $A \in \mathsf{Alg}(\mathcal{C}^{\times \mathbb{N}})$. Let $\mathbb{N}\text{-}\mathsf{GrMod}_A(\mathcal{C})$ be the category whose objects consist of the data:

(a) a collection of objects $\{M_j\}_{j \in \mathbb{N}}$ in $\mathcal{C}$,

(b) a collection of **(left) action morphisms** $\{\triangleright_{i,j} : A_i \otimes M_j \to M_{i+j}\}_{i,j \in \mathbb{N}}$ in $\mathcal{C}$,

satisfying the associativity and unitality constraints, respectively,

- $\triangleright_{i+j,k}(m_{i,j} \otimes \mathrm{id}_{A_k}) = \triangleright_{i,j+k}(\mathrm{id}_{A_i} \otimes \triangleright_{j,k})$, for all $i, j, k \in \mathbb{N}$,

- $\triangleright_{0,i}(u_0 \otimes \mathrm{id}_{M_i}) = \mathrm{id}_{M_i}$, for all $i \in \mathbb{N}$.

Morphisms $(\{M_j\}_j, \{\triangleright_{i,j}\}_{i,j}) \to (\{M'_j\}_j, \{\triangleright'_{i,j}\}_{i,j})$ are tuples $(\phi_j : M_j \to M'_j)_{j \in \mathbb{N}}$ of morphisms in $\mathcal{C}$, where $\triangleright'_{i,j}(\mathrm{id}_{A_i} \otimes \phi_j) = \phi_{i+j}\triangleright_{i,j}$ for all $i, j \in \mathbb{N}$.

Now the following result follows similarly to the proof of Proposition 4.48.

Proposition 4.52. *Given an $\mathbb{N}$-graded algebra A in $\mathcal{C}$, we have a category isomorphism:*

$$A\text{-}\mathsf{Mod}(\mathcal{C}^{\times \mathbb{N}}) \cong \mathbb{N}\text{-}\mathsf{GrAlg}_A(\mathcal{C}).$$

$\square$

With this, we interchange objects and morphisms $A\text{-}\mathsf{Mod}(\mathcal{C}^{\times \mathbb{N}})$ and $\mathbb{N}\text{-}\mathsf{GrAlg}_A(\mathcal{C})$, and call these the **category of $\mathbb{N}$-graded left A-modules in $\mathcal{C}$.**

Likewise, we can define categories of $\mathbb{N}$-graded right modules and of $\mathbb{N}$-graded bimodules over A in $\mathcal{C}$ to yield:

$$\mathsf{Mod}\text{-}A(\mathcal{C}^{\times \mathbb{N}}) \cong \mathbb{N}\text{-}\mathsf{GrAlg}(\mathcal{C})_A \quad \text{and} \quad (B_1, B_2)\text{-}\mathsf{Mod}(\mathcal{C}^{\times \mathbb{N}}) \cong \mathbb{N}\text{-}\mathsf{GrAlg}_{B_1}(\mathcal{C})_{B_2}.$$

We leave it to the reader to explore this, and we will revisit these categories briefly later when discussing graded Morita equivalence in monoidal categories in §4.7.3.

§4.7. Morita equivalence in the monoidal setting

Now we extend the notion of Morita equivalence of $\Bbbk$-algebras introduced §2.4.4 to the monoidal setting for algebras in $\mathcal{C}$. But first, we generalize the Eilenberg-Watts Theorem, namely parts (a')-(c') of Theorem 2.51 for $\mathcal{C} = \mathsf{FdVec}$, to a more general monoidal setting.

Standing hypothesis. Assume that $\mathcal{C}$ is abelian monoidal as in §3.1.3, and that $(X \otimes -)$ and $(- \otimes X)$ are right exact, for each object $X \in \mathcal{C}$ (see §2.8.1). See Remark 4.14, along with Figure 3.15, about the latter condition.

Consider the following lemma.

Lemma 4.53. *Let A and B be algebras in $\mathcal{C}$. Then, the functor*

$$Q \otimes_A - : A\text{-}\mathsf{Mod}(\mathcal{C}) \to B\text{-}\mathsf{Mod}(\mathcal{C})$$

is right exact, for any $Q \in (B, A)\text{-}\mathsf{Bimod}(\mathcal{C})$.

Proof. We will sketch the proof, and leave the details to the reader as part of Exercise 4.46. By Proposition 2.49(a), it suffices to show that $Q \otimes_A -$ preserves cokernels. In particular, the cokernel of any morphism $\phi : M \to N$ in $A\text{-}\mathsf{Mod}(\mathcal{C})$,

$$(\mathrm{coker}(\phi),\ \alpha : N \to \mathrm{coker}(\phi)),$$

exists in $A\text{-}\mathsf{Mod}(\mathcal{C})$ by Proposition 4.31(d). We want that as left B-modules in $\mathcal{C}$:

$$Q \otimes_A \mathrm{coker}(\phi) \cong \mathrm{coker}(\mathrm{id}_Q \otimes_A \phi).$$

Now $\alpha \phi = \vec{0}_{M,\mathrm{coker}(\phi)}$, so $(\mathrm{id}_Q \otimes_A \alpha)(\mathrm{id}_Q \otimes_A \phi) = \vec{0}_{Q \otimes_A M, Q \otimes_A \mathrm{coker}(\phi)}$. Hence, we get a unique morphism

$$\gamma_1 : \mathrm{coker}(\mathrm{id}_Q \otimes_A \phi) \to Q \otimes_A \mathrm{coker}(\phi)$$

by the universal property of cokernels. It can be shown that γ_1 is a morphism in $B\text{-}\mathsf{Mod}(\mathcal{C})$ using the assumption that $(B \otimes -)$ is right exact (and thus preserves cokernels by Proposition 2.49(a)). On the other hand, we can use the standing hypothesis and Proposition 2.49(a) to get that $(Q \otimes -)$ preserves cokernels. This, in turn, with the universal property of coequalizers, yields a unique morphism

$$\gamma_2 : Q \otimes_A \mathrm{coker}(\phi) \to \mathrm{coker}(\mathrm{id}_Q \otimes_A \phi).$$

The morphisms γ_1 and γ_2 are then mutually inverse. $\qquad\square$

§4.7.1. Generalized Eilenberg-Watts Theorem

We begin by reintroducing a category from §3.6.4 attached to right C-module categories $\mathcal{M}$ and $\mathcal{N}$.

- $\text{Rex}_{\text{Mod-}C}(\mathcal{M}, \mathcal{N})$: objects are right exact, right C-module functors (F, t) from $\mathcal{M}$ to $\mathcal{N}$, and morphisms are natural transformations ϕ between such functors that are compatible with the right C-actions. That is, for $\phi : F \Rightarrow F'$, we have that $(\phi_M \triangleleft \text{id}_X) t_{M,X} = t'_{M,X} \phi_{M \triangleleft X}$, for all $X \in C$ and $M \in \mathcal{M}$.

Theorem 4.54 (Generalized Eilenberg-Watts Theorem). *Let A and B be algebras in C. Take the right C-module actions $\triangleleft_A$ on A-Mod(C) and $\triangleleft_B$ on B-Mod(C) given in Proposition 4.30. Then, we have the equivalence of categories below.*

$$\text{Rex}_{\text{Mod-}C}\, (A\text{-Mod}(C), B\text{-Mod}(C)) \;\simeq\; (B, A)\text{-Bimod}(C)$$

$$[F : A\text{-Mod}(C) \to B\text{-Mod}(C)] \xrightarrow{\;\Phi\;} F({}_A A_{\text{reg}})$$

$$[Q \otimes_A - : A\text{-Mod}(C) \to B\text{-Mod}(C)] \xleftarrow{\;\Psi\;} {}_B Q_A$$

In particular, $F({}_A A_{\text{reg}}) \in \text{Mod-}A(C)$ via:

$$\triangleleft^A_{F(A)} : \; F(A) \otimes A \;=\; F(A) \triangleleft_B A \;\cong\; F(A \triangleleft_A A) \;=\; F(A \otimes A) \xrightarrow{F(m_A)} F(A).$$

Proof. We will sketch the proof, and leave the details to the reader as part of Exercise 4.46. First, it is straightforward to check that $F({}_A A_{\text{reg}}) \in (B, A)\text{-Bimod}(C)$. With this, the functor Φ is well-defined. Towards defining the functor Ψ, we have that $Q \otimes_A M \in B\text{-Mod}(C)$ for $M \in A\text{-Mod}(C)$ by Proposition 4.42, which uses the right exactness of $(B \otimes -)$. Moreover, $(Q \otimes_A -)$ is a right C-module functor with structure morphisms $t_{M,X}$, for $M \in A\text{-Mod}(C)$ and $X \in C$, given by:

$$t_{M,X} : Q \otimes_A (M \triangleleft_A X) \;=\; Q \otimes_A (M \otimes X) \;\cong\; (Q \otimes_A M) \otimes X \;=\; (Q \otimes_A M) \triangleleft_B X.$$

For the iso, see Example 4.43 and modify Lemma 4.44(a). We also have that $(Q \otimes_A -)$ is right exact by Lemma 4.53. So, the functor Ψ is well-defined.

Next, we obtain that $\Phi\Psi \cong \text{Id}_{(B,A)\text{-Bimod}(C)}$ due to the computation below:

$$\Phi\Psi({}_B Q_A) \;=\; \Phi(Q \otimes_A -) \;=\; Q \otimes_A {}_A A_{\text{reg}} \;\cong\; Q.$$

The last iso holds by Lemma 4.44(b).

Finally, we verify that $\Psi\Phi(F) \cong F$, for all $F \in \text{Rex}_{\text{Mod-}C}\, (A\text{-Mod}(C), B\text{-Mod}(C))$. That is, for each $M \in A\text{-Mod}(C)$, we need to show that $F({}_A A_{\text{reg}}) \otimes_A M \cong F(M)$ as left B-modules in C. Since F is a right C-module functor, we get the first two vertical isos in the commutative diagram in $\mathcal{C}$ below.

$$
\begin{array}{ccccc}
F(A) \otimes A \otimes M & \underset{\mathrm{id}\,\otimes\,\triangleright}{\overset{\triangleleft\,\otimes\,\mathrm{id}}{\rightrightarrows}} & F(A) \otimes M & \longrightarrow & F(A) \otimes_A M \\
\cong \Big\updownarrow & & \cong \Big\uparrow & & \Big\updownarrow \\
F(A \otimes A \otimes M) & \underset{F(\mathrm{id}\,\otimes\,\triangleright)}{\overset{F(\triangleleft\,\otimes\,\mathrm{id})}{\rightrightarrows}} & F(A \otimes M) & \longrightarrow & F(A \otimes_A M) \cong F(M)
\end{array}
$$

The vertical dashed arrows arise by the universal property of coequalizers. But for the upward dashed arrow, we also use the assumption that F is right exact, so it preserves coequalizers [Proposition 2.49(a), Exercise 2.13(d)]. These vertical dashed arrows are mutually inverse morphisms in $B\text{-Mod}(\mathcal{C})$, as required. $\qquad\square$

Remark 4.55. With a similar argument, we also obtain a version of the Generalized Eilenberg-Watts Theorem for right modules in $\mathcal{C}$:

$$\mathsf{Rex}_{\mathcal{C}\text{-Mod}}\left(\mathsf{Mod}\text{-}A(\mathcal{C}), \mathsf{Mod}\text{-}B(\mathcal{C})\right) \simeq (A, B)\text{-Bimod}(\mathcal{C}).$$

Here, the correspondences are given by $\Phi(F) := F((A_{\mathrm{reg}})_A)$ and $\Psi(_A P_B) = - \otimes_A P$.

§4.7.2. Morita equivalence of algebras

We say that $A, B \in \mathsf{Alg}(\mathcal{C})$ are **Morita equivalent** if $A\text{-Mod}(\mathcal{C})$ and $B\text{-Mod}(\mathcal{C})$ are equivalent as right $\mathcal{C}$-module categories, with the right $\mathcal{C}$-module actions $\triangleleft_A$ on $A\text{-Mod}(\mathcal{C})$ and $\triangleleft_B$ on $B\text{-Mod}(\mathcal{C})$ given in Proposition 4.30.

Morita equivalence here is an equivalence relation for algebras in $\mathcal{C}$ [Exercise 2.28]. Next, we provide a useful characterization of this notion below, which generalizes Theorem 2.18.

Theorem 4.56 (Generalized Morita's Theorem). *Take $A, B \in \mathsf{Alg}(\mathcal{C})$. Then, the following statements are equivalent.*

(a) *$(A\text{-Mod}(\mathcal{C}), \triangleleft_A) \simeq (B\text{-Mod}(\mathcal{C}), \triangleleft_B)$, as right $\mathcal{C}$-module categories.*

(b) *There exist bimodules $P \in (A, B)\text{-Bimod}(\mathcal{C})$ and $Q \in (B, A)\text{-Bimod}(\mathcal{C})$ such that $P \otimes_B Q \cong A_{\mathrm{reg}}$ as A-bimodules in C and $Q \otimes_A P \cong B_{\mathrm{reg}}$ as B-bimodules in C.*

Bimodules P and Q satisfying the conditions in part (b) are said to be **invertible**.

Proof. Note that (b) $\Rightarrow$ (a) is the more elementary direction. To prove this, consider the functors F and G below, which are defined by Proposition 4.42:

$$F := (_B Q_A) \otimes_A - : \; A\text{-Mod}(\mathcal{C}) \to B\text{-Mod}(\mathcal{C}),$$

$$G := (_A P_B) \otimes_B - : \; B\text{-Mod}(\mathcal{C}) \to A\text{-Mod}(\mathcal{C}).$$

Now for $M \in A\text{-Mod}(\mathcal{C})$, we obtain that:

$$GF(M) = P \otimes_B (Q \otimes_A M) \cong (P \otimes_B Q) \otimes_A M \cong A_{\mathrm{reg}} \otimes_A M \cong M.$$

Here, the first iso holds by a modification of Lemma 4.44(a), and the last iso holds by Lemma 4.44(b). Therefore, $GF \cong \mathrm{Id}_{A\text{-Mod}(\mathcal{C})}$. Likewise, $FG \cong \mathrm{Id}_{B\text{-Mod}(\mathcal{C})}$. So, the functors F and G imply that $A\text{-Mod}(\mathcal{C}) \simeq B\text{-Mod}(\mathcal{C})$ as categories. This is updated to an equivalence of right $\mathcal{C}$-module categories via the structures morphisms $t_{M,X}$ for F defined below, for $M \in A\text{-Mod}(\mathcal{C})$ and $X \in \mathcal{C}$:

$$t_{M,X} : F(M \triangleleft_A X) = F(M \otimes X) = Q \otimes_A (M \otimes X)$$
$$\cong (Q \otimes_A M) \otimes X = F(M) \otimes X = F(M) \triangleleft_B X.$$

For the iso, see Example 4.43 and modify Lemma 4.44(a). Thus, we established that (b) $\Rightarrow$ (a).

Now for (a) $\Rightarrow$ (b), take quasi-inverse functors $F : A\text{-Mod}(\mathcal{C}) \to B\text{-Mod}(\mathcal{C})$ and $G : B\text{-Mod}(\mathcal{C}) \to A\text{-Mod}(\mathcal{C})$, which are both equivalences of categories (see, e.g., Exercise 2.28). So, F and G have right adjoints by Proposition 2.25(b), and thus are right exact by Proposition 2.49(b). So by the Generalized Eilenberg Watts Theorem [Theorem 4.54], there exist bimodules $P \in (A, B)\text{-Bimod}(\mathcal{C})$ and $Q \in (B, A)\text{-Bimod}(\mathcal{C})$ such that $F \cong (Q \otimes_A -)$ and $G \cong (P \otimes_B -)$. We will proceed with showing that $P \otimes_B Q \cong A_{\mathrm{reg}}$ as A-bimodules in $\mathcal{C}$, and leave the other required isomorphism to the reader. By the assumption, we have a natural isomorphism $\Phi : GF \xRightarrow{\sim} \mathrm{Id}_{A\text{-Mod}(\mathcal{C})}$, with component,

$$\Phi_A := \Phi_{A_{\mathrm{reg}}} : P \otimes_B Q \cong GF(A_{\mathrm{reg}}) \xrightarrow{\sim} A_{\mathrm{reg}},$$

being an isomorphism of left A-modules in $\mathcal{C}$. Here, the right A-action on $P \otimes_B Q$ is given as follows:

$$\triangleleft^A_{P \otimes_B Q} : (P \otimes_B Q) \otimes A \cong P \otimes_B ((Q \otimes_A A) \otimes A) \cong P \otimes_B (F(A) \otimes A) \cong P \otimes_B F(A \otimes A)$$
$$\xrightarrow{GF(m_A)} P \otimes_B F(A) \cong P \otimes_B (Q \otimes_A A) \cong P \otimes_B Q.$$

For the above, note that $F(A \otimes A) = F(A \triangleleft_A A) \cong F(A) \triangleleft_B A = F(A) \otimes A$ since F is a right $\mathcal{C}$-module functor. Finally, Φ_A is a morphism of right A-modules in $\mathcal{C}$ since:

$$\Phi_A \circ \triangleleft^A_{P \otimes_B Q} = \Phi_A \circ GF(m_A) = m_A \circ \Phi_{A \otimes A} = m_A \circ (\Phi_A \otimes \mathrm{id}_A).$$

The second equation holds by the naturality of Φ, and the third equation holds since F, G are right $\mathcal{C}$-module functors. So, $P \otimes_B Q \cong A_{\mathrm{reg}}$ as A-bimodules in $\mathcal{C}$. $\square$

To use the characterization above in practice, one first produces morphisms $P \otimes_B Q \to A$ and $Q \otimes_A P \to B$ using the universal property of coequalizers. But obtaining morphisms in the reverse direction is trickier. It may be helpful to use the result below, which is a generalization of a standard fact in Morita theory; see Lemma 4.5.2 of Cohn [2003]. This result is Proposition 2.20 of Morales et al. [2022], and we leave the proof as Exercise 4.47 for a curious reader.

Proposition 4.57. *Take algebras $A, B \in C$, and take bimodules $P \in (A, B)\text{-Bimod}(C)$ and $Q \in (B, A)\text{-Bimod}(C)$. If there exist epis*

$$\phi_A : P \otimes_B Q \to A_{\mathrm{reg}} \in A\text{-Bimod}(C), \qquad \phi_B : Q \otimes_A P \to B_{\mathrm{reg}} \in B\text{-Bimod}(C)$$

that make the following diagrams commute,

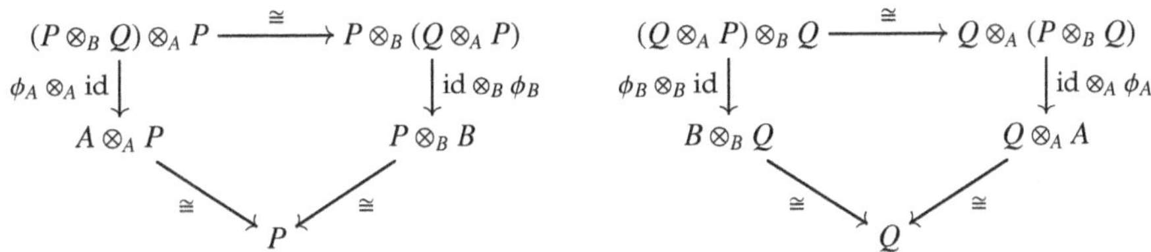

then ϕ_A and ϕ_B are isomorphisms of bimodules in C. $\quad\square$

The data $(A, B, P, Q, \phi_A, \phi_B)$ in the hypotheses of Proposition 4.57, with ϕ_A and ϕ_B not necessarily epic, is called a **Morita context**.

Example 4.58. If X is a left rigid object in C and $\mathbb{1}$ is a simple object in C (e.g., if C is a tensor category as from §3.10.1), then the C-endomorphism algebra $X \otimes X^*$ from Example 4.9(a) is Morita equivalent to the unit algebra $\mathbb{1}$ from Example 4.1(a). This specializes to the fact that $\mathrm{Mat}_n(\mathbb{k})$ is Morita equivalent to $\mathbb{k}$ as $\mathbb{k}$-algebras [Example 2.21]. Verifying the details here and more is Exercise 4.48.

See also Exercises 4.49 and 4.50 for additional practice.

Remark 4.59. Note that everything in §4.7 can be stated and proved in terms of right modules in C via the version of the Generalized Eilenberg-Watts Theorem in Remark 4.55. Therefore, one can obtain that algebras A and B in C are Morita equivalent if and only if $\mathrm{Mod}\text{-}A(C) \simeq \mathrm{Mod}\text{-}B(C)$ as left C-module categories. This implies that Morita equivalence in C is a left-right symmetric condition.

§4.7.3. On Morita equivalence of graded algebras

Here, we consider when two graded algebras in a monoidal category have equivalent module categories consisting of graded modules.

Note that in §4.6.1 morphisms between graded objects preserve degree. That is, for objects $X, Y \in C^{\times \mathbb{N}}$, components of a morphism $\phi : X \to Y$ sends a component X_i of X to the component Y_i of Y, for all $i \in \mathbb{N}$. However, traditional **graded Morita equivalence** allows for morphisms of any degree. Namely, ϕ is of **degree** d if X_i sent to Y_{i+d} for all $i \in \mathbb{N}$. See, for instance, Theorem 3.2(6) in the work of Boisen [1994]. But note that a category with such morphisms is not additive as the sum of morphisms of different degrees is no longer graded. We like abelian categories here, so we work in categories of graded objects with morphisms in degree 0, such as the product category $C^{\times \mathbb{N}}$ used in §4.6.

Another framework for studying the Morita equivalence of graded algebras (in categories with morphisms of degree 0) was set up in the work of Zhang [1996] for $\Bbbk$-algebras. There, *twisting systems* of graded algebras were introduced, and it was shown that $\mathbb{N}$-graded $\Bbbk$-algebras A and B have equivalent categories of graded modules over $\Bbbk$ with morphisms in degree 0 when A is a *twist* of B. The converse statement holds when the algebras satisfy the condition that $A_0 = B_0 = \Bbbk$.

This was generalized to the monoidal/enriched setting in the work of Liu Lopez and Walton [2023] by first generalizing the twisting systems of Zhang [1996]. Then, it was shown that if $A, B \in \mathbb{N}\text{-GrAlg}(\mathcal{C})$ and A is a *twist* of B, we have that

$$\mathbb{N}\text{-GrMod}(\mathcal{C})_A \simeq^{\mathcal{C}} \mathbb{N}\text{-GrMod}(\mathcal{C})_B.$$

(See §3.11.2 for details about enriched equivalence.) The equivalence of graded modules here is called **Zhang-Morita equivalence**. The converse statement also holds under certain conditions.

§4.8. Internal Homs and Ends

In the last section, we discussed how algebras in $\mathcal{C}$ are considered the 'same' if their categories of modules *in* $\mathcal{C}$ are equivalent as module categories *over* $\mathcal{C}$. In this section, we will explore when these internal and external notions of module categories coincide. Recall Proposition 4.30, and consider the terminology below.

A left (resp., right) $\mathcal{C}$-module category $\mathcal{M}$ is **represented** by $A \in \mathsf{Alg}(\mathcal{C})$ if

$$\mathcal{M} \sim \text{Mod-}A(\mathcal{C}) \quad \text{as left } \mathcal{C}\text{-module categories}$$
$$(\text{resp., } \mathcal{M} \sim A\text{-Mod}(\mathcal{C}) \quad \text{as right } \mathcal{C}\text{-module categories}).$$

The algebra representatives of $\mathcal{C}$-module categories constructed here will arise as *internal Ends* as in §3.11.4. We will stick with left $\mathcal{C}$-module categories in the discussion below, but it can be translated for right $\mathcal{C}$-module categories in a straightforward way.

§4.8.1. Internal End algebras

Recall from §3.11.4 that a left $\mathcal{C}$-module category $(\mathcal{M}, \rhd, m, p)$ is **closed** if, for each $M \in \mathcal{M}$, the functor $(- \rhd M) : \mathcal{C} \to \mathcal{M}$ has a right adjoint, which we denote by

$$\underline{\text{Hom}}(M, -) : \mathcal{M} \to \mathcal{C}.$$

Here, we have a bijection for each $Z \in \mathcal{C}$ and $N \in \mathcal{M}$:

$$\zeta := \zeta_{Z,N} : \text{Hom}_{\mathcal{M}}(Z \rhd M, N) \xrightarrow{\sim} \text{Hom}_{\mathcal{C}}(Z, \underline{\text{Hom}}(M, N)). \tag{4.60}$$

The object $\underline{\mathrm{Hom}}(M, N) \in \mathcal{C}$ is the **internal Hom** of $M, N \in \mathcal{M}$, and we call the object

$$\underline{\mathrm{End}}(M) := \underline{\mathrm{Hom}}(M, M) \in \mathcal{C}$$

the **internal End** of $M \in \mathcal{M}$.

To give internal Homs and internal Ends algebraic structure, first consider the following morphism derived from the adjunction above, for each $M, N \in \mathcal{M}$:

$$\mathrm{ev}_{M,N} := \zeta^{-1}(\mathrm{id}_{\underline{\mathrm{Hom}}(M,N)}) : \ \underline{\mathrm{Hom}}(M, N) \rhd M \to N. \qquad (4.61)$$

Moreover, consider the composition of morphisms, $\mathrm{ev}_{M,N,P}$, for each $M, N, P \in \mathcal{M}$:

$$
\begin{array}{ccc}
\left(\underline{\mathrm{Hom}}(N, P) \otimes \underline{\mathrm{Hom}}(M, N)\right) \rhd M & \xrightarrow{\quad\mathrm{ev}_{M,N,P}\quad} & P \\
{\scriptstyle m_{\underline{\mathrm{Hom}}(N,P),\underline{\mathrm{Hom}}(M,N),M}} \Big\downarrow & & \Big\uparrow {\scriptstyle \mathrm{ev}_{N,P}} \\
\underline{\mathrm{Hom}}(N, P) \rhd \left(\underline{\mathrm{Hom}}(M, N) \rhd M\right) & \xrightarrow{\ \mathrm{id} \,\rhd\, \mathrm{ev}_{M,N}\ } & \underline{\mathrm{Hom}}(N, P) \rhd N.
\end{array}
$$

One then gets the morphisms below in $\mathcal{C}$, for each $M, N, P \in \mathcal{M}$:

$$\mathrm{comp}_{M,N,P} := \zeta(\mathrm{ev}_{M,N,P}) : \ \underline{\mathrm{Hom}}(N, P) \otimes \underline{\mathrm{Hom}}(M, N) \to \underline{\mathrm{Hom}}(M, P).$$

Also, for $M \in \mathcal{M}$, take:

$$\mathrm{coev}_{\mathbb{1},M} := \zeta(p_M) : \ \mathbb{1} \to \underline{\mathrm{Hom}}(M, M).$$

Now with these morphisms, internal Ends and internal Homs arise, resp., as algebras and modules in $\mathcal{C}$ as described below. We leave the proof as Exercise 4.51.

Proposition 4.62. *Suppose that* $(\mathcal{M}, \rhd, m, p)$ *is a closed left $\mathcal{C}$-module category. Then, the following statements hold.*

(a) *For any $M \in \mathcal{M}$, we obtain that $\underline{\mathrm{End}}(M)$ is an algebra in $\mathcal{C}$ with*

$$m_{\underline{\mathrm{End}}(M)} := \mathrm{comp}_{M,M,M} : \underline{\mathrm{End}}(M) \otimes \underline{\mathrm{End}}(M) \to \underline{\mathrm{End}}(M),$$

$$u_{\underline{\mathrm{End}}(M)} := \mathrm{coev}_{\mathbb{1},M} : \mathbb{1} \to \underline{\mathrm{End}}(M),$$

(b) $\underline{\mathrm{Hom}}(M, N)$ *is a right $\underline{\mathrm{End}}(M)$-module in $\mathcal{C}$, for any $M, N \in \mathcal{M}$, with right action:*

$$\vartriangleleft_{\underline{\mathrm{Hom}}(M,N)} := \mathrm{comp}_{M,M,N} : \underline{\mathrm{Hom}}(M, N) \otimes \underline{\mathrm{End}}(M) \to \underline{\mathrm{Hom}}(M, N).$$

(c) $\underline{\mathrm{Hom}}(N, M)$ *is a left $\underline{\mathrm{End}}(M)$-module in $\mathcal{C}$, for any $M, N \in \mathcal{M}$, with left action:*

$$\vartriangleright_{\underline{\mathrm{Hom}}(M,N)} := \mathrm{comp}_{N,M,M} : \underline{\mathrm{End}}(M) \otimes \underline{\mathrm{Hom}}(N, M) \to \underline{\mathrm{Hom}}(N, M). \qquad \square$$

We refer to the algebras constructed in part (a) above as **internal End algebras**, and refer to the modules constructed in parts (b,c) as **internal Hom modules**.

Example 4.63. Recall from Example 3.70 that if C is right closed monoidal [§3.11.3], then the regular left C-module category C_{reg} is closed. Here, $\underline{\mathrm{Hom}}(-,-)$ is the left internal Hom from the right closed monoidal condition on C. In particular by Exercise 3.40, when C is left rigid, we get for $Y, Z \in C_{\mathrm{reg}}$ that

$$\underline{\mathrm{Hom}}(Y, Z) := Z \otimes Y^*$$

as an object in C. The bijections $\zeta_{X,Z} : \mathrm{Hom}_C(X \otimes Y, Z) \xrightarrow{\sim} \mathrm{Hom}_C(X, Z \otimes Y^*)$ are given in Proposition 3.33(d). Using these maps, we obtain that

$$\underline{\mathrm{End}}(Y) = Y \otimes Y^* \in \mathsf{Alg}(C).$$

is the C-endomorphism algebra from Example 4.9. See Exercise 4.52.

Upgrading the example above (when C is rigid and abelian) for the left C-module category $M = \mathrm{Mod}\text{-}A(C)$ comprises Exercise 4.53.

Example 4.64. Take a group G, and consider the monoidal category $(G\text{-Mod}, \otimes_{\Bbbk}, \Bbbk)$. Recall from Example 3.19 that the (strong monoidal) forgetful functor $G\text{-Mod} \to \mathsf{Vec}$ equips Vec with the structure of a left $(G\text{-Mod})$-module category. Namely, for $(X, \triangleright) \in G\text{-Mod}$ and $V \in \mathsf{Vec}$ we have the action:

$$(X, \triangleright) \triangleright V := X \otimes_{\Bbbk} V \in \mathsf{Vec}.$$

The internal End algebra of the object $\Bbbk \in \mathsf{Vec}$ exists, and we claim that $\underline{\mathrm{End}}(\Bbbk)$ is the dual group algebra $(\Bbbk G)^* \in \mathsf{Alg}(G\text{-Mod})$ from Exercise 4.2. Indeed, consider the bijection that $\underline{\mathrm{End}}(\Bbbk)$ must satisfy:

$$\zeta : \mathrm{Hom}_{\mathsf{Vec}}(X, \Bbbk) = \mathrm{Hom}_{\mathsf{Vec}}((X, \triangleright) \triangleright \Bbbk, \Bbbk) \xrightarrow{\sim} \mathrm{Hom}_{G\text{-Mod}}((X, \triangleright), \underline{\mathrm{End}}(\Bbbk)).$$

Taking $\underline{\mathrm{End}}(\Bbbk) = (\Bbbk G)^*$ works via the assignments below:

$$\zeta(\phi : X \to \Bbbk) = [\zeta(\phi) : (X, \triangleright) \to (\Bbbk G)^*], \quad \text{for } \zeta(\phi)(x) = [\Bbbk G \to \Bbbk, \ g' \mapsto \phi(g' \triangleright x)]$$

$$\zeta^{-1}(\psi : (X, \triangleright) \to (\Bbbk G)^*) = [\zeta^{-1}(\psi) : X \to \Bbbk], \quad \text{for } \zeta^{-1}(\psi)(x) = \psi(x)(e_G),$$

for all linear maps $\phi : X \to \Bbbk$, all left G-module maps $\psi : (X, \triangleright) \to (\Bbbk G)^*$, and all elements $x \in X$, $g' \in G$. Completing the details here is part of Exercise 4.54.

Upgrading the example above for the module category $H\text{-Mod}$ from Example 3.19, with H a subgroup of G, is also part of Exercise 4.54.

Moreover, Exercises 4.55 and 4.56 use other module categories to produce examples of internal End algebras and internal Hom modules.

We end this part with a useful lemma.

Lemma 4.65. *If C is left rigid, and $\mathcal{M}$ is a closed left C-module category, then*

$$\underline{\mathrm{Hom}}(M, X \rhd N) \cong X \otimes \underline{\mathrm{Hom}}(M, N)$$

for all $X \in C$ and $M, N \in \mathcal{M}$.

Proof. Take an arbitrary object $Y \in C$. Then, consider the computation below:

$$
\begin{aligned}
\mathrm{Hom}_C(Y, \underline{\mathrm{Hom}}(M, X \rhd N)) &\overset{(4.60)}{\cong} \mathrm{Hom}_{\mathcal{M}}(Y \rhd M, \ X \rhd N) \\
&\overset{\text{Prop. 3.45(b)}}{\cong} \mathrm{Hom}_{\mathcal{M}}(X^* \rhd (Y \rhd M), \ N) \\
&\overset{\text{mod. assoc.}}{\cong} \mathrm{Hom}_{\mathcal{M}}((X^* \otimes Y) \rhd M, \ N) \\
&\overset{(4.60)}{\cong} \mathrm{Hom}_C(X^* \otimes Y, \ \underline{\mathrm{Hom}}(M, N)) \\
&\overset{\text{Prop. 3.33(b)}}{\cong} \mathrm{Hom}_C(Y, \ X \otimes \underline{\mathrm{Hom}}(M, N)).
\end{aligned}
$$

The result then holds by the contravariant Yoneda's lemma, Lemma 2.33. $\qquad\square$

§4.8.2. Ostrik's Theorem

Here, we will discuss how, after imposing certain conditions on C, the internal End algebras from Proposition 4.62(a) serve as Morita equivalence class representatives of algebras in C. This is due to the work of Ostrik [2003c].

First, recall that internal Homs do not exist in general. But when C is a finite multitensor category and $\mathcal{M}$ is a left C-module category, then $\mathcal{M}$ is closed [Corollary 3.72]; thus, internal Homs exist in this case. So, we impose the hypothesis below for the rest of this section.

Standing hypothesis.   Assume that C is a finite multitensor category here.

To proceed, recall the assumptions on C-module categories introduced in §3.10.3. Now we collect some results from Douglas et al. [2019] about C-module categories in this setting. The reader is encouraged to explore the proof of the result below before consulting the aforementioned reference.

Lemma 4.66. *Let $\mathcal{M}$ be a left nonzero C-module category.*

(a) *If M is a projective object in $\mathcal{M}$, then the functor $\underline{\mathrm{Hom}}(M, -)$ is right exact.*

(b) *If $\mathcal{M}$ is semisimple, then $\underline{\mathrm{Hom}}(M, -)$ is exact for all $M \in \mathcal{M}$.*

(c) *The morphism $\mathrm{ev}_{M,N} : \underline{\mathrm{Hom}}(M, N) \rhd M \to N$ from (4.61) is epic for each $N \in \mathcal{M}$ if and only if the functor $\underline{\mathrm{Hom}}(M, -) : \mathcal{M} \to C$ is faithful.*

(d) *If $\mathcal{M}$ is semisimple and indecomposable, then part (c) holds for all nonzero $M \in \mathcal{M}$.*

Proof. Parts (a) and (c) are in the extended preprint version of the article by Douglas et al. [2019]. Namely, see Lemma 2.22 and the proof of Lemma 2.25 of Douglas et al. [2018].

Part (b) then follows from part (a), Proposition 2.49(b), and Corollary 2.56.

For part (d), note that if $\mathcal{M}$ is semisimple and indecomposable, then $\mathrm{Hom}_{\mathcal{M}}(M, -)$ is faithful as the objects $N \in \mathcal{M}$ for which $\mathrm{Hom}_{\mathcal{M}}(M, N) = 0$ forms a $\mathcal{C}$-module subcategory of $\mathcal{M}$. $\qquad\square$

This brings us to the main result of the section. Recall the assumptions on $\mathcal{C}$-module categories in §3.9.4 when $\mathcal{C}$ is semisimple (e.g., when $\mathcal{C}$ is multifusion).

Theorem 4.67 (Ostrik's Theorem). *Suppose that $\mathcal{C}$ is semisimple, i.e., $\mathcal{C}$ is multifusion. If $\mathcal{M}$ is a nonzero, indecomposable left $\mathcal{C}$-module category, then for any nonzero object $M \in \mathcal{M}$, we obtain that*

$$\mathcal{M} \simeq \mathrm{Mod}\text{-}(\underline{\mathrm{End}}(M))(\mathcal{C}) \quad \textit{as left } \mathcal{C}\text{-module categories}$$

Here, $\underline{\mathrm{End}}(M) \in \mathrm{Alg}(\mathcal{C})$ via Proposition 4.62(a).

Proof. First, $\mathcal{M}$ is closed by Corollary 3.72, so internal Homs and internal Ends for $\mathcal{M}$ exist. Then, for any nonzero object $M \in \mathcal{M}$, denote

$$A := \underline{\mathrm{End}}(M),$$

and, again, note that $A \in \mathrm{Alg}(\mathcal{C})$ by Proposition 4.62(a). Further by Proposition 4.62(b), we have the functor below:

$$F : \mathcal{M} \to \mathrm{Mod}\text{-}A(\mathcal{C}), \quad N \mapsto \underline{\mathrm{Hom}}(M, N).$$

In particular, $\underline{\mathrm{Hom}}(M, -)$ is covariant, so for a morphism $f : N \to N'$ in $\mathcal{M}$, we have that $F(f)$ is a morphism $\underline{\mathrm{Hom}}(M, N) \to \underline{\mathrm{Hom}}(M, N')$ in $\mathrm{Mod}\text{-}A(\mathcal{C})$. We aim to show that F is an equivalence of left $\mathcal{C}$-module categories.

To proceed, note that F is a left $\mathcal{C}$-module functor due to Lemma 4.65. Namely, for the left $\mathcal{C}$-module categories $(\mathcal{M}, \rhd)$ and $(\mathrm{Mod}\text{-}A(\mathcal{C}), \rhd_A)$ [Proposition 4.30], and for all $X \in \mathcal{C}$ and $N \in \mathcal{M}$, we have that

$$F(X \rhd N) = \underline{\mathrm{Hom}}(M, X \rhd N) \overset{\text{Lem. 4.65}}{\cong} X \otimes \underline{\mathrm{Hom}}(M, N) = X \rhd_A F(N).$$

Next, we will show that F is fully faithful, that is,

$$F_{N,N'} : \mathrm{Hom}_{\mathcal{M}}(N, N') \to \mathrm{Hom}_{\mathrm{Mod}\text{-}A(\mathcal{C})}(F(N), F(N')), \quad f \mapsto F(f).$$

is an isomorphism. This is true when $N \in \mathcal{M}$ is an object of the form:

$$N = X \rhd M, \quad \text{for some } X \in \mathcal{C}. \tag{$\dagger$}$$

To see this, observe that $F(N) = \underline{\mathrm{Hom}}(M, X \rhd M) \cong X \otimes A$ by Lemma 4.65, which is a free object in Mod-$A(\mathcal{C})$. Now under the assumption (†) above, F is fully faithful as follows:

$$
\begin{aligned}
\mathrm{Hom}_{\mathrm{Mod}\text{-}A(\mathcal{C})}(F(N), F(N')) &\cong \mathrm{Hom}_{\mathrm{Mod}\text{-}A(\mathcal{C})}(X \otimes A, F(N')) \\
&\cong \mathrm{Hom}_{\mathcal{C}}(X, \mathrm{Forg}(F(N'))) \quad &\text{(Exer. 4.25(c) for Mod-}A(\mathcal{C})) \\
&= \mathrm{Hom}_{\mathcal{C}}(X, \underline{\mathrm{Hom}}(M, N')) \quad &\text{(definition of } F) \\
&\cong \mathrm{Hom}_{\mathcal{M}}(X \rhd M, N') \quad &(\mathcal{M} \text{ closed}) \\
&= \mathrm{Hom}_{\mathcal{M}}(N, N'). \quad &(†)
\end{aligned}
$$

In general, we can remove the assumption (†) towards faithfulness. Namely, by Lemma 4.66(d), the composition of F with the forgetful functor Mod-$A(\mathcal{C}) \to \mathcal{C}$ is faithful. Thus, F is faithful. This means that $F_{L,N'}$ is monic, for all $L \in \mathcal{M}$.

Now to remove the assumption (†) for fullness, note that again by Lemma 4.66(d), the morphism $\mathrm{ev}_{M,N}$ is epic for each $N \in \mathcal{M}$. Thus, we obtain the exact sequence in $\mathcal{C}$ below, for $X := \underline{\mathrm{Hom}}(M, N)$ and $K := \ker(\mathrm{ev}_{M,N})$:

$$
0 \longrightarrow K \longrightarrow X \rhd M \longrightarrow N \longrightarrow 0.
$$

Next, by Lemma 4.66(b), we get the exact sequence in $\mathcal{C}$ below:

$$
0 \longrightarrow F(K) \longrightarrow F(X \rhd M) \longrightarrow F(N) \longrightarrow 0.
$$

Apply the left exact, contravariant functors, $\mathrm{Hom}_{\mathcal{M}}(-, N')$, $\mathrm{Hom}_{\mathrm{Mod}\text{-}A(\mathcal{C})}(-, F(N'))$, respectively to the sequences above to yield the commutative diagram below.

$$
\begin{array}{ccccccc}
0 & \longrightarrow & \mathrm{Hom}_{\mathcal{M}}(N, N') & \longrightarrow & \mathrm{Hom}_{\mathcal{M}}(X \rhd M, N') & \longrightarrow & \mathrm{Hom}_{\mathcal{M}}(K, N') \\
& & \downarrow{\scriptstyle F_{0,0}} & & \downarrow{\scriptstyle F_{N,N'}} & & \downarrow{\scriptstyle F_{(X \rhd M),N'}} \quad \downarrow{\scriptstyle F_{K,N'}} \\
0 & \longrightarrow & \mathrm{Hom}_{\mathrm{Mod}\text{-}A(\mathcal{C})}(F(N), F(N')) & \longrightarrow & \mathrm{Hom}_{\mathrm{Mod}\text{-}A(\mathcal{C})}(F(X \rhd M), F(N')) & \longrightarrow & \mathrm{Hom}_{\mathrm{Mod}\text{-}A(\mathcal{C})}(F(K), F(N'))
\end{array}
$$

The first and third vertical maps are isos, trivially and as shown under the assumption (†), respectively. Moreover, $F_{K,N'}$ is monic due to the faithfulness of F. Hence, $F_{N,N'}$ is epic by a Four Lemma [Lemma 2.48(b)]. Together with the faithfulness of F, we get that $F_{N,N'}$ is an isomorphism [Proposition 2.4]. Thus, F is fully faithful.

To obtain that F is essentially surjective, consider the adjunction

$$
\zeta_{X,Y} : \mathrm{Hom}_{\mathrm{Mod}\text{-}A(\mathcal{C})}(X \otimes A, Y) \xrightarrow{\sim} \mathrm{Hom}_{\mathcal{C}}(X, \mathrm{Forg}(Y));
$$

see Exercise 4.25 adapted for right modules. Take $Z \in \mathrm{Mod}\text{-}A(\mathcal{C})$. Then, we get that $\zeta^{-1}_{\mathrm{Forg}(Z),Z}(\mathrm{id}_Z) : Z \otimes A \to Z$ is an epi in Mod-$A(\mathcal{C})$ (using the counit for Free-Forget adjunction here and Exercise 2.41(b)). Take X to be the kernel of this morphism, and by composing with the epi $\zeta^{-1}_{\mathrm{Forg}(X),X}(\mathrm{id}_X)$, we get the sequence of morphisms,

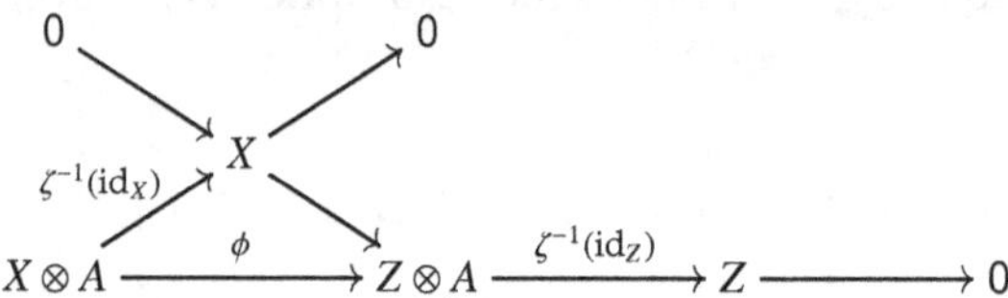

in Mod-$A(\mathcal{C})$, for $A := \underline{\mathrm{End}}(M)$. We then get that

$$\mathrm{Hom}_{\mathrm{Mod}\text{-}A(\mathcal{C})}(X \otimes A, Z \otimes A) \cong \mathrm{Hom}_{\mathrm{Mod}\text{-}A(\mathcal{C})}(F(X \rhd M), F(Z \rhd M)) \qquad \text{(Lemma 4.65)}$$

$$\cong \mathrm{Hom}_{\mathcal{M}}(X \rhd M, Z \rhd M). \qquad \text{(F is fully-faithful)}$$

Take $\phi' : X \rhd M \to Z \rhd M$ in $\mathcal{M}$ to be the image of ϕ under the isomorphism above. Then, $F(\mathrm{coker}(\phi')) \cong \mathrm{coker}(F(\phi'))$ since F is right exact [Lemma 4.66(b)]. Further, the cokernel of $F(\phi')$, which is isomorphic to the cokernel of ϕ, is in turn isomorphic to Z. Thus, F is essentially surjective since $F(\mathrm{coker}(\phi')) \cong Z$. $\qquad \square$

Note that the semisimplicity of $\mathcal{C}$ (which implies that $\mathcal{M}$ is semisimple by definition), along with the indecomposability of $\mathcal{M}$, were used to employ Lemma 4.66.

Remark 4.68. The hypothesis of Ostrik's Theorem [Theorem 4.67] can be weakened to achieve generalizations, as shown in the works mentioned below.

(a) In Theorem 7.10.1 of the textbook by Etingof et al. [2015], the authors achieve the conclusion of Theorem 4.67 without imposing semisimplicity. Instead, $\mathcal{C}$ is assumed to be finite tensor, and $\mathcal{M}$ is exact and indecomposable.

(b) The work of Douglas et al. [2018] (namely, their Theorem 2.24) achieves the conclusion of Theorem 4.67 with neither semisimplicity nor rigidity imposed. There, the conditions in Lemma 4.66(a,c) for $\underline{\mathrm{Hom}}(M, -)$ are still imposed.

Remark 4.69. We have shown that internal End algebras serve as representatives of Morita equivalence classes of algebras in numerous monoidal categories, but in practice one may want to work with explicit algebra representatives.

For instance, take the fusion category Vec_G^ω from Exercise 3.35. Then, the twisted group algebras $\Bbbk L_\psi$ from Exercise 4.4 (and Exercise 4.50) serve as representatives of Morita equivalence classes of algebras in Vec_G^ω. This is due to the works of Ostrik [2003b] and of Natale [2017]. Exercise 4.57 compares these two collections of algebra representatives of Morita equivalence classes in $\mathrm{Alg}(\mathrm{Vec}_G^\omega)$.

§4.9. Properties of algebras in monoidal categories

Now we turn our attention to various properties for algebras A in monoidal categories $\mathcal{C}$, building on those introduced in Chapter 1 for algebras over a field. We will also highlight when these properties satisfy the following conditions.

- If a property P of $A \in \mathsf{Alg}(\mathcal{C})$ is given in terms of the underlying object and the structure morphisms of A, then we say that P is **intrinsic**.

- If a property P of $A \in \mathsf{Alg}(\mathcal{C})$ is given in terms of a category of A-modules in $\mathcal{C}$, then we say that P is **module-theoretic**.

Module-theoretic properties are often **Morita invariant**, i.e., if an algebra has such a property, then so does any algebra that is Morita equivalent to it (see §4.7.2).

For ease, we set the following hypotheses.

Standing hypothesis. Assume that $\mathcal{C}$ is a tensor category as in §3.10; see also §3.1.3 and Remark 4.14, along with Figure 3.15, for related conditions.

One useful observation is the following.

Remark 4.70. Under the standing assumption, the category A-$\mathsf{Mod}(\mathcal{C})$ is abelian for any algebra A in $\mathcal{C}$, due to Propositions 3.68 and 4.31(d).

Here, we will discuss the following properties, implications, and question below.

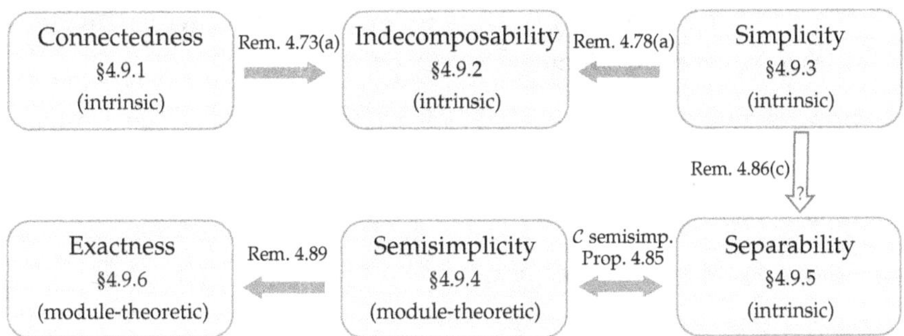

Figure 4.12: Properties for algebras in monoidal categories.

The properties in Figure 4.12 are illustrated here via the algebras below.

- The unit algebra $\mathbb{1}$ from Example 4.1(a), discussed throughout the section.

- The $\mathcal{C}$-endomorphism algebra $X \otimes X^*$ from Example 4.9, discussed throughout.

- Monads from §4.3.2, specifically towards separability, in Exercise 4.60.

- The twisted group algebras mentioned Remark 4.69, explored in Exercise 4.61.

§4.9.1. Connected algebras

One intrinsic property that is used often in the literature is the following: we say that $A \in \mathsf{Alg}(\mathcal{C})$ is **connected** or **haploid** if $\dim_{\Bbbk} \mathsf{Hom}_{\mathcal{C}}(\mathbb{1}, A) = 1$.

Connectedness is a strong condition on $A \in \mathsf{Alg}(\mathcal{C})$. For instance, when $\mathcal{C} = \mathsf{Vec}$, we have that A is connected if and only if $\dim_{\Bbbk} A = \dim_{\Bbbk} \mathsf{Hom}_{\mathsf{Vec}}(\Bbbk, A) = 1$.

We also see that connectedness is not Morita invariant (so, there is little hope that connectedness could be module-theoretic). Namely, take $\mathcal{C} = \mathsf{Vec}$, and recall that $\mathsf{Mat}_n(\Bbbk)$ is Morita equivalent to $\Bbbk$ for all $n \geq 1$ [Example 2.21]. Now, $\Bbbk$ is connected, yet $\mathsf{Mat}_n(\Bbbk)$ is connected only when $n = 1$.

Next, we introduce our running examples for the section.

Example 4.71. Consider the unit algebra $\mathbb{1}$ from Example 4.1(a). Also, consider the $\mathcal{C}$-endomorphism algebra $X \otimes X^*$, for X a nonzero object of $\mathcal{C}$, from Example 4.9(a).

(a) The algebra $\mathbb{1}$ is connected. Indeed, $\mathsf{Hom}_{\mathcal{C}}(\mathbb{1}, \mathbb{1}) \cong \Bbbk$ since $\mathbb{1} \in \mathcal{C}$ is assumed to be absolutely simple.

(b) The algebra $X \otimes X^*$ is connected if and only if X is an absolutely simple object of $\mathcal{C}$. Indeed, $\mathsf{Hom}_{\mathcal{C}}(\mathbb{1}, X \otimes X^*) \cong \mathsf{Hom}_{\mathcal{C}}(X, X)$ by Proposition 3.33(d). Therefore, by Proposition 2.60, $X \otimes X^*$ is connected if X is a simple object of $\mathcal{C}$, and the converse holds when $\mathcal{C}$ is semisimple.

(c) As a special case of part (a), take $\mathcal{C} = \mathsf{FdVec}$ and recall from Example 4.9(b) that $X \otimes X^* \cong \mathsf{Mat}_n(\Bbbk)$ as $\Bbbk$-algebras, for $n = \dim_{\Bbbk} X$. Now X is an (absolutely) simple object of $\mathcal{C}$ if and only if $X \cong \Bbbk$. So, we recover the precise condition for $\mathsf{Mat}_n(\Bbbk)$ to be connected (namely, $n = 1$) discussed above.

Exploring connectedness when $\mathcal{C} = G\text{-}\mathsf{Mod}$, for a group G, is Exercise 4.58.

One useful characterization of connectedness is given as follows; this expands on Example 1.23, and on Exercise 1.26(a), in the case when $\mathcal{C} = \mathsf{Vec}$.

Proposition 4.72. *Take an algebra* (A, m_A, u_A) *in* $\mathcal{C}$*, and consider the following triple:*

$$\left(\mathsf{Hom}_{\mathcal{C}}(\mathbb{1}, A), \quad m_{\mathsf{Hom}_{\mathcal{C}}(\mathbb{1},A)}(f \otimes f') := m_A\,(f \otimes f')\,\ell_{\mathbb{1}}^{-1}, \quad u_{\mathsf{Hom}_{\mathcal{C}}(\mathbb{1},A)}(f) := u_A \right).$$

Here, f and f' are morphisms in the $\Bbbk$-vector space $\mathsf{Hom}_{\mathcal{C}}(\mathbb{1}, A)$.

(a) *We have that* $(\mathsf{Hom}_{\mathcal{C}}(\mathbb{1}, A), m_{\mathsf{Hom}_{\mathcal{C}}(\mathbb{1},A)}, u_{\mathsf{Hom}_{\mathcal{C}}(\mathbb{1},A)})$ *is a $\Bbbk$-algebra.*

(b) *We have that as $\Bbbk$-algebras:*

$$\mathsf{Hom}_{\mathcal{C}}(\mathbb{1}, A) \cong \mathsf{End}_{\mathsf{Mod}\text{-}A(\mathcal{C})}((A_{\mathrm{reg}})_A), \qquad \mathsf{Hom}_{\mathcal{C}}(\mathbb{1}, A)^{\mathrm{op}} \cong \mathsf{End}_{A\text{-}\mathsf{Mod}(\mathcal{C})}({}_A(A_{\mathrm{reg}})).$$

(c) *A is connected if and only if any of the $\Bbbk$-algebras in part (b) is 1-dimensional.*

Proof. Most of the details are reserved as Exercise 4.59, including the verification of part (a). Towards part (b), the first isomorphism is given below:

$$\phi : \mathsf{Hom}_{\mathcal{C}}(\mathbb{1}, A) \to \mathsf{End}_{\mathsf{Mod}\text{-}A(\mathcal{C})}((A_{\mathrm{reg}})_A), \quad f \mapsto m_A\,(f \otimes \mathrm{id}_A)\,\ell_A^{-1}.$$

The inverse ψ of ϕ is defined by $\psi(g) := g\,u_A$, for $g \in \mathsf{End}_{\mathsf{Mod}\text{-}A(\mathcal{C})}((A_{\mathrm{reg}})_A)$. Part (c) follows immediately from part (b). $\square$

§4.9.2. Indecomposable algebras

Recall from §4.5.1i that an algebra A in $\mathcal{C}$ is **indecomposable** if it is not isomorphic as algebras to a biproduct of nonzero algebras in $\mathcal{C}$.

Remark 4.73. (a) Connected algebras are indecomposable. Indeed, if $A \cong A_1 \,\square\, A_2$, for nonzero algebras A_1, A_2 in $\mathcal{C}$, then the morphisms u_{A_1}, u_{A_2} yield morphisms $u_i : \mathbb{1} \to A_i \to A$, for $i = 1, 2$, that are not scalar multiples of each other.

(b) Indecomposable algebras are not necessarily connected. For instance, for each $n \geq 1$, $\mathrm{Mat}_n(\Bbbk)$ is indecomposable as a $\Bbbk$-algebra since $Z(\mathrm{Mat}_n(\Bbbk)) \cong \Bbbk$ [Proposition 1.17]. But $\mathrm{Mat}_n(\Bbbk)$ is not connected when $n > 1$ [Example 4.71(c)].

Next, we continue our running examples of the section.

Example 4.74. Consider the unit algebra $\mathbb{1}$ [Example 4.1(a)]. Also, consider the $\mathcal{C}$-endomorphism algebra $X \otimes X^*$, for X a nonzero object of $\mathcal{C}$ [Example 4.9(a)].

(a) The algebra $\mathbb{1}$ is indecomposable by Example 4.71(a) and Remark 4.73(a).

(b) The algebra $X \otimes X^*$ is indecomposable by Example 4.80(b) and Remark 4.78(a) in §4.9.3 below.

Even though the connected condition is not Morita invariant, indecomposability is module-theoretic due to the result below (cf. Exercise 3.14).

Proposition 4.75. *Take $A, A_1, A_2 \in \mathrm{Alg}(\mathcal{C})$, and recall Proposition 4.30. Then, the following statements hold.*

(a) *$(A_1 \,\square\, A_2)$-$\mathrm{Mod}(\mathcal{C}) \simeq A_1$-$\mathrm{Mod}(\mathcal{C}) \times A_2$-$\mathrm{Mod}(\mathcal{C})$ as right $\mathcal{C}$-module categories.*

(b) *Mod-$(A_1 \,\square\, A_2)(\mathcal{C}) \simeq \mathrm{Mod}$-$A_1(\mathcal{C}) \times \mathrm{Mod}$-$A_2(\mathcal{C})$ as left $\mathcal{C}$-module categories.*

(c) *If A-$\mathrm{Mod}(\mathcal{C})$ (or, $\mathrm{Mod}(\mathcal{C})$-$A$) is indecomposable as a right (or, left) $\mathcal{C}$-module category, then A is an indecomposable algebra in $\mathcal{C}$.*

Proof. The equivalence in part (a) holds via the functors below,

$$F : (A_1 \,\square\, A_2)\text{-}\mathrm{Mod}(\mathcal{C}) \longrightarrow A_1\text{-}\mathrm{Mod}(\mathcal{C}) \times A_2\text{-}\mathrm{Mod}(\mathcal{C}),$$

$$G : A_1\text{-}\mathrm{Mod}(\mathcal{C}) \times A_2\text{-}\mathrm{Mod}(\mathcal{C}) \longrightarrow (A_1 \,\square\, A_2)\text{-}\mathrm{Mod}(\mathcal{C}),$$

where $F(M, \triangleright : (A_1 \square A_2) \otimes M \to M) := ((M, \triangleright \alpha_1), (M, \triangleright \alpha_2))$. Here, we use Lemma 3.4, and the morphisms below that come from the universal property of the biproduct:

$$\alpha_1 : A_1 \otimes M \to (A_1 \otimes M) \,\square\, (A_2 \otimes M), \qquad \alpha_2 : A_2 \otimes M \to (A_1 \otimes M) \,\square\, (A_2 \otimes M).$$

Also, $G((M_1, \triangleright_1), (M_2, \triangleright_2)) := (M_1 \,\square\, M_2, \triangleright')$, where by way of Lemma 3.4, we define:

$$\triangleright' := \triangleright_1 \,\square\, {}_{A_1 \otimes M_2}\vec{0} \,\square\, {}_{A_2 \otimes M_1}\vec{0} \,\square\, \triangleright_2 : (A_1 \,\square\, A_2) \otimes (M_1 \,\square\, M_2) \to M_1 \,\square\, M_2.$$

We leave it to the reader to complete the details. Part (b) follows likewise. Part (c) then follows from parts (a,b). $\qquad \square$

Exploration 4.76. The reader is encouraged to investigate a monoidal generalization of Propositions 1.17 and 1.22 in Chapter 1. Those results provide characterizations of indecomposability for algebras and modules in Vec in terms of idempotent elements.

§4.9.3. Simple algebras

Next, we generalize the notion of simplicity for algebras, introduced in §1.5.2 for algebras over a field. We say that an algebra $A \in \mathsf{Alg}(\mathcal{C})$ is **simple** if the only ideals of A are the zero ideal and itself; see §4.2.1 and Example 4.10.

Simple algebras in the monoidal context have not been explored deeply in the literature to date, so many of the results in §§1.5.2, 1.5.3 are open for generalization. In particular, the classification of finite-dimensional simple $\Bbbk$-algebras is achieved via the Cartan-Wedderburn Theorem [Theorem 1.37], but its generalization to the monoidal context is an open problem.

Research Problem 4.77. (a) Establish a generalization of the Cartan-Wedderburn Theorem [Theorem 1.37] for algebras in $\mathcal{C}$. A finiteness hypothesis should be imposed since finite-dimensionality is needed for the setting of $\Bbbk$-algebras.

(b) Provide an example of an 'infinite' simple algebra in $\mathcal{C}$, outside of the classification of simple algebras in part (a). Compare to Remark 1.39.

Towards resolving this problem and towards understanding simple algebras in $\mathcal{C}$ in general, consider the following results.

Remark 4.78. (a) A simple algebra in $\mathcal{C}$ is indecomposable. Indeed, if $A \cong A_1 \square A_2$ is a decomposable algebra in $\mathcal{C}$, then $(A_1)_{\mathrm{reg}} \square 0$ is a proper ideal of A in $\mathcal{C}$.

(b) Indecomposable algebras are not necessarily simple. Recall from Remark 1.33 that $\Bbbk[v]$ is an indecomposable $\Bbbk$-algebra that is not a simple $\Bbbk$-algebra.

Next, we will see that simplicity is Morita invariant.

Proposition 4.79. *Let A and B be Morita equivalent algebras in $\mathcal{C}$. If B is simple as an algebra in $\mathcal{C}$, then so is A.*

Proof. We proceed in the strict case via the Strictification Theorem [Theorem 3.26]. Since A and B are Morita equivalent, there are bimodules $P \in (A, B)\text{-}\mathsf{Bimod}(\mathcal{C})$ and $Q \in (B, A)\text{-}\mathsf{Bimod}(\mathcal{C})$ such that $P \otimes_B Q \cong A_{\mathrm{reg}}$ as A-bimodules and $Q \otimes_A P \cong B_{\mathrm{reg}}$ as B-bimodules, due to the Generalized Morita Theorem [Theorem 4.56]. For an ideal I of A in $\mathcal{C}$, we have that $Q \otimes_A I \otimes_A P$ is isomorphic to an ideal of B in $\mathcal{C}$ via the morphism below:

$$\iota : Q \otimes_A I \otimes_A P \xrightarrow{\;\mathrm{id}_Q \,\otimes_A\, \iota_I^A \,\otimes_A\, \mathrm{id}_P\;} Q \otimes_A A \otimes_A P \cong Q \otimes_A P \cong B.$$

(We leave the details to the reader; see also Lemma 4.44.) Now if B is simple, then $Q \otimes_A I \otimes_A P$ is isomorphic to 0, or to B, as B-bimodules in $\mathcal{C}$. This implies that

$$I \cong A \otimes_A I \otimes_A A \cong P \otimes_B Q \otimes_A I \otimes_A P \otimes_B Q$$

is isomorphic to $P \otimes_B 0 \otimes_B Q \cong 0$, or to $P \otimes_B B \otimes_B Q \cong A$, as A-bimodules in $\mathcal{C}$. This implies that A is a simple algebra in $\mathcal{C}$. $\qquad\qquad\square$

Example 4.80. Take the unit algebra $\mathbb{1}$ [Example 4.1(a)], and the $\mathcal{C}$-endomorphism algebra $X \otimes X^*$, for X a nonzero object of $\mathcal{C}$ [Example 4.9(a)].

(a) The algebra $\mathbb{1}$ is simple since $\mathbb{1}$ as an object is simple.

(b) The algebra $X \otimes X^*$ is simple via Proposition 4.79 and part (a) because it is Morita equivalent to the unit algebra $\mathbb{1}$ [Example 4.58].

§4.9.4. Semisimple algebras

As one would expect, the study of semisimple algebras in the monoidal context prompts more questions than answers at this point. The definition of a semisimple $\mathbb{k}$-algebra in §1.6 is module-theoretic, but we obtain by the Artin-Wedderburn theorem [Theorem 1.44] an intrinsic characterization of such algebras.

Following the approach in §1.6, we say that $A \in \mathsf{Alg}(\mathcal{C})$ is **semisimple** if A-Mod($\mathcal{C}$) is a semisimple category.

Note that semisimplicity for algebras in $\mathcal{C}$ is Morita invariant.

Example 4.81. Take the unit algebra $\mathbb{1}$ from Example 4.1(a), and consider the $\mathcal{C}$-endomorphism algebra $X \otimes X^*$, for X a nonzero object of $\mathcal{C}$, from Example 4.9(a).

(a) The algebra $\mathbb{1}$ is semisimple if and only if $\mathcal{C}$ is semisimple because $\mathbb{1}$-Mod($\mathcal{C}$) $\cong \mathcal{C}$ [Example 4.29].

(b) The algebra $X \otimes X^*$ is semisimple if and only if $\mathcal{C}$ is semisimple, due to part (a) and $X \otimes X^*$ being Morita equivalent to $\mathbb{1}$ [Example 4.58].

Following up with Research Problem 4.77, we also propose the task below.

Research Problem 4.82. Establish a version of the Artin-Wedderburn theorem [Theorem 1.44] in the (semisimple) monoidal context, thereby providing an intrinsic description of semisimple algebras in $\mathcal{C}$.

One useful (module-theoretic) characterization is given below, which partially generalizes Proposition 2.57 in §2.8.3.

Proposition 4.83. *An algebra A in $\mathcal{C}$ is semisimple if and only if each object in A-Mod($\mathcal{C}$) (or in Mod-$A(\mathcal{C})$) is projective.*

Proof. We will discuss the arguments for left modules; the arguments for right modules hold similarly. If A is a semisimple algebra in $\mathcal{C}$, then A-Mod($\mathcal{C}$) is a semisimple, abelian category by the definition and by Remark 4.70. Then, each object in A-Mod($\mathcal{C}$) is projective by Corollary 2.56.

Conversely, suppose that each object in A-Mod($\mathcal{C}$) is projective. Then, similar to the proof of Proposition 2.57, it suffices to show that every left ideal I of A has a complement in $\mathcal{C}$. Indeed, the short exact sequence $0 \to I \to A \to A/I \to 0$ splits in A-Mod($\mathcal{C}$). So, $I \,\square\, (A/I) \cong A$ in A-Mod($\mathcal{C}$), as desired. $\qquad\square$

§4.9.5. Separable algebras

Now we generalize the notion of separability from §1.7.

An algebra (A, m, u) in $\mathcal{C}$ is said to be **separable** if there exists a right-inverse $\phi : A \to A \otimes A$ of m in A-Bimod($\mathcal{C}$). Here, $m\phi = \mathrm{id}_A$, and $\phi \in A$-Bimod($\mathcal{C}$) with the A-actions on A given by $\triangleright_A = m = \triangleleft_A$, and with the A-actions on $A \otimes A$ given by $\triangleright_{A \otimes A} = (m \otimes \mathrm{id}_A)\, a_{A,A,A}^{-1}$ and $\triangleleft_{A \otimes A} = (\mathrm{id}_A \otimes m)\, a_{A,A,A}$.

Let us continue the running example of this section.

Example 4.84. Consider the unit algebra $\mathbb{1}$ from Example 4.1(a). Also, consider the $\mathcal{C}$-endomorphism algebra $X \otimes X^*$, for X a nonzero object of $\mathcal{C}$, from Example 4.9(a).

(a) The algebra $(\mathbb{1}, m_{\mathbb{1}} := \ell_{\mathbb{1}}, u_{\mathbb{1}} := \mathrm{id}_{\mathbb{1}})$ in $\mathcal{C}$ is separable by using $\phi_{\mathbb{1}} = \ell_{\mathbb{1}}^{-1}$.

(b) Suppose that $\mathcal{C}$ is pivotal [§3.7], and take $\mathcal{C}$ to be strict via Theorem 3.26. Moreover, take an object $X \in \mathcal{C}$ with invertible pivotal dimension, $\dim_j X$. (Recall that $\dim_j X$ is the morphism $\mathrm{ev}_X\, (\mathrm{id}_{X^\vee} \otimes j_X^{-1})\, \mathrm{coev}_{X^\vee} : \mathbb{1} \to \mathbb{1}$.) Then, the algebra $A := X \otimes X^\vee$ is separable. Here, $m_A := \mathrm{id}_X \otimes \mathrm{ev}_X \otimes \mathrm{id}_{X^\vee}$, and we use

$$\phi_A := (\dim_j X)^{-1} [\mathrm{id}_X \otimes (\mathrm{id}_{X^\vee} \otimes j_X^{-1})\, \mathrm{coev}_{X^\vee} \otimes \mathrm{id}_{X^\vee}].$$

Indeed, $m_A \phi_A = \mathrm{id}_A$. Moreover, $\phi_A \in A$-Mod($\mathcal{C}$) since

$$
\begin{aligned}
\phi_A \circ \triangleright_A &= (\dim_j X)^{-1} [\mathrm{id}_X \otimes (\mathrm{id}_{X^\vee} \otimes j_X^{-1})\, \mathrm{coev}_{X^\vee} \otimes \mathrm{id}_{X^\vee}] [\mathrm{id}_X \otimes \mathrm{ev}_X \otimes \mathrm{id}_{X^\vee}] \\
&= (\dim_j X)^{-1} [\mathrm{id}_X \otimes \mathrm{ev}_X \otimes \mathrm{id}_{X^\vee \otimes X \otimes X^\vee}] [\mathrm{id}_{X \otimes X^\vee \otimes X} \otimes (\mathrm{id}_{X^\vee} \otimes j_X^{-1})\, \mathrm{coev}_{X^\vee} \otimes \mathrm{id}_{X^\vee}] \\
&= \triangleright_{A \otimes A} \circ (\mathrm{id}_A \otimes \phi_A).
\end{aligned}
$$

Likewise, $\phi_A \in $ Mod-A($\mathcal{C}$). Therefore, $\phi_A \in A$-Bimod($\mathcal{C}$).

An exercise on separable monads [§4.3.2] is provided in Exercise 4.60.

Next, we see that when $\mathcal{C}$ is semisimple, separability is a module-theoretic property. This result first appeared as Proposition 2.7 of Davydov et al. [2013].

Proposition 4.85. *Assume that $\mathcal{C}$ is a multifusion category. Then, the following statements are equivalent for $A \in \mathrm{Alg}(\mathcal{C})$.*

(a) *A is separable.*

(b) *A-Mod($\mathcal{C}$) is semisimple.*

(c) *Mod-$A(\mathcal{C})$ is semisimple.*

(d) *A-Bimod($\mathcal{C}$) is semisimple.*

Proof. For part (a) to imply part (b), suppose that A is a separable algebra. Then, by Lemma 4.2, we get the short exact sequence in A-Bimod($\mathcal{C}$):

$$0 \longrightarrow \ker(m_A) \longrightarrow A \otimes A \xrightarrow{m_A} A \longrightarrow 0.$$

An argument similar to the proof of Proposition 4.13 shows why $K := \ker(m_A)$ is in A-Bimod($\mathcal{C}$). Next, Proposition 2.45, along with A being separable, implies that $A \otimes A \cong A \,\square\, K$ in A-Bimod($\mathcal{C}$). Then, for any $M \in A$-Mod($\mathcal{C}$) in $\mathcal{C}$, we obtain that:

$$\begin{aligned}
A \otimes M &\cong (A \otimes A) \otimes_A M \;\cong\; (A \,\square\, K) \otimes_A M \\
&\cong (A \otimes_A M) \,\square\, (K \otimes_A M) \;\cong\; M \,\square\, (K \otimes_A M),
\end{aligned}$$

as left A-modules in $\mathcal{C}$. The isos above follow from Lemmas 4.44 and 3.4, and using $\otimes_A$ as the monoidal product of A-Bimod($\mathcal{C}$). Now the adjunctions in Exercise 4.25(c) and Proposition 3.33(d) imply that for all $Z \in A$-Mod($\mathcal{C}$):

$$\begin{aligned}
\mathrm{Hom}_{A\text{-Mod}(\mathcal{C})}(A \otimes M, Z) &\cong \mathrm{Hom}_{\mathcal{C}}(M, \mathrm{Forg}(Z)) \\
&\cong \mathrm{Hom}_{\mathcal{C}}(\mathbb{1} \otimes M, \mathrm{Forg}(Z)) \;\cong\; \mathrm{Hom}_{\mathcal{C}}(\mathbb{1}, \mathrm{Forg}(Z) \otimes M^*).
\end{aligned}$$

The functor Forg is exact, and the functor $- \otimes M^*$ is exact by Proposition 3.68. Also, $\mathcal{C}$ is semisimple, so $\mathbb{1}$ is projective by Corollary 2.56; thus, $\mathrm{Hom}_{\mathcal{C}}(\mathbb{1}, -)$ is exact [Propositions 2.52(a) and 2.53(a)]. The composition of these functors is then exact, and via the isomorphism above, $\mathrm{Hom}_{A\text{-Mod}(\mathcal{C})}(A \otimes M, -)$ is exact. Hence, $A \otimes M$ is a projective object in A-Mod($\mathcal{C}$) by Proposition 2.53(a). Since $A \otimes M \cong M \,\square\, (K \otimes_A M)$ in A-Mod($\mathcal{C}$), Corollary 2.55 implies that M is a projective object in A-Mod($\mathcal{C}$). So, A-Mod($\mathcal{C}$) is semisimple by Proposition 4.83.

Next, part (b) implies part (d) as follows. Note that A-Bimod($\mathcal{C}$) is equivalent to $\mathrm{Rex}_{\mathrm{Mod}\text{-}\mathcal{C}}(A\text{-Mod}(\mathcal{C}), A\text{-Mod}(\mathcal{C}))$ by the Generalized Eilenberg-Watts Theorem [Theorem 4.54]. The latter category is then semisimple by part (b) and by the powerful result, Theorem 2.18 of Etingof et al. [2005]: $\mathrm{Rex}_{\mathrm{Mod}\text{-}\mathcal{C}}(\mathcal{M}_1, \mathcal{M}_2)$ is semisimple when the right $\mathcal{C}$-module categories $\mathcal{M}_1$ and $\mathcal{M}_2$ are semisimple. Indeed, semisimplicity is preserved across an equivalence of categories.

Likewise, part (a) implies part (c), and part (c) implies part (d).

Part (d) implies part (a) due to Proposition 2.46 applied to the epi $m_A : A \otimes A \to A$ in A-Bimod($\mathcal{C}$); see Lemma 4.2. $\qquad\square$

Remark 4.86. Regarding the proof of Proposition 4.85, observe the following.

(a) Separability and semisimplicity are interchangeable for algebras in multifusion categories. Semisimplicity is both right- and left-module theoretic here.

(b) Since FdVec is multifusion, Proposition 4.85 recovers Proposition 1.57. Namely, algebras in FdVec are semisimple if and only if they are separable.

(c) Resolving Research Problem 4.77 (Generalized Cartan-Wedderburn Theorem) should imply that 'finite' simple algebras are separable; see Example 1.54. One may want to first consider this problem in the semisimple setting.

(d) But separability and semisimplicity are not interchangeable outside of the semisimple setting. Namely, the unit algebra $\mathbb{1}$ is always separable [Example 4.84(a)], but it is not necessarily semisimple [Example 4.81(a)].

§4.9.6. Exact algebras

Separable and semisimple algebras are best behaved in semisimple categories; see, e.g., Proposition 4.85 and Remark 4.86. But for the nonsemisimple setting, the following module-theoretic property for algebras is often imposed.

We say that an algebra A in C is **exact** if the category A-$\mathrm{Mod}(C)$ is exact as a right C-module category; that is, for all projective $P \in C$ and all $M \in A$-$\mathrm{Mod}(C)$, we have that $M \otimes P$ is a projective object in A-$\mathrm{Mod}(C)$ [§3.10.4, Proposition 4.30].

As defined here, exactness is left-module theoretic, but sometimes a right-module theoretic definition is used in the literature. To resolve this ambiguity (as done for separable/semisimple algebras in the semisimple setting), and towards furthering the applications of exact algebras/modules, many researchers as of the date of publication have asked the question below.

Research Question 4.87. Is there an intrinsic notion of an exact algebra in C?

Towards resolving this question, characterizations of exact module categories (and, thus, of exact algebras) are useful; one from Sections 7.6 and 7.9 of Etingof et al. [2015] is given below.

Proposition 4.88. *Suppose that C is a finite tensor category. A left C-module category $\mathcal{M}$ is exact if and only if any left C-module functor from $\mathcal{M}$ is exact. A similar statement holds for right module categories.*

Proof. Suppose that $\mathcal{M}$ is exact, and take a left C-module functor $F : \mathcal{M} \to \mathcal{N}$. Suppose there exists a short exact sequence $0 \to M' \to M \to M'' \to 0$ in $\mathcal{M}$, but $0 \to F(M') \to F(M) \to F(M'') \to 0$ is not exact in $\mathcal{N}$, by way of contradiction. Now for any projective $P \in C$, we get that $0 \to P \rhd M' \to P \rhd M \to P \rhd M'' \to 0$ is exact as $(P \rhd -)$ is exact [Propositions 2.49(b), 3.45], and is split as $P \rhd M''$ is projective

[Proposition 2.53]. Thus, $0 \to F(P \rhd M') \to F(P \rhd M) \to F(P \rhd M'') \to 0$ is exact by Proposition 2.49(c). Since F is a module category functor, we obtain that

$$0 \to P \rhd F(M') \to P \rhd F(M) \to P \rhd F(M'') \to 0$$

is exact. This contradicts the non-exactness of $0 \to F(M') \to F(M) \to F(M'') \to 0$, since $(P \rhd -)$ is exact. Thus, F is an exact functor.

Conversely, suppose that any left $\mathcal{C}$-module functor $F : \mathcal{M} \to \mathcal{N}$ is exact. Then, the internal Hom functor $\underline{\mathrm{Hom}}(M, -) : \mathcal{M} \to \mathcal{C}$ is exact for any $M \in \mathcal{M}$. (Internal Homs exist here by Corollary 3.72.) Therefore, by Proposition 2.53,

$$\mathrm{Hom}_{\mathcal{M}}(P \rhd M, -) \cong \mathrm{Hom}_{\mathcal{C}}(P, \underline{\mathrm{Hom}}(M, -))$$

is exact, for any projective object $P \in \mathcal{C}$. Therefore, $P \rhd M$ is projective in this case, and again, by Proposition 2.53, $\mathcal{M}$ is exact. $\qquad\square$

Another characterization is given in terms of *relative Serre functors* when $\mathcal{C}$ is a finite tensor category; see Proposition 4.24 of Fuchs et al. [2020] for details.

Returning to exact algebras, consider the following observations and examples.

Remark 4.89. Semisimple algebras in $\mathcal{C}$ are exact. Indeed, if A is semisimple, then A-Mod($\mathcal{C}$) is semisimple and all left A-modules in $\mathcal{C}$ are projective [Proposition 4.83]. Therefore, for any projective $P \in \mathcal{C}$ and any $M \in A$-Mod($\mathcal{C}$), we obtain that $M \otimes P$ is a projective left A-module in $\mathcal{C}$, as required.

Example 4.90. Take the unit algebra $\mathbb{1}$ from Example 4.1(a), and consider the $\mathcal{C}$-endomorphism algebra $X \otimes X^*$, for X a nonzero object of $\mathcal{C}$, from Example 4.9(a).

(a) The unit algebra $\mathbb{1}$ is always exact due to Example 4.29 and Proposition 3.69.

(b) Exactness for algebras is Morita invariant. So, the algebra $X \otimes X^*$ is exact by part (a) and Example 4.58.

§4.10. Bimodules and beyond

The final bit of material that we will introduce in this book is on bimodules over algebras in monoidal categories– on their categorical structure, on a related notion of sameness, and briefly, on their higher category theory.

Standing hypothesis. Assume that $\mathcal{C}$ is a tensor category as in §3.10; see also §3.1.3 and Remark 4.14, along with Figure 3.15, for related conditions.

---

§4.10.1. On the monoidal category of bimodules

The main result of the section is on how close a monoidal category of bimodules (from §4.5.2iii) is to being (multi)tensor or (multi)fusion. Compare to Figure 3.15. See also Section 7.11 of Etingof et al. [2015].

Theorem 4.91. *Take $A \in \mathrm{Alg}(\mathcal{C})$. Then, the following statements hold for the monoidal category $(A\text{-Bimod}(\mathcal{C}), \otimes_A, A_{\mathrm{reg}})$.*

(a) *It is always abelian, linear, locally finite, and $\otimes_A$ is bilinear on morphisms.*

(b) *It has an absolutely simple monoidal unit when A is connected.*

(c) *It is finite when $\mathcal{C}$ is finite.*

(d) *It is semisimple when $\mathcal{C}$ is semisimple and A is separable.*

(e) *It is rigid when $\mathcal{C}$ is finite and A is exact.*

As a consequence, $(A\text{-Bimod}(\mathcal{C}), \otimes_A, A_{\mathrm{reg}})$ is a finite tensor (resp., fusion) category when $\mathcal{C}$ is finite tensor (resp., fusion), A is connected, and A is exact (resp., separable).

Proof. (a) The category $A\text{-Bimod}(\mathcal{C})$ is abelian and $\Bbbk$-linear by Proposition 4.33. Moreover, we leave it to the reader to consider how $A\text{-Bimod}(\mathcal{C})$ inherits the locally finite condition from $\mathcal{C}$, and how $\otimes_A$ is bilinear on morphisms as this holds for $\otimes$.

(b) Since $\mathrm{End}_{A\text{-Bimod}(\mathcal{C})}(A_{\mathrm{reg}})$ contains the identity map, it is a nonzero subspace of $\mathrm{End}_{\mathrm{Mod}\text{-}A(\mathcal{C})}(A_{\mathrm{reg}})$. The latter is 1-dimensional since A is connected [Proposition 4.72(c)]. Thus, $\dim_{\Bbbk}\mathrm{End}_{A\text{-Bimod}(\mathcal{C})}(A_{\mathrm{reg}}) = 1$ as required.

(c) We have that $A\text{-Bimod}(\mathcal{C})$ is locally finite from part (a). Also, $A\text{-Bimod}(\mathcal{C})$ has finitely many isoclasses of simple objects as $\mathcal{C}$ does. Next, take $M \in A\text{-Bimod}(\mathcal{C})$, and (by abusing notation) consider the object $M \in \mathcal{C}$ after forgetting the A-actions. Since $\mathcal{C}$ has enough projectives, there exists an epi $\pi_M : P(M) \to M$ in $\mathcal{C}$, with $P(M)$ a projective object in $\mathcal{C}$. Consider the morphism:

$$\pi_M^A : (A \otimes P(M)) \otimes A \xrightarrow{\ \mathrm{id}\,\otimes\,\pi_M\,\otimes\,\mathrm{id}\ } (A \otimes M) \otimes A \xrightarrow{\ \blacktriangleleft(\triangleright\,\otimes\,\mathrm{id})\ } M.$$

Now $(A \otimes P(M)) \otimes A \in A\text{-Bimod}(\mathcal{C})$ is projective via Propositions 2.53 and 3.33 as

$$\mathrm{Hom}_{A\text{-Bimod}(\mathcal{C})}((A \otimes P(M)) \otimes A, -) \cong \mathrm{Hom}_{\mathcal{C}}(P(M), {}^*A \otimes (- \otimes A^*))$$

is (right) exact. Namely, the functors $-\otimes A^*$ and ${}^*A \otimes -$ are exact by Proposition 3.68, and the functor $\mathrm{Hom}_{\mathcal{C}}(P(M), -)$ is exact by Proposition 2.53. Moreover, $\blacktriangleleft(\triangleright\,\otimes\,\mathrm{id})$ is epic (cf. Lemma 4.2), and so is $\mathrm{id}\,\otimes\,\pi_M\,\otimes\,\mathrm{id}$. Thus, π_M^A is epic. Thus, $A\text{-Bimod}(\mathcal{C})$ has enough projectives, and is finite as a result.

(d) This holds by Proposition 4.85.

(e) By the Generalized Eilenberg-Watts Theorem [Theorem 4.54], it suffices to show that $\mathrm{Rex}_{\mathrm{Mod}\text{-}\mathcal{C}}(A\text{-Mod}(\mathcal{C}), A\text{-Mod}(\mathcal{C}))$ is rigid. Duals in categories of endofunctors are adjoint functors [Example 3.41]. Now the result holds by Proposition 4.88, Corollary 2.62, and part (c). $\qquad\square$

Remark 4.92. Observe from Examples 3.37 and 3.38 that the monoidal category $(A\text{-Bimod}(\mathcal{C}), \otimes_A, A_{\mathrm{reg}})$ is often not rigid, even when $\mathcal{C} = \mathsf{FdVec}$. The exactness of the algebra A plays a key role in rigidity.

Remark 4.93. When $\mathcal{C}$ is a *braided* tensor category, one can form a *commutative* algebra in $\mathcal{C}$; this is discussed in a future volume. One can then construct a monoidal category $\mathrm{Rep}_{\mathcal{C}}(A)$ of A-bimodules in $\mathcal{C}$, where the right A-action on an object in $\mathcal{C}$ is induced by the left A-action (just as a left module over a commutative algebra A over $\Bbbk$ forms an A-bimodule over $\Bbbk$). If, further, $\mathcal{C}$ is finite and A is separable, then the monoidal category $(\mathrm{Rep}_{\mathcal{C}}(A), \otimes_A, A_{\mathrm{reg}})$ is rigid due to Proposition 3.11 and Lemma 4.20 in the work of Laugwitz and Walton [2023]. It is expected that this result holds when the separability of A is replaced by exactness.

Remark 4.94. A generalized notion of rigidity can also be used to examine monoidal categories of bimodules: namely, *Grothendieck-Verdier (GV-)duality*. This was introduced by Boyarchenko and Drinfeld [2013]. Here, the anti-equivalence $(-)^*$ is replaced with a more general duality functor, such that its quasi-inverse specializes to $^*(-)$. This circumvents the need for strong hypotheses on A, while achieving many results that one would get if A-Bimod$(\mathcal{C})$ is rigid. For instance, see the work of Fuchs et al. [2023] where it is shown that A-Bimod(FdVec) satisfies GV-duality. See also the work of Allen et al. [2021], where GV-duality is used to study the representation theory of *vertex operator algebras*.

Example 4.95. One vital collection of monoidal categories of bimodules are those formed by taking the twisted group algebras $\Bbbk L_\psi$ in the monoidal category Vec_G^ω; see Exercises 4.4 and 3.35. The resulting category, $(\Bbbk L_\psi)$-Bimod(Vec_G^ω) is called a **group-theoretical fusion category**, and is denoted by $\mathcal{C}(G, \omega, L, \psi)$. Verifying that $\mathcal{C}(G, \omega, L, \psi)$ is indeed a fusion category is Exercise 4.62. Group-theoretical fusion categories often serve as a test case for a result about fusion categories.

§4.10.2. Categorical Morita equivalence

By replacing algebras and their categories of modules with tensor categories and their module categories, we can consider a notion of sameness for tensor categories in the context of Morita equivalence. That is, for tensor categories $\mathcal{C}$ and $\mathcal{D}$, we want a way of saying that:

$$\mathcal{C}\text{-Mod} \simeq \mathcal{D}\text{-Mod}, \tag{4.96}$$

for these collections of module categories. To do this properly, one needs to go 'up' a categorical level as discussed in §4.10.3 below. Namely, we want to keep track module categories, module functors between them, and we want to compare such structures with transformations. But for now, we proceed as follows.

By Theorem 2.18 and Remark 2.20, we have that two $\Bbbk$-algebras A and B are Morita equivalent precisely when one of the equivalent conditions below holds.

($\star$) There exist bimodules $_AP_B$ and $_BQ_A$ such that $P \otimes_B Q \cong A_{\mathrm{reg}}$ as A-bimodules and $Q \otimes_A P \cong B_{\mathrm{reg}}$ as B-bimodules.

($\star\star$) We have that $B^{\mathrm{op}} \cong \mathrm{End}_{A\text{-Mod}}(M)$ as algebras, for some finitely generated, projective generator M of A-Mod.

Motivated by the characterization ($\star\star$), we say that tensor categories $\mathcal{C}$ and $\mathcal{D}$ are **categorically Morita equivalent** if there exists an exact left $\mathcal{C}$-module category $\mathcal{M}$ such that, as tensor categories:

$$\mathcal{D}^{\otimes\mathrm{op}} \overset{\otimes}{\simeq} (\mathrm{Rex}_{\mathcal{C}\text{-Mod}}(\mathcal{M}, \mathcal{M}), \circ, \mathrm{Id}_{\mathcal{M}}).$$

Categorical Morita equivalence is indeed an equivalence relation. See Section 7.12 of the book by Etingof et al. [2015] for details about this fact and more.

Remark 4.97. When the setting of (a generalized) Ostrik's Theorem holds [Theorem 4.67, Remark 4.68], we get that $\mathcal{M} \simeq \mathrm{Mod}\text{-}A(\mathcal{C})$, for some $A \in \mathrm{Alg}(\mathcal{C})$. Here,

$$\mathrm{Rex}_{\mathcal{C}\text{-Mod}}(\mathcal{M}, \mathcal{M}) \overset{\otimes}{\simeq} (A\text{-Bimod}(\mathcal{C}))^{\otimes\mathrm{op}}$$

as tensor categories by way of the Generalized Eilenberg-Watts Theorem [Remark 4.55]. If $F_1, F_2 \in \mathrm{Rex}_{\mathcal{C}\text{-Mod}}(\mathcal{M}, \mathcal{M})$ corresponds to $M_1, M_2 \in A\text{-Bimod}(\mathcal{C})$, respectively, then $F_1 \circ F_2 \in \mathrm{Rex}_{\mathcal{C}\text{-Mod}}(\mathcal{M}, \mathcal{M})$ corresponds to $M_2 \otimes_A M_1 \in A\text{-Bimod}(\mathcal{C})$. Therefore, $\mathcal{D}$ is categorically Morita equivalent to $\mathcal{C}$ in this setting if and only if there exists an exact algebra A in $\mathcal{C}$ such that, as tensor categories:

$$\mathcal{D} \overset{\otimes}{\simeq} (A\text{-Bimod}(\mathcal{C}), \otimes_A, A_{\mathrm{reg}}).$$

Example 4.98. Take a finite group G. We have that the tensor categories, G-Mod and Vec_G, are categorically Morita equivalent. Indeed, the forgetful functor $\mathrm{Vec}_G \to \mathrm{Vec}$ is strong monoidal (check this), so Vec is a left module category over Vec_G [Example 3.18]. This module category is exact since Vec is a semisimple, abelian category [§§2.7.3, 2.2.2iii, Corollary 2.56]. Now:

$$\mathrm{Rex}_{\mathrm{Vec}_G\text{-Mod}}(\mathrm{Vec}, \mathrm{Vec}) \overset{\otimes}{\simeq} (G\text{-Mod})^{\otimes\mathrm{op}},$$

given by sending a module functor $(F, s) \in \mathrm{Rex}_{\mathrm{Vec}_G\text{-Mod}}(\mathrm{Vec}, \mathrm{Vec})$ to the left G-module $F(\Bbbk) =: V$, with action $g \triangleright v := s_{\delta_g, \Bbbk}(v)$. Here, δ_g is the simple object of Vec_G from Exercise 3.35(a). Completing the details here is Exercise 4.63.

Next, we discuss briefly Morita equivalence in the context of $(\star)$ above.

Remark 4.99. Take tensor categories $\mathcal{C}$ and $\mathcal{D}$, along with a $(\mathcal{C}, \mathcal{D})$-bimodule category $\mathcal{P}$ and a $(\mathcal{D}, \mathcal{C})$-bimodule category $\mathcal{Q}$ (which are abelian as in §3.3.4).

- The **Deligne product** of $\mathcal{P}$ and $\mathcal{Q}$ is an abelian category $\mathcal{P} \boxtimes \mathcal{Q}$ consisting of objects $P \boxtimes Q$ for $P \in \mathcal{P}$ and $Q \in \mathcal{Q}$, and morphisms $f \boxtimes g$ for $f \in \mathcal{P}$ and $g \in \mathcal{Q}$.

- A functor $F : \mathcal{P} \boxtimes \mathcal{Q} \to \mathcal{Z}$ is $\mathcal{D}$-**balanced** if there is a natural isomorphism $\{F((P \lhd Y) \boxtimes Q) \xrightarrow{\sim} F(P \boxtimes (Y \rhd Q))\}_{P \in \mathcal{P}, Q \in \mathcal{Q}, Y \in \mathcal{D}}$ satisfying an associativity axiom.

- The **relative** (or **balanced**) **tensor product** of $\mathcal{P}$ and $\mathcal{Q}$ is an abelian category $\mathcal{P} \boxtimes_{\mathcal{D}} \mathcal{Q}$, equipped with a right exact $\mathcal{D}$-balanced functor $\mathcal{P} \boxtimes \mathcal{Q} \to \mathcal{P} \boxtimes_{\mathcal{D}} \mathcal{Q}$ that is universal amongst all right exact $\mathcal{D}$-balanced functors from $\mathcal{P} \boxtimes \mathcal{Q}$.

- Then, $\mathcal{P} \boxtimes_{\mathcal{D}} \mathcal{Q}$ is a $\mathcal{C}$-bimodule category. Also, $\mathcal{Q} \boxtimes_{\mathcal{C}} \mathcal{P}$ is a $\mathcal{D}$-bimodule category.

Now by Proposition 4.2 of Etingof et al. [2010], $\mathcal{C}$ and $\mathcal{D}$ are categorically Morita equivalent if and only if there exists $\mathcal{P}, \mathcal{Q}$ as above, where, as bimodule categories:

$$\mathcal{P} \boxtimes_{\mathcal{D}} \mathcal{Q} \simeq \mathcal{C}_{\mathrm{reg}}, \qquad \mathcal{Q} \boxtimes_{\mathcal{C}} \mathcal{P} \simeq \mathcal{D}_{\mathrm{reg}}.$$

Here, $\mathcal{P}$ and $\mathcal{Q}$ are said to be **invertible bimodule categories**.

Check out the article by Etingof et al. [2010], along with the work of Greenough [2013], for further details. See also the work of Douglas et al. [2019] on how to apply Ostrik's Theorem [Theorem 4.67] to this setting; compare to Remark 4.97.

§4.10.3. On the higher category theory of bimodules

When determining if two structures have the same representation theory, it is best to work in a framework that is increased by one categorical level. Indeed, two $\Bbbk$-algebras (at "categorical level 0") have the same representation theory precisely when their categories of modules (at "categorical level 1") are equivalent. Also, two tensor categories $\mathcal{C}$ and $\mathcal{D}$ (at "categorical level 1") have the same representation theory precisely when (4.96) holds, and this condition resides at "categorical level 2" in a sense. So, the goal of this section is to discuss briefly higher category theory. In particular, this framework captures the Morita theory of algebras and categorical Morita theory, both involving bimodules.

A *higher categorical structure* is an entity that is comprised of objects, morphisms between objects, morphisms between morphisms (that is, *2-morphisms*), and possibly 'higher' morphisms. How, precisely, these entities contain identities, compositions, and possess compatibilities between operations, vary from one higher categorical structure to another. For instance, 2-categorical structures involve objects, morphisms, and 2-morphisms, and these include *(strict) 2-categories, bicategories*, and *double categories*, just to mention a few. But we will only peek into this Pandora's box and not crack it wide open here...for there are beasts inside.

A **2-category** (also called a **strict 2-category**) $\mathcal{E}$ consists of the data (a)-(h) below.

(a) A collection of **0-cells**, $\mathcal{E}_0$.

(b) A collection of **1-cells**, $\mathcal{E}_1(X, Y)$, for any $X, Y \in \mathcal{E}_0$.

$$X \longrightarrow Y$$

(c) A collection of **2-cells**, $\mathcal{E}_2(f, f') := \mathcal{E}_2^{X,Y}(f, f')$, for any $f, f' \in \mathcal{E}_1(X, Y)$.

(d) An **identity 1-cell**, $\mathrm{id}_X \in \mathcal{E}_1(X, X)$, for any $X \in \mathcal{E}_0$.

(e) An **identity 2-cell**, $\mathrm{Id}_f := \mathrm{Id}_f^{X,Y} \in \mathcal{E}_2(f, f)$, for any $f \in \mathcal{E}_1(X, Y)$.

(f) A **horizontal composition of 1-cells**, $gf \in \mathcal{E}_1(W, Y)$, for $f \in \mathcal{E}_1(W, X), g \in \mathcal{E}_1(X, Y)$.

$$gf : W \xrightarrow{\ f\ } X \xrightarrow{\ g\ } Y$$

(g) A **vertical composition of 2-cells**, $F' \circ^{\mathrm{ver}} F \in \mathcal{E}_2^{X,Y}(f, f'')$, for any $F \in \mathcal{E}_2^{X,Y}(f, f')$ and $F' \in \mathcal{E}_2^{X,Y}(f', f'')$, with $f, f', f'' \in \mathcal{E}_1(X, Y)$.

(h) A **horizontal composition of 2-cells**, $G \circ^{\mathrm{hor}} F \in \mathcal{E}_2^{W,Y}(gf, g'f')$, which exists for any $F \in \mathcal{E}_2^{W,X}(f, f')$ and $G \in \mathcal{E}_2^{X,Y}(g, g')$, with $f, f' \in \mathcal{E}_1(W, X)$ and $g, g' \in \mathcal{E}_1(X, Y)$.

This data must satisfy the compatibility axioms (i)-(viii) below.

(i) **(1-cell horizontal composition associativity)**

$$(hg)f = h(gf) \quad \text{in } \mathcal{E}_1(W, Z),$$

for $f \in \mathcal{E}_1(W, X), g \in \mathcal{E}_1(X, Y), h \in \mathcal{E}_1(Y, Z)$.

(ii) **(1-cell horizontal composition unitality)**

$$\mathrm{id}_X f = f \quad \text{in } \mathcal{E}_1(W, X), \qquad g\, \mathrm{id}_X = g \quad \text{in } \mathcal{E}_1(X, Y),$$

for $f \in \mathcal{E}_1(W, X)$ and $g \in \mathcal{E}_1(X, Y)$.

(iii) **(2-cell vertical composition associativity)**

$$(F'' \circ^{\text{ver}} F') \circ^{\text{ver}} F \;=\; F'' \circ^{\text{ver}} (F' \circ^{\text{ver}} F) \quad \text{in } \mathcal{E}_2(f, f'''),$$

for $F \in \mathcal{E}_2(f, f')$, $F' \in \mathcal{E}_2(f', f'')$, $F'' \in \mathcal{E}_2(f'', f''')$.

(iv) **(2-cell vertical composition unitality)**

$$\mathrm{Id}_{f'} \circ^{\text{ver}} F = F \;\; \text{in } \mathcal{E}_2(f, f'), \qquad F' \circ^{\text{ver}} \mathrm{Id}_{f'} = F' \;\; \text{in } \mathcal{E}_2(f', f''),$$

for $F \in \mathcal{E}_2(f, f')$ and $F' \in \mathcal{E}_2(f', f'')$.

(v) **(2-cell horizontal composition associativity)**

$$(H \circ^{\text{hor}} G) \circ^{\text{hor}} F \;=\; H \circ^{\text{hor}} (G \circ^{\text{hor}} F) \;\; \text{in } \mathcal{E}_2(hgf, h'g'f'),$$

for $F \in \mathcal{E}_2(f, f')$, $G \in \mathcal{E}_2(g, g')$, $H \in \mathcal{E}_2(h, h')$.

(vi) **(2-cell horizontal composition unitality)**

$$\mathrm{Id}_{\mathrm{id}_X} \circ^{\text{hor}} F = F \;\; \text{in } \mathcal{E}_2^{W,X}(f, f'), \qquad G \circ^{\text{hor}} \mathrm{Id}_{\mathrm{id}_X} = G \;\; \text{in } \mathcal{E}_2^{X,Y}(g, g'),$$

for $F \in \mathcal{E}_2^{W,X}(f, f')$ and $G \in \mathcal{E}_2^{X,Y}(g, g')$.

(vii) **(interchange law for 2-cell identities)**

$$\mathrm{Id}_g \circ^{\text{hor}} \mathrm{Id}_f = \mathrm{Id}_{gf},$$

for $f \in \mathcal{E}_1(W, X)$ and $g \in \mathcal{E}_1(X, Y)$.

(viii) **(interchange law for 2-cell compositions)** As illustrated below,

$$(G' \circ^{\text{ver}} G) \circ^{\text{hor}} (F' \circ^{\text{ver}} F) \;=\; (G' \circ^{\text{hor}} F') \circ^{\text{ver}} (G \circ^{\text{hor}} F).$$

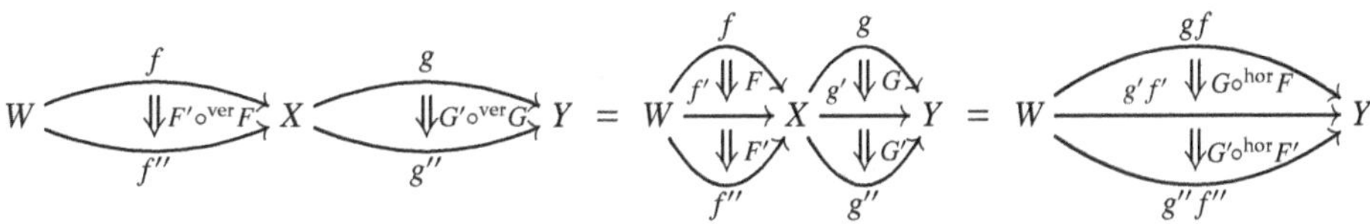

Remark 4.100. There are many underlying categories within the data above.

- The data (a), (b) (d), (f), along with axioms (i) and (ii), forms a traditional category as defined in §2.1.1.

- The data (b), (c), (e), (g), along with axioms (iii) and (iv), also forms a category.

- Likewise, the data (b), (c), (e), (h), with axioms (v) and (vi), forms a category.

Example 4.101. The **2-category of categories**, $\mathfrak{Cat}$, is defined by taking:

(a) $\mathfrak{Cat}_0$ to be the collection of all categories;

(b) $\mathfrak{Cat}_1(\mathcal{A}, \mathcal{B})$ to be the collection of functors from a category $\mathcal{A}$ to a category $\mathcal{B}$;

(c) $\mathfrak{Cat}_2^{\mathcal{A},\mathcal{B}}(F, F')$ to be the collection of natural transformations from a functor $F : \mathcal{A} \to \mathcal{B}$ to a functor $F' : \mathcal{A} \to \mathcal{B}$.

Verifying the details of this example is Exercise 4.64.

Returning to the theme of the section, let us consider some higher categorical data pertaining to bimodules. This is motivated by the Generalized Morita's Theorem [Theorem 4.56], which states that two algebras A and B in $\mathcal{C}$ are Morita equivalent if and only if there exists an invertible (A, B)-bimodule in $\mathcal{C}$. The 2-categorical structure that arises in this context is beyond the scope of strict 2-categories because one needs to consider invertibility using isomorphisms of bimodules. Equalities are too restrictive. Now consider the following notion.

A **bicategory** (also called a **weak 2-category**) consists of the data (a)–(h) from the definition of a strict 2-category, subject to the same axioms (iii) and (iv) for 2-cell vertical composition. However, the axioms (v) and (vi) for 2-cell horizontal composition are replaced with an *associator* and two *unitor* natural isomorphisms, subject to a pentagon axiom and a triangle axiom. As a result, a weaker version of axioms (i) and (ii) for 1-cell horizontal composition arise in this construction, and the interchange axioms (vii) and (viii) still hold. We refer the reader to Chapter 2 of the textbook by Johnson and Yau [2021] for more details.

Example 4.102. The bicategory of bimodules over algebras in $\mathcal{C}$, denoted by $\mathfrak{Bim}(\mathcal{C})$, is given by the following data.

(a) $\mathfrak{Bim}(\mathcal{C})_0$ as the collection of algebras in $\mathcal{C}$.

(b) $\mathfrak{Bim}(\mathcal{C})_1(A, B)$ as the collection of (A, B)-bimodules in $\mathcal{C}$, for algebras A, B in $\mathcal{C}$.

(c) $\mathfrak{Bim}(\mathcal{C})_2^{A,B}(M, N)$ as the morphisms $M \to N$ of (A, B)-bimodules in $\mathcal{C}$.

(d) Regular bimodules $_A(A_{\mathrm{reg}})_A$ as the identity 1-cells.

(e) Identity bimodule morphisms in $\mathcal{C}$ as the identity 2-cells.

(f) Tensor product $\otimes_A$ of bimodules in $\mathcal{C}$ as the horizontal composition of 1-cells.

(g) Bimodule morphism composition in $\mathcal{C}$ as the vertical composition of 2-cells.

(h) Tensor product $\otimes_A$ of bimodule morphisms in $\mathcal{C}$ as the horizontal composition of 2-cells.

Here, $\mathfrak{Bim}(\mathcal{C})$ fails to be a strict 2-category. Namely, axiom (i) fails as the tensor product of bimodules over algebras in $\mathcal{C}$ is not associative, nor unital, on the nose; this only holds up to isomorphism [Lemma 4.44]. But with $\mathfrak{Bim}(\mathcal{C})$ being a bicategory, the Morita equivalence of algebras in $\mathcal{C}$ is captured by the invertibility of 1-cells. For this reason, $\mathfrak{Bim}(\mathcal{C})$ is sometimes referred to as a *Morita 2-category*.

On a similar note, recall Remark 4.99, and consider the bicategory below.

Example 4.103. The bicategory of bimodule categories over tensor categories, denoted by $\mathfrak{BimCat}$, is given by:

(a) $\mathfrak{BimCat}_0$ as the collection of tensor categories;

(b) $\mathfrak{BimCat}_1(\mathcal{C}, \mathcal{D})$ as the collection of $(\mathcal{C}, \mathcal{D})$-bimodule categories, for tensor categories $\mathcal{C}$ and $\mathcal{D}$;

(c) $\mathfrak{BimCat}_2^{\mathcal{C}, \mathcal{D}}(\mathcal{M}, \mathcal{N})$ consisting of $(\mathcal{C}, \mathcal{D})$-bimodule functors $\mathcal{M} \to \mathcal{N}$.

Now, two tensor categories $\mathcal{C}$ and $\mathcal{D}$ are categorically Morita equivalent if and only if there exists an invertible 1-cell in the bicategory $\mathfrak{BimCat}_1(\mathcal{C}, \mathcal{D})$.

But this is getting quite fancy. So, we should end here.

References for further exploration

Here are some nice references on the higher category theory of bimodules.

- As mentioned in §4.10.2, the articles, "Fusion categories and homotopy theory" by Etingof et al. [2010] and "Monoidal 2-structure of bimodule categories" by Greenough [2013], are great references for the details of Example 4.103.

- The work in the articles, "A trace for bimodule categories" by Fuchs et al. [2017] and "State sum models with defects based on spherical fusion categories" by Meusburger [2023], use higher categorical structures involving bimodule categories to produce topological invariants. For the latter reference, see:

  Catherine Meusburger's 2023 SwissMAP Research Station workshop talk on "Turaev-Viro-Barrett-Westbury state sums with defects".
  `https://youtu.be/u0nLxahjQGU`

- In a different direction, the results in "Cell 2-representations of finitary 2-categories" by Mazorchuk and Miemietz [2011] introduce a way of generalizing the representation theory of $\Bbbk$-algebras within the framework of 2-categorical structures. The role of bimodules is prominent throughout their theory. See:

  Vanessa Miemietz's 2023 International Centre for Mathematical Sciences workshop talk on "Higher Representation Theory".
  `https://youtu.be/B1Xoh1ZS0qA`

§4.11. Summary

The goal of this final chapter was to upgrade the constructions and results in abstract algebra presented in Chapter 1 to the monoidal setting developed in Chapters 2 and 3. See Figure 4.13. We began with the definition of an algebra in a monoidal category, which boiled down to algebras over a field $\Bbbk$ when in the monoidal category $(\text{Vec}, \otimes_{\Bbbk}, \Bbbk)$. Monads are key examples of algebras; they arise in the monoidal category of endofunctors. General constructions of algebras are images of monoidal functors. In fact, right adjoints of strong monoidal functors are monoidal, so we can build algebras in monoidal categories via adjunction.

We also studied subalgebras, ideals, and quotient algebras here, in which case we needed to require that monoidal categories were abelian to work with kernels and cokernels of morphisms. Moreover, we examined the further hypotheses needed to form quotient algebras, and saw that rigidity suffices.

Next, we studied modules over algebras in the monoidal setting. Key construction included Eilenberg-Moore categories for monads (and this is not the entire category of modules over a monad). Operations on, and graded structures of, algebras and modules were discussed.

The three capstone results of the book, (1) the Generalized Eilenberg-Watts Theorem, (2) the Generalized Morita's Theorem, and (3) Ostrik's theorem were presented here. The first theorem is a generalization of a Chapter 2 result, on how right exact functors between categories of modules are given by tensoring with a bimodule. The second result is also a generalization of a result in Chapter 2, on when categories of modules are equivalent as module categories. The last result is new to the monoidal setting: it provides conditions when a module category over a monoidal category $\mathcal{C}$ is realized as a category of modules in $\mathcal{C}$. That is, Ostrik's theorem shows how external representations of $\mathcal{C}$ are realized as internal representations in $\mathcal{C}$; the algebras that resolve this problem are internal Ends.

We then examined several properties attached to algebras in monoidal categories. Some of these generalize properties in Chapter 1, and some are new to the monoidal setting. Moreover, some are intrinsic to the structure of the algebra, and others are defined by their categories of modules (pertaining to Morita theory). But there are still many research questions to consider here, like cooking up an extensive generalization of the Artin-Wedderburn Theorem from Chapter 1.

Lastly, we circled back to bimodules and studied the properties of their monoidal categorical structure, chatted about their higher categorical structure, and used these notions to examine a type of Morita equivalence for tensor categories.

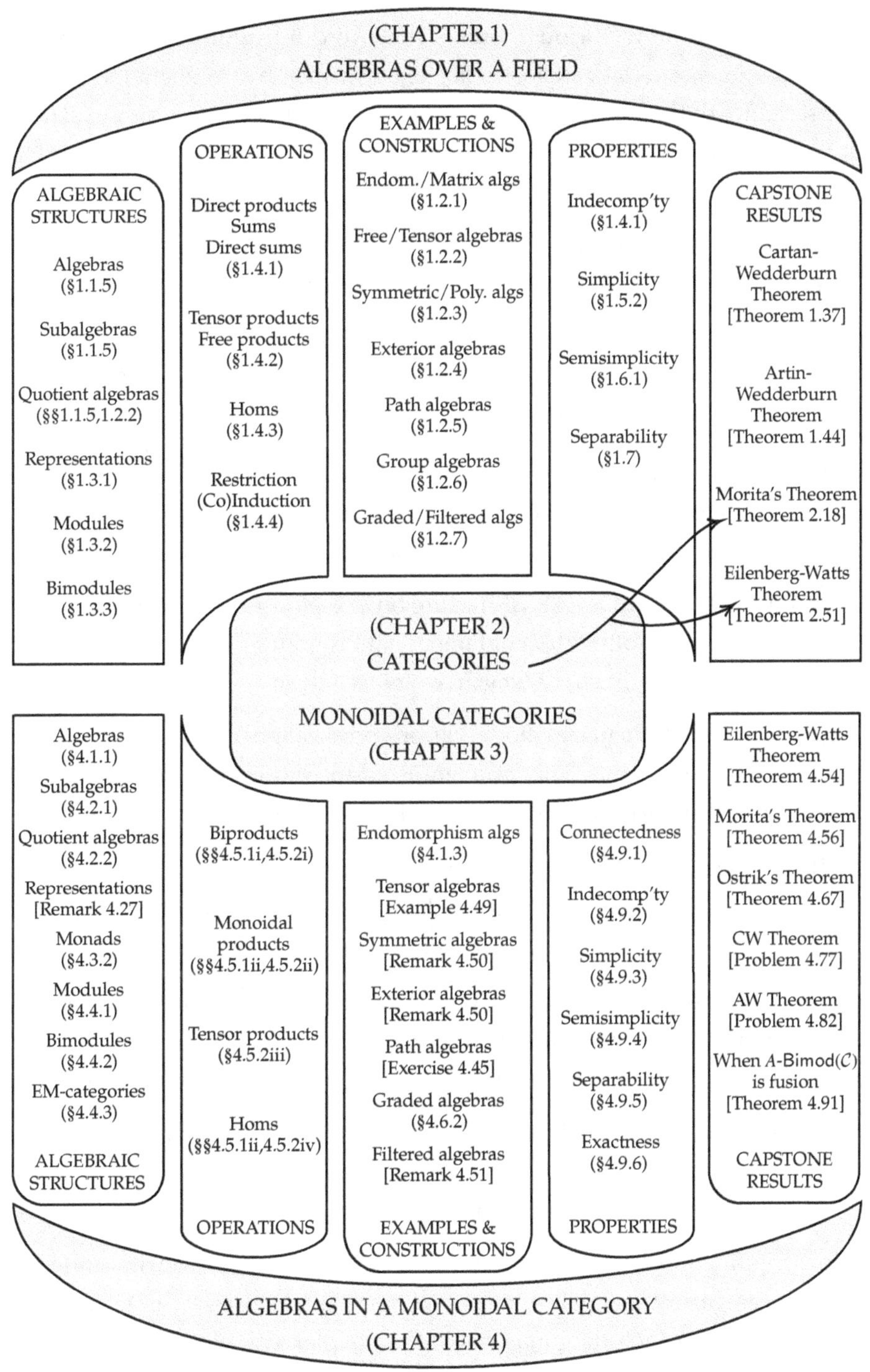

Figure 4.13: Summary of how Chapter 4 upgrades Chapter 1.

§4.12. Modern applications

We now illustrate how various notions that were introduced in this chapter on algebras in monoidal categories are used in modern mathematics. A full understanding of the resources here is not expected. Instead, we aim to put the chapter's material into context by offering videos and content to casually explore.

Group actions on algebras (that is, $A \in \text{Alg}(G\text{-Mod})$) is the original setting for Symmetries of Algebras. The friendly talk below discusses classical results on invariant subalgebras, and how they can be expanded in general **invariant theory**.

Ellen Kirkman's 2020 Simons Laufer Mathematical Sciences Institute lecture on
"Invariants of actions of Artin-Schelter regular algebras"
https://vimeo.com/908089874/949b3cf63c

Here is an accessible talk on monads in **functional programming**. It is through the lens of not getting bogged down in mathematical details (quite different than this book, of course). Still, it is good to see monads 'out in the wild'.

César Tron-Lozai's 2022 Devoxx UK lecture on "No Nonsense Monad & Functor -
The foundation of Functional Programming"
https://youtu.be/e6tWJD5q8uw

Algebras in fusion categories model various physical phenomena, so it is worth classifying such structures. This problem is addressed in the enlightening talk below for **group-theoretical fusion categories**, up to Morita equivalence.

Ana Ros Camacho's 2021 Rocky Mountain Representation Theory seminar talk
on "Algebra objects in group-theoretical fusion categories"
https://youtu.be/o59jvTXSAS4

The talk below is a sneak peek of how algebraic structures in monoidal categories are used fruitfully in **subfactor theory**, a subfield of the study of von Neumann algebras. The video is a tad blurry, but connections to this chapter are quite clear.

Noah Snyder's 2022 Hausdorff Center for Mathematics lecture on
"An algebraic version of the small index subfactor classification"
https://youtu.be/35Q4zhZ9Rb8

The intriguing talk below introduces a way of developing a **Morita context** in the higher categorical setting by using exact module categories (or exact algebras). If you enjoyed the concluding material in this chapter, you will enjoy this talk.

César Galindo's 2022 Centre de recherches mathématiques lecture on
"Spherical Morita contexts and relative Serre functors"
https://youtu.be/7IjTxYTfP6Q

§4.13. References for further exploration

- The series of articles by Bodo Pareigis from the late 1970s on "Non-additive ring and module theory" are must-reads if you like the material in this chapter.

  - In the first installment, "I. General theory of monoids" by Pareigis [1977a], algebras (or, there, *monoids*) and A-modules (or, there, *A-objects*) in a monoidal category are introduced. *Commutative* algebras in *symmetric* monoidal categories are also discussed; these will make an appearance in a future volume. Monads, tensor products over algebras, and internal Homs, are also studied.

  - The second installment, "II. C-categories, C-functors and C-morphisms" by Pareigis [1977b], develops the theory of C-module categories and their functors, especially for categories of modules over algebras in C.

  - The third installment, "III. Morita equivalences" by Pareigis [1978] introduces Morita equivalence in the same manner as it is done in this chapter; there, it is called C-*equivalence*. A monoidal version of the fact that Morita equivalent $\Bbbk$-algebras have isomorphic centers [Proposition 2.22] is also presented.

- If you are curious about the formal theory of monads and their Eilenberg-Moore categories, check out the foundational article by Street [1972]. Interesting duality theorems are explored in this work, where the author uses opposites of the various underlying categories of the 2-category $\mathfrak{Cat}$. Follow-up work was provided more recently by Lack and Street [2002].

- An important study of algebras in the monoidal categories G-Mod and Vec_G, for G a finite group, is established by Cohen and Montgomery [1984]. Their duality theorems start with such an algebra A, and uses its G-symmetry to build certain extensions of A, that eventually yield the matrix algebra $\mathrm{Mat}_n(A)$ for $n = |G|$. In short, they show how G-symmetries can produce Morita contexts. As stated in its MathSciNet review, the article "introduces a machine that really works".

- Vital results on algebras on monoidal categories are presented in the work of Ostrik [2003c], which was partly covered in §4.8. In this work, algebras in G-Mod, and in a *Lie-theoretic* monoidal category C_ℓ of $\widehat{\mathfrak{sl}}_2$-*modules at level* $\ell \in \mathbb{Z}_{>0}$, are classified up to Morita equivalence. The latter is used to produce invariants of *modular tensor categories*; these categories are discussed in a future volume.

- The work by Fuchs et al. [2002] employs certain kinds of algebras in monoidal categories to understand *rational conformal field theories* in mathematical physics. Here, *Frobenius algebras* (examined in a future volume) in modular tensor categories are the key structures at play, and bimodules over such algebras are used to model the boundaries of the physical theories of interest.

§4.14. Exercises

4.1 Recall the definition of an algebra in a monoidal category from §4.1.1 and various examples of monoidal categories from §3.1.2.

(a) Verify that algebras in $(\mathsf{Vec}, \otimes_{\Bbbk}, \Bbbk)$ are $\Bbbk$-algebras.

(b) Verify that algebras in $(\mathsf{Set}, \times, \{\cdot\})$ are monoids.

(c) Describe the algebras in the following monoidal categories.

$\qquad$ (i) $(\mathsf{Ab}, \otimes_{\mathbb{Z}}, \mathbb{Z})$ $\qquad\qquad$ (ii) $(\mathsf{Cat}, \times, 1)$

4.2 Recall the notion of an algebra in a monoidal category from §4.1.1, and take a group G. A *G-module algebra* (or a *G-algebra*) is, by definition, an algebra in the monoidal category $(G\text{-Mod}, \otimes_{\Bbbk}, \Bbbk)$ (see §3.1.2i). This is a left G-module $(A, \triangleright_A : G \times A \to A)$ as in §1.3.4, equipped with left G-module morphisms, $m_A : A \otimes_{\Bbbk} A \to A$, $u_A : \Bbbk \to A$, that satisfy associativity and unitality axioms.

(a) Show that a $\Bbbk$-vector space A is a G-module algebra if and only if A is a $\Bbbk$-algebra, where for $g \in G$ and $a, b \in A$:

$$g \triangleright (ab) = (g \triangleright a) \otimes_{\Bbbk} (g \triangleright b), \qquad g \triangleright 1_A = 1_A.$$

Here, $A := (A, m, u)$, with $ab := m(a \otimes_{\Bbbk} b)$ and $1_A := u(1_{\Bbbk})$.

(b) Take G to be a finite group. Consider the $\Bbbk$-vector space $\mathrm{Hom}_{\Bbbk}(\Bbbk G, \Bbbk)$, which has basis elements $p_g : G \to \Bbbk$, with $p_g(g') = \delta_{g,g'} 1_{\Bbbk}$, for all $g, g' \in G$. Show that this is a G-module algebra, where for $g, g', g_1, g_2 \in G$:

$$g \triangleright p_{g'} := p_{g'g}, \quad m(p_{g_1} \otimes_{\Bbbk} p_{g_2}) := \delta_{g_1,g_2} p_{g_1}, \quad u(1_{\Bbbk}) := \textstyle\sum_{g \in G} p_g,$$

extended $\Bbbk$-linearly to elements of $\mathrm{Hom}_{\Bbbk}(\Bbbk G, \Bbbk)$. We call this a *$G$-dual group algebra*, and denote it by $(\Bbbk G)^*$.

4.3 Recall the definition of an algebra in a monoidal category from §4.1.1 and the monoidal category $(\mathsf{Vec}_G, \otimes_{\Bbbk}, \Bbbk)$, for a group G, from §3.1.2i.

(a) Describe the algebras in Vec_G. These are called *G-graded algebras*.

(b) Take a subgroup L of G, and consider the $\Bbbk$-vector space

$$\Bbbk L \;\cong\; \bigoplus\nolimits_{l \in L} \delta_l,$$

where δ_l is a 1-dimensional G-graded vector space, with $(\delta_l)_l = \Bbbk$, and $(\delta_l)_k = 0$ for $k \neq l$; see Exercise 3.35. Show that this is a G-graded algebra, where for $l_1, l_2 \in L$:

$$m(\delta_{l_1} \otimes_{\Bbbk} \delta_{l_2}) := \delta_{l_1 l_2}, \qquad u(1_{\Bbbk}) := \delta_e.$$

This is a *(G-graded) group algebra*, denoted by $G\text{-}\Bbbk L$, or by $\Bbbk L$ for short.

4.4 Recall the definition of an algebra in a monoidal category from §4.1.1. For a group G and 3-cocycle ω on G, also recall the monoidal category $(\mathrm{Vec}_G^\omega, \otimes_\Bbbk, \Bbbk)$ from Exercise 3.35. Take the following data:

(i) A subgroup L of G such that $\omega|_{L \times L \times L}$ is cohomologically trivial;

(ii) A $\Bbbk^\times$-**valued 2-cocycle** ψ **on** G **with** $d^3\psi = \omega|_{L \times L \times L}$, which is a function $\psi : L \times L \to \Bbbk^\times$ that satisfies

$$\psi(l_1, l_2 l_3)\, \psi(l_2, l_3) \;=\; \omega(l_1, l_2, l_3)\, \psi(l_1 l_2, l_3)\, \psi(l_1, l_2),$$

for all $l_1, l_2, l_3 \in L$.

Show that the $\Bbbk$-vector space $\Bbbk L \cong \bigoplus_{l \in L} \delta_l$ as in Exercise 4.3(b) is an algebra in Vec_G^ω, where for $l_1, l_2 \in L$, we define:

$$m(\delta_{l_1} \otimes_\Bbbk \delta_{l_2}) := \psi(l_1, l_2)^{-1}\, \delta_{l_1 l_2}, \qquad u(1_\Bbbk) := \psi(e, e)\, \delta_e.$$

We call this algebra in Vec_G^ω a **twisted (G-graded) group algebra**, and denote it by $G\text{-}\Bbbk L_\psi$, or by $\Bbbk L_\psi$, for short.

4.5 Complete the details of Example 4.1 in §4.1.1 in verifying the associativity and unitality axioms for the unit algebra $(\mathbb{1},\ \ell_1,\ \mathrm{id}_1)$, and for the zero algebra $(0,\ _{0 \otimes 0}\vec{0},\ _1\vec{0})$, in $\mathcal{C}$. Assume that $\mathcal{C}$ has a zero object 0 for the latter task.

4.6 Verify that the collection of $\mathrm{Alg}(\mathcal{C})$ of algebras in a monoidal category $\mathcal{C}$, along with algebra morphisms, from §4.1.1 forms a category.

4.7 Complete the proof of Proposition 4.3 in §4.1.1 on how monoidal functors send algebras to algebras, functorially.

Hint. By The Strictification Theorem [Theorem 3.26], it suffices to establish the result in the strict case. To show that $m_{F(A)}$ is associative and left unital, with unit $u_{F(A)}$, complete and justify the commutative diagrams below.

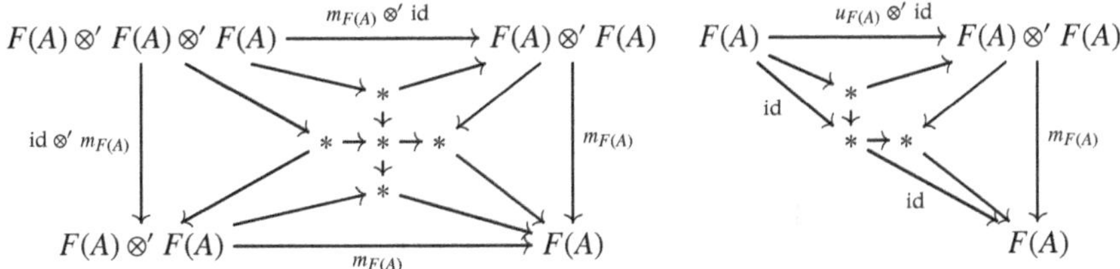

4.8 Here, we continue Exercise 4.2 on G-algebras, for a group G.

(a) Use the monoidal isomorphisms from Exercise 3.6(b,c) to describe each of the categories, and to verify category isomorphisms, below:

$$\mathrm{Alg}(G\text{-Mod}) \;\cong\; \mathrm{Alg}(\mathrm{Rep}(G)) \;\cong\; \mathrm{Alg}(\mathrm{Rep}(\Bbbk G)) \;\cong\; \mathrm{Alg}(\Bbbk G\text{-Mod}).$$

(b) For a subgroup L of G, describe the image of the G-dual group algebra, $(\Bbbk L)^*$, under the isomorphisms above.

(c) Now take another subgroup H of G.

    (i) Construct a restriction functor $\mathrm{Res}^G_H : G\text{-Mod} \to H\text{-Mod}$ that is strong monoidal; cf. Exercise 3.6(d).

    (ii) For a subgroup L of G, describe the H-algebra, $\mathrm{Res}^G_H((\Bbbk L)^*)$. In particular, describe $\mathrm{Res}^G_{\langle e\rangle}((\Bbbk L)^*) \in \mathsf{Alg}(\mathsf{Vec})$.

4.9 [Open-ended] Explore examples of algebras in various monoidal categories in §3.1.2 (other than what is done in the exercises above), say by:

(a) Constructing algebras directly, and by

(b) Using Proposition 4.3 and examples of monoidal functors from §3.2.3.

4.10 Take $\mathcal{A}_\sqcap := (\mathcal{A}, \sqcap, T)$ to be cartesian monoidal category; see §§2.2.1i,ii, 3.1.2ii. Also, recall the definition of an algebra in $\mathcal{A}_\sqcap$ presented in §4.1.1, and consider it as a "monoid" in $\mathcal{A}_\sqcap$ (in name) for the purpose of this exercise.

(a) Define the category: $\mathsf{Group}(\mathcal{A}_\sqcap)$. Objects should be tuples (G, m, u, i), where G is an object of $\mathcal{A}$, and $m : G \sqcap G \to G, u : T \to G, i : G \to G$ are morphisms in $\mathcal{A}$ (**group objects**) satisfying group compatibility conditions. Then, obtain the isomorphism of categories below:

$$\mathsf{Group}((\mathsf{Set}, \times, \{\cdot\})) \cong \mathsf{Group}.$$

(b) Recall the Eckmann-Hilton Principle from §4.1.2. Verify that

$$\mathsf{Group}((\mathsf{Group}, \times, \{e\})) \cong \mathsf{Ab}.$$

(c) Show that an object G admits the structure of a group object in $\mathcal{A}_\sqcap$ if and only if, for each object X of $\mathcal{A}$, there exists a contravariant functor:

$$\mathcal{A} \to \mathsf{Group}, \ X \mapsto \mathrm{Hom}_{\mathcal{A}}(X, G).$$

An aside. Examples of groups in monoidal categories include *topological groups* (resp., *Lie groups, algebraic groups*) in the monoidal category of topological spaces (resp., of smooth manifolds, of affine varieties).

4.11 Recall the notion of an endomorphism algebra of an object in an enriched (e.g., rigid) category from §4.1.3. For $V \in \mathsf{FdVec}$, show that $\mathrm{End}_{\mathsf{FdVec}}(V)$ and $\mathrm{End}_{\Bbbk}(V)$ are isomorphic as algebras in FdVec, as stated in Example 4.9(b).

4.12 Complete the details of Example 4.10 in §4.2.1 in showing that, for an algebra A in $\mathcal{C}$, both A_{obj} and the zero object 0 are ideals of A in $\mathcal{C}$.

4.13 Prove Corollary 4.12 in §4.2.1 on showing that if I is an ideal that 'contains' the unit of an algebra A in $\mathcal{C}$, then $I \cong A$ as algebras in $\mathcal{C}$.

Hint. Show that a right inverse ψ of ι_I^A is a two-sided inverse, since $\iota_I^A : I \to A$ is monic. Use a unit axiom and an ideal condition to construct ψ. Do not forget to establish an isomorphism of algebras, not just of objects.

4.14 Complete the proof of Proposition 4.13 in §4.2.1 in showing that the kernel of an algebra morphism $\phi : A \to A'$ in $\mathcal{C}$ is an ideal of A in $\mathcal{C}$.

4.15 Recall from §§4.2.1, 4.2.2 material on subalgebras, ideals, and quotient algebras of algebras in a monoidal category. Take a group G with subgroup L. Continuing Exercises 4.2, 4.3, 4.4 above, construct a non-trivial example of:

    (i) a subalgebra,   (ii) an ideal,   (iii) a quotient algebra,

of each of the following algebras in monoidal categories.

(a) A G-dual group algebra, $(\Bbbk G)^*$, in $\mathsf{Alg}(G\text{-Mod})$.

(b) A G-graded group algebra, $\Bbbk L$, in $\mathsf{Alg}(\mathsf{Vec}_G)$.

(c) A twisted G-graded group algebra, $\Bbbk L_\psi$, in $\mathsf{Alg}(\mathsf{Vec}_G^\omega)$.

Hint. For a group G, one may want to use (normal) subgroups, or use the $\Bbbk$-vector space $(\Bbbk G)^+$ with basis $\{g - e\}_{g \in G}$. The latter generates the **augmentation ideal** of the group algebra $\Bbbk G$ over $\Bbbk$.

4.16 Recall the proof of part (a) of Theorem 4.22, the special case of Doctrinal Adjunction in §4.3.1.

(a) Verify that the diagram in Figure 4.7 commutes. In particular, provide details to show that each region is a commutative diagram.

(b) If part (a) is fun for you, construct the analogue of Figure 4.7 to verify the monoidal right unitality axiom.

(c) If part (a) is really fun for you, verify that the diagram in Figure 4.8 commutes. In particular, provide details to show that each region in Figure 4.8 is a commutative diagram.

4.17 Prove parts (b,c) of Theorem 4.22, the special case of Doctrinal Adjunction in §4.3.1. Recall that part (a) states the right adjoint G of a strong monoidal functor F is monoidal. Part (b) claims that the unit and counit of $F \dashv G$ are monoidal natural transformations, and part (c) claims the components of the unit and counit at algebras are algebra morphisms.

Hint. For part (b), use Exercise 3.4, along with the triangle identities for adjunction (see §2.5.1), and several instances of naturality (see §2.3.4).

4.18 For a morphism $\phi : H \to G$ of finite groups, recall the coinduced algebra:

$$\mathrm{Coind}_H^G(A) := \mathrm{Hom}_{H\text{-Mod}}(\Bbbk G, A) \in \mathrm{Alg}(G\text{-Mod}),$$

for $A \in \mathrm{Alg}(H\text{-Mod})$ from Example 4.23 in §4.3.1.

(a) Write down the unit η and counit ε of adjunction, $\mathrm{Res}_H^G \dashv \mathrm{Coind}_H^G$.

(b) Use the solution of Exercise 4.8(c.i) and the special case of Doctrinal Adjunction, Theorem 4.22(a), to give Coind_H^G a monoidal structure.

(c) Use part (b) and Proposition 4.3 to endow $\mathrm{Coind}_H^G(A)$ with a multiplication morphism and unit morphism to get $\mathrm{Coind}_H^G(A) \in \mathrm{Alg}(G\text{-Mod})$.

(d) Now take $H = \langle e \rangle$, and $A = \Bbbk$ to be the unit algebra in G-Mod. Describe the coinduced algebra: $\mathrm{Coind}_{\langle e \rangle}^G(\Bbbk)$ (cf. Exercise 4.2(b)).

4.19 [Open-ended] Recall the Doctrinal Adjunction from Theorem 4.22(a) in §4.3.1: For an adjunction $(F : \mathcal{C} \to \mathcal{D}) \dashv (G : \mathcal{D} \to \mathcal{C})$, with F strong monoidal, we get that G is monoidal. Use this to construct examples of algebras $G(A) \in \mathrm{Alg}(\mathcal{C})$, for $A \in \mathrm{Alg}(\mathcal{D})$. Do this for boring and interesting right adjoint functors G, besides what is presented in Exercise 4.18.

4.20 Verify the details of Example 4.25 in §4.3.2. Namely, given an object $A \in \mathcal{C}$, for $\mathcal{C}$ strict, we obtain that algebra structures on A in $\mathcal{C}$ are in bijective correspondence with monad structures on the endofunctor $(A \otimes -)$ on $\mathcal{C}$.

4.21 Recall the notion of a monad on a category from §4.3.2. Consider the endofunctor $P : \mathrm{Set} \to \mathrm{Set}$, that sends a set X to its power set $P(X)$ (the set whose elements are all subsets of X). Construct natural transformations $\mu : P \circ P \Rightarrow P$ and $\eta : \mathrm{Id}_{\mathrm{Set}} \Rightarrow P$, such that $(P, \mu, \eta) \in \mathrm{Monad}(\mathrm{Set})$. We refer to this as a **power set monad**.

4.22 Let $\mathcal{C}$ be an additive monoidal category with infinite biproducts $\square$. Consider the endofunctor $T : \mathcal{C} \to \mathcal{C}$, that sends X to $\square_{n \geq 0} X^{\otimes n}$. Endow T with the structure of a monad on $\mathcal{C}$ (see §4.3.2). We call this a **free algebra monad** (or **tensor algebra monad**) on $\mathcal{C}$.

4.23 [Open-ended] Explore more examples of monads on categories in the literature, other than what is presented in the last two exercises. For instance, explore special cases of Example 4.26 in §4.3.2.

4.24 Recall the notion of a monad from §4.3.2.

   (a) Verify the details of Example 4.26: Given a pair of adjoint functors $(F : \mathcal{A} \to \mathcal{B}) \dashv (G : \mathcal{B} \to \mathcal{A})$, we obtain that $(GF,\ G\varepsilon F,\ \eta) \in \mathsf{Monad}(\mathcal{A})$.

   (b) Show that the identity monad in Example 4.24 arises as a special case of the adjunction monad in part (a).

   (c) Show that the monad $(A \otimes -)$ in Example 4.25 arises as a special case of the adjunction monad in part (a) (after reading §4.4.1).

4.25 Take an algebra A in $\mathcal{C}$, and recall from Example 4.28(c) in §4.4.1i the free left A-module $A \otimes X$ on an object X in $\mathcal{C}$.

   (a) Verify that $(A \otimes X,\ \triangleright_{A\otimes X} := (m \otimes \mathrm{id}_X)\, a^{-1}_{A,A,X}) \in A\text{-}\mathsf{Mod}(\mathcal{C})$.

   (b) Show that Free: $\mathcal{C} \to A\text{-}\mathsf{Mod}(\mathcal{C})$, $X \mapsto (A \otimes X, \triangleright_{A\otimes X})$, is a functor.

   (c) Consider the forgetful functor, Forg: $A\text{-}\mathsf{Mod}(\mathcal{C}) \to \mathcal{C}$, $(M,\triangleright) \mapsto M$. Establish Free-Forget adjunction (see §2.5.3) in this context:

$$(\text{Free} : \mathcal{C} \to A\text{-}\mathsf{Mod}(\mathcal{C})) \dashv (\text{Forg} : A\text{-}\mathsf{Mod}(\mathcal{C}) \to \mathcal{C}).$$

4.26 Recall the notion of a (bi)module in a monoidal category from §§4.4.1i, 4.4.2i. Take the $\mathcal{V}$-endomorphism algebra $A := \mathrm{End}_\mathcal{V}(X)$ from §4.1.3, and construct examples of objects in the categories of (bi)modules below:

$$\text{(a) } A\text{-}\mathsf{Mod}(\mathcal{V}), \qquad \text{(b) } \mathsf{Mod}\text{-}A(\mathcal{V}), \qquad \text{(c) } A\text{-}\mathsf{Bimod}(\mathcal{V}).$$

Illustrate these examples in the setting of Example 4.9 from §4.1.3.

4.27 Take an algebra A in $\mathcal{C}$, and a left A-module $(M,\triangleright)$ in $\mathcal{C}$. Recall from §4.4.1ii the discussion about left quotient modules over A. Verify that if the endofunctor $A\otimes -$ on $\mathcal{C}$ is right exact, then there exists a morphism $\triangleright_{M/N} : A\otimes M/N \to M/N$ in $\mathcal{C}$ that makes $\pi^M_N : M \to M/N$ a morphism of left A-modules in $\mathcal{C}$.

Hint. Use Remark 4.14(d)$\Rightarrow$(e) and part of the proof of Proposition 4.18.

4.28 Take $A \in \mathsf{Alg}(\mathcal{C})$. Prove Proposition 4.30 from §4.4.1iii verifying that $A\text{-}\mathsf{Mod}(\mathcal{C})$ (resp., $\mathsf{Mod}\text{-}A(\mathcal{C})$) is a right (resp., left) $\mathcal{C}$-module category as in §3.3.1.

4.29 For $A \in \mathsf{Alg}(\mathcal{C})$, complete the proof of Proposition 4.31 from §4.4.1iv on various properties of $A\text{-}\mathsf{Mod}(\mathcal{C})$ and $\mathsf{Mod}\text{-}A(\mathcal{C})$ inherited from $\mathcal{C}$ and A.

4.30 Verify the details of Examples 4.34 and 4.35 in §4.4.3ii. Namely,

   (a) For the identity monad on Id on $\mathcal{A}$, explain why $\mathcal{A}^{\mathrm{Id}} \cong \mathcal{A}$.

   (b) Given a monad $(A \otimes -)$ on $\mathcal{C}$, show that its Eilenberg-Moore category $\mathcal{C}^{(A\otimes -)}$ is isomorphic to $A\text{-}\mathsf{Mod}(\mathcal{C})$.

4.31 [Open-ended] Recall the power set monad P on Set from Exercise 4.21. Now, pertaining to §4.4.3ii, the Eilenberg-Moore objects of P are *sup-complete (semi)lattices*. Investigate this in the literature and discuss such constructions.

4.32 Recall the free algebra monad T on an additive monoidal category C (with infinite biproducts) from Exercise 4.22. Pertaining to §4.4.3ii, what are the Eilenberg-Moore objects of T?

4.33 Complete the details of the proof of Theorem 4.37 in §4.4.3ii in establishing that every monad T on a category $\mathcal{A}$ yields an adjunction:

$$(\mathrm{Free}^T : \mathcal{A} \to \mathcal{A}^T) \dashv (\mathrm{Forg}^T : \mathcal{A}^T \to \mathcal{A}),$$

and that T is recovered as the adjunction monad: $\mathrm{Forg}^T \circ \mathrm{Free}^T$.

4.34 Complete the details of the proof of Theorem 4.39 in §4.4.3iv in establishing the contravariant correspondence between monad morphisms and functors between their EM-categories: $(T \Rightarrow T') \leftrightsquigarrow (\mathcal{A}^{T'} \to \mathcal{A}^T)$.

4.35 In the manner of Example 4.41 in §4.4.3iv, take a group G with subgroup H, and illustrate Corollary 4.40 for the group algebras $A = \Bbbk H$ and $A' = \Bbbk G$ in the monoidal category $C = (\mathsf{Vec}, \otimes_\Bbbk, \Bbbk)$.

4.36 Complete the details of §4.5.1i in showing that the biproduct $\square$ of algebras in C is an algebra in C, and that $\mathsf{Alg}(C)_\square := (\mathsf{Alg}(C), \square, 0)$ is a monoidal category.

4.37 Complete the details of the proof of Proposition 4.42 in §4.5.2iii in showing that when $(B_1 \otimes -)$ and $(- \otimes B_2)$ are right exact endofunctors of C, and when M is a (B_1, A)-bimodule in C and N is a (A, B_2)-bimodule in C, we obtain that $M \otimes_A N$ is a (B_1, B_2)-bimodule in C.

4.38 Following up with Example 4.43 in §4.5.2iii, suppose that $(B_1 \otimes -)$ and $(- \otimes B_2)$ are right exact endofunctors of C, and take bimodules $M \in (B_1, \mathbb{1})$-$\mathsf{Bimod}(C)$ and $N \in (\mathbb{1}, B_2)$-$\mathsf{Bimod}(C)$, for the unit algebra $\mathbb{1}$. Verify that

$$M \otimes_\mathbb{1} N \cong M \otimes N$$

as (B_1, B_2)-bimodules in C. Here, $M \otimes N \in (B_1, B_2)$-$\mathsf{Bimod}(C)$ as in §4.5.2ii.

4.39 Complete the proof of Lemma 4.44 in §4.5.2iii in showing that, for A-bimodules M, N, P in C, we have the following identities in A-$\mathsf{Bimod}(C)$.

(a) $(M \otimes_A N) \otimes_A P \cong M \otimes_A (N \otimes_A P)$.

(b) $M \otimes_A A_{\mathrm{reg}} \cong M \cong A_{\mathrm{reg}} \otimes_A M$.

4.40 [Open-ended] Here, we follow up with Remark 4.45 in §4.5.2iv. It is shown in Section 3 of Liu Lopez and Walton [2023] that if $\mathcal{C}$ is a right closed monoidal category as in §3.11.3, then Mod-$A(\mathcal{C})$ is enriched over $\mathcal{C}$. Explore this result and other related results in this reference.

4.41 Recall Proposition 4.46 from §4.5.2iv on constructing duals of modules in (rigid) monoidal categories.

    (a) Prove Proposition 4.46 (in the strict case) using graphical calculus.

    (b) Formulate and prove a non-strict version of Proposition 4.46 using equations or commutative diagrams. Namely, show that when given actions $\triangleright : A \otimes M \to M$ and $\triangleleft : N \otimes A \to N$, with M left rigid and N right rigid, we obtain induced actions $\triangleleft : M^* \otimes A \to M^*$ and $\triangleright : A \otimes {}^*N \to {}^*N$.

4.42 Recall the product category $\mathcal{C}^{\times \mathbb{N}}$ examined in §4.6.1.

    (a) Verify that $\mathcal{C}^{\times \mathbb{N}} \cong \mathsf{Fun}(\underline{\mathbb{N}}, \mathcal{C})$ as categories.

    (b) Establish part of Proposition 4.47(a): If $\mathcal{C}$ is additive, then so is $\mathcal{C}^{\times \mathbb{N}}$.

    (c) Verify the rest of Proposition 4.47(a): If $\mathcal{C}$ is abelian, then so is $\mathcal{C}^{\times \mathbb{N}}$.

    (d) Prove that if $\mathcal{C}$ is additive, then $\mathcal{C}^{\times \mathbb{N}}$ has the Cauchy monoidal structure given in Proposition 4.47(b).

4.43 Complete the proof of Proposition 4.48 in §4.6.2 in establishing that $\mathsf{Alg}(\mathcal{C}^{\times \mathbb{N}})$ is isomorphic to $\mathbb{N}$-$\mathsf{GrAlg}(\mathcal{C})$ as categories.

4.44 Let $X \in \mathcal{C}$, and recall the $\mathcal{C}$-tensor algebra $T(X)$ in $\mathsf{Alg}(\mathcal{C}^{\times \mathbb{N}}) \cong \mathbb{N}$-$\mathsf{GrAlg}(\mathcal{C})$ described in Example 4.49 from §4.6.2. Complete the following tasks.

    (a) Show that $I(X) := (0, X, X^{\otimes 2}, X^{\otimes 3}, \dots)$ is an ideal of $T(X)$ in $\mathsf{Alg}(\mathcal{C}^{\times \mathbb{N}})$ (or in $\mathbb{N}$-$\mathsf{GrAlg}(\mathcal{C})$); see §4.2.1.

    (b) Describe the quotient algebra $T(X)/I(X)$ in $\mathsf{Alg}(\mathcal{C}^{\times \mathbb{N}})$ (or in $\mathbb{N}$-$\mathsf{GrAlg}(\mathcal{C})$); see §4.2.2.

    (c) Take the object $I(X)^r := (0, \dots, 0, X^{\otimes r}, X^{\otimes (r+1)}, X^{\otimes (r+2)}, \dots)$, and show that this is an ideal of $T(X)$. Moreover, describe the quotient algebra $T(X)/I(X)^r$ in $\mathsf{Alg}(\mathcal{C}^{\times \mathbb{N}})$ (or in $\mathbb{N}$-$\mathsf{GrAlg}(\mathcal{C})$).

4.45 [Open-ended] Recall the $\mathcal{C}$-tensor algebras in $\mathsf{Alg}(\mathcal{C}^{\times \mathbb{N}})$ from Example 4.49 in §4.6.2, which generalize the tensor algebras over $\Bbbk$ from §1.2.2. Define $\mathcal{C}$-*path algebras* in $\mathsf{Alg}(\mathcal{C}^{\times \mathbb{N}})$ ($\cong \mathbb{N}$-$\mathsf{GrAlg}(\mathcal{C})$) that generalize the path algebras over $\Bbbk$ from §1.2.5. Such $\mathcal{C}$-path algebras should specialize to $\mathcal{C}$-tensor algebras.

4.46 In this exercise, we examine the details of preparatory results towards proving the Generalized Morita's Theorem in §4.7.2

    (a) Complete the details of the proof of Lemma 4.53 in §4.7.

    (b) Complete the details of the proof of the Generalized Eilenberg-Watts Theorem [Theorem 4.54] in §4.7.1.

4.47 Establish Proposition 4.57 in §4.7.2, which provides another characterization of when two algebras in C are Morita equivalent.

Hint. It suffices to show that the morphisms ϕ_A and ϕ_B are monos since A-Bimod(C) and B-Bimod(C) are abelian [Proposition 4.33]. To do so for ϕ_A, take $f_1, f_2 : W \to P \otimes_B Q \in A$-Bimod($C$) with $\phi_A f_1 = \phi_A f_2$. Next, show that the diagram below is commutative (suppressing associativity for brevity).

$$
\begin{array}{ccccc}
W \otimes_A P \otimes_B Q & \xrightarrow{\;\mathrm{id}\,\otimes_A\,\phi_A\;} & W \otimes_A A & \xrightarrow{\;f_1\,\otimes_A\,\mathrm{id},\; f_2\,\otimes_A\,\mathrm{id}\;} & P \otimes_B Q \otimes_A A \\
{\scriptstyle f_1\,\otimes_A\,\mathrm{id}\,\otimes_B\,\mathrm{id},}\ \downarrow & & & & \downarrow{\scriptstyle \cong} \\
{\scriptstyle f_2\,\otimes_A\,\mathrm{id}\,\otimes_B\,\mathrm{id}} & & & & \\
P \otimes_B Q \otimes_A P \otimes_B Q & \xrightarrow{\;\phi_A\,\otimes_A\,\mathrm{id}\,\otimes_B\,\mathrm{id}\;} & A \otimes_A P \otimes_B Q & \xrightarrow{\;\cong\;} & P \otimes_B Q
\end{array}
$$

This will imply that $f_1 \otimes_A \mathrm{id}_A = f_2 \otimes_A \mathrm{id}_A$, which yields $f_1 = f_2$, as required.

4.48 Recall the notion of Morita equivalence from §4.7.2, with Example 4.58.

    (a) Show that if $\mathbb{1} \in C$ is a simple object and if X is a left rigid object in C, then the algebra $\mathrm{End}_C(X) = X \otimes X^*$ from Example 4.9(a) is Morita equivalent to the algebra $\mathbb{1}$ from Example 4.1(a).

    (b) Show that part (a) specializes to the statement from Example 2.21: $\mathrm{Mat}_n(\mathbb{k})$ is Morita equivalent to $\mathbb{k}$ as $\mathbb{k}$-algebras.

    (c) Recall the full result of Example 2.21: $\mathrm{Mat}_n(A)$ is Morita equivalent to A, for any $\mathbb{k}$-algebra A. Formulate and prove a generalization of this fact in the setting of §4.7.2.

Hint. Use Proposition 4.57. Moreover, for part (a), the assumption that $\mathbb{1} \in C$ is simple implies that evaluation morphisms in C are epic.

4.49 Recall Exercise 4.3 and §4.7.2, and take the monoidal category $(\mathrm{Vec}_{C_2}, \otimes_{\mathbb{k}}, \mathbb{k})$, for the cyclic group C_2 of order 2. Are the group algebras $\mathbb{k}\langle e \rangle$ and $\mathbb{k}C_2$ Morita equivalent as algebras in Vec_{C_2}?

4.50 [Open-ended] Recall Exercise 4.4, along with §4.7.2. Here, we work in $(\mathrm{Vec}^{\omega}_{C_2 \times C_2}, \otimes_{\mathbb{k}}, \mathbb{k})$, for the Klein-four group $C_2 \times C_2$. Note that

$$
H^2(C_2 \times C_2, \mathbb{k}^{\times}) \cong C_2, \qquad H^3(C_2 \times C_2, \mathbb{k}^{\times}) \cong C_2 \times C_2 \times C_2.
$$

(a) Are there non-cohomologous 2-cocycles ψ and ψ' of $C_2 \times C_2$ such that the twisted group algebras $\Bbbk(C_2 \times C_2)_\psi$ and $\Bbbk(C_2 \times C_2)_{\psi'}$ are Morita equivalent as algebras in $\mathrm{Vec}^1_{C_2 \times C_2}$? Here, the superscript "1" denotes the trivial 3-cocycle on $C_2 \times C_2$.

(b) Pick a 3-cocycle ω on $C_2 \times C_2$ that is not cohomologically trivial. Explore whether you can construct non-cohomologous 2-cocycles ψ and ψ' of $C_2 \times C_2$ such that the twisted group algebras $\Bbbk(C_2 \times C_2)_\psi$ and $\Bbbk(C_2 \times C_2)_{\psi'}$ are Morita equivalent as algebras in $\mathrm{Vec}^\omega_{C_2 \times C_2}$.

4.51 Verify Proposition 4.62 in §4.8.1, which claims that internal Ends arise as algebras in $\mathcal{C}$, and internal Homs arise as modules over such algebras in $\mathcal{C}$.

4.52 Here, we follow up on Example 4.63 in §4.8.1. Assume that $\mathcal{C}$ is left rigid and take $\mathcal{M}$ to be the regular left $\mathcal{C}$-module category $\mathcal{C}_{\mathrm{reg}}$.

(a) For $Y \in \mathcal{C}_{\mathrm{reg}}$, show that the internal End algebra $\underline{\mathrm{End}}(Y) \in \mathcal{C}$ in Proposition 4.62(a) is the $\mathcal{C}$-endomorphism algebra from Example 4.9 in §4.1.3.

(b) If you previously completed Exercise 4.26, is one of your examples for part (b) of that exercise the same as the internal Hom module from Proposition 4.62(b)?

4.53 Assume that $\mathcal{C}$ is rigid and abelian. Upgrade the previous exercise for the left $\mathcal{C}$-module category $\mathcal{M} = \mathrm{Mod}\text{-}A(\mathcal{C})$, for an algebra $A \in \mathcal{C}$. Namely, recall Proposition 4.62 in §4.8.1 and complete the following tasks.

(a) Given $M, N \in \mathrm{Mod}\text{-}A(\mathcal{C})$, compute the internal Hom object $\underline{\mathrm{Hom}}(M, N)$.

(b) Describe the multiplication morphism and the unit morphism for the internal End algebra $\underline{\mathrm{End}}(M) \in \mathcal{C}$ from Proposition 4.62(a). Show that the multiplication is indeed associative and unital.

(c) Describe the action morphism for the right $\underline{\mathrm{End}}(M)$-module $\underline{\mathrm{Hom}}(M, N)$ in $\mathcal{C}$ from Proposition 4.62(b), and verify that it is associative and unital.

(d) Describe the action morphism for the left $\underline{\mathrm{End}}(M)$-module $\underline{\mathrm{Hom}}(N, M)$ in $\mathcal{C}$ from Proposition 4.62(c), and verify that it is associative and unital.

Hint. For part (a), note that the internal Hom object $\underline{\mathrm{Hom}}(X, Y) := Y \otimes X^*$ in Exercise 4.52 is isomorphic to the object $(X \otimes {}^*Y)^*$ in $\mathcal{C}$.

4.54 Recall the notion of an internal End algebra from §4.8.1.

(a) Complete the details of Example 4.64 on showing that when $\mathcal{C}$ is the monoidal category $(G\text{-}\mathrm{Mod}, \otimes_\Bbbk, \Bbbk)$, for a group G, and $\mathcal{M} = \mathrm{Vec}$ is the left $\mathcal{C}$-module category via the forgetful functor [Example 3.19], then $\underline{\mathrm{End}}(\Bbbk)$ is the dual group algebra $(\Bbbk G)^* \in \mathrm{Alg}(G\text{-}\mathrm{Mod})$ from Exercise 4.2.

(b) Take a subgroup H of G, and let $\mathcal{M}$ be the left $\mathcal{C}$-module category H-Mod from Example 3.19 via the restriction functor Res_H^G. Then, provide an example of an internal End algebra in G-Mod, with respect to H-Mod.

4.55 Recall from Example 3.21 that $\mathcal{C} := (A\text{-Bimod}, \otimes_A, A)$ is a monoidal category and $\mathcal{M} := (A, B)\text{-Bimod}$ is a left $\mathcal{C}$-module category, for $\Bbbk$-algebras A, B. Also recall the material from §4.8.1.

(a) Provide an example of an internal End algebra in $\mathcal{C}$ with respect to $\mathcal{M}$.

(b) Provide an example of an internal Hom module in $\mathcal{C}$ over the internal End algebra in part (a).

4.56 [Open-ended] Recall the notions of an internal End algebra and an internal Hom module from §4.8.1. Produce examples of such structures using module categories (say from §3.3.2) different than those in Examples 4.63, 4.64, and in Exercises 4.54, 4.55.

4.57 Continuing Exercise 4.53 and Remark 4.69 in §4.8.2, do the twisted group algebras $\Bbbk L_\psi \in \mathrm{Vec}_G^\omega$ from Exercise 4.4 arise as internal End algebras in Vec_G^ω?

4.58 For a group G, recall the definition of a G-module algebra $A \in \mathrm{Alg}(G\text{-Mod})$ from Exercise 4.2. Consider its **invariant subalgebra**:

$$A^G = \{a \in A \mid g \triangleright a = a, \ \ \forall g \in G\}.$$

Show that $A \in \mathrm{Alg}(G\text{-Mod})$ is connected as in §4.9.1 if and only if $\dim_\Bbbk A^G = 1$.

Hint. Show there exists a $\Bbbk$-algebra structure on $\mathrm{Hom}_{G\text{-Mod}}(\Bbbk, A)$ and that it is isomorphic to A^G as $\Bbbk$-algebras.

4.59 Complete the details of the proof of Proposition 4.72 in §4.9.1 about the $\Bbbk$-algebra $\mathrm{Hom}_\mathcal{C}(\mathbb{1}, A)$ and its relationship with endomorphism algebras.

4.60 Let $\mathcal{A}$ be a category, and recall from §4.3.2 that a monad on $\mathcal{A}$ is an algebra $(T : \mathcal{A} \to \mathcal{A}, \mu : T^2 \Rightarrow T, \eta : \mathrm{Id}_\mathcal{A} \Rightarrow T)$ in $(\mathrm{End}(\mathcal{A}), \circ, \mathrm{Id}_\mathcal{A})$. Moreover, for a pair of adjoint functors $(F : \mathcal{A} \to \mathcal{B}) \dashv (G : \mathcal{B} \to \mathcal{A})$ with unit η and counit ε, recall that $(GF, G\varepsilon F, \eta)$ is a monad on $\mathcal{A}$ [Example 4.26]. Here, we study the separability property of the monad GF; see §4.9.5.

(a) A monad on $\mathcal{A}$ is said to be **separable** if it is separable as an algebra in $\mathrm{End}(\mathcal{A})$. Write down the precise conditions of this definition.

(b) A functor $G : \mathcal{B} \to \mathcal{A}$ is **separable** if for any pair of objects $X, Y \in \mathcal{B}$ there exists an assignment, $G'_{X,Y} : \mathrm{Hom}_\mathcal{A}(G(X), G(Y)) \to \mathrm{Hom}_\mathcal{B}(X, Y)$, that is natural in X and Y, such that $G'_{X,Y}(G(f)) = f$ for any $f : X \to Y \in \mathcal{B}$.

Show that when G is a right adjoint of a functor $F : \mathcal{A} \to \mathcal{B}$ with counit $\varepsilon : FG \Rightarrow \mathrm{Id}_\mathcal{B}$, then G is separable if and only if there exists a natural transformation $\theta : \mathrm{Id}_\mathcal{B} \Rightarrow FG$ such that $\varepsilon \circ^{\mathrm{ver}} \theta = \mathrm{Id}_\mathcal{B}$.

(c) For the adjunction $F \dashv G$, verify that the monad GF is separable (as in part (a)) when the functor G is separable (as in part (b)).

(d) For an algebra A in a monoidal category $\mathcal{C}$, illustrate part (c) for the adjunction from Example 4.28(c) and Exercise 4.25:

$$(\text{Free} : \mathcal{C} \to A\text{-Mod}(\mathcal{C})) \dashv (\text{Forg} : A\text{-Mod}(\mathcal{C}) \to \mathcal{C}).$$

Does the converse statement of part (c) also hold? That is, if $\text{Forg} \circ \text{Free}$ is a separable monad on $\mathcal{C}$, is the functor Forg separable?

4.61 Consider the twisted group algebras $\Bbbk L_\psi$ in Vec_G^ω from Exercise 4.4, and recall the various algebraic properties discussed in §§4.9.1–4.9.6. Here, G is a finite group, and ω is a $\Bbbk$-valued 3-cocycle on G. Moreover, L is a subgroup of G, and ψ is a $\Bbbk$-valued 2-cocycle on G with $d^3\psi = \omega|_{L \times L \times L}$.

Discuss when $\Bbbk L_\psi$ possesses each of the following properties:

(a) connected; (c) simple; (e) separable;

(b) indecomposable; (d) semisimple; (f) exact.

Hint. Try special cases at first, say with trivial cocycles ω and ψ.

4.62 Continuing Exercise 4.61, consider the monoidal category of bimodules

$$\mathcal{C}(G, \omega, L, \psi) := (\Bbbk L_\psi)\text{-Bimod}(\mathrm{Vec}_G^\omega)$$

as discussed in Example 4.95 in §4.10.1. Verify that $\mathcal{C}(G, \omega, L, \psi)$ is always a fusion category, justifying its name: a **group-theoretical fusion category**.

4.63 Recall the discussion of categorical Morita equivalence from §4.10.2. Complete the details of Example 4.98 in verifying that, for a finite group G, the tensor categories G-Mod and Vec_G are categorically Morita equivalent.

4.64 Recall the discussion of 2-categories from §4.10.3. Verify that $\mathfrak{Cat}$ from Example 4.101 is a (strict) 2-category.

BIBLIOGRAPHY

Marcelo Aguiar and Swapneel Mahajan. *Monoidal functors, species and Hopf algebras*, volume 29 of *CRM Monograph Series*. American Mathematical Society, Providence, RI, 2010.

Robert Allen, Simon Lentner, Christoph Schweigert, and Simon Wood. Duality structures for module categories of vertex operator algebras and the Feigin Fuchs boson. *ArXiv preprint math.QA/2107.05718v2. Retrieved October 6, 2023*, 2021.

Jorge Almeida and Emanuele Rodaro. Semisimple synchronizing automata and the Wedderburn-Artin theory. *Internat. J. Found. Comput. Sci.*, 27(2):127–145, 2016.

Paolo Aluffi. *Algebra: Chapter 0*, volume 104 of *Graduate Studies in Mathematics*. American Mathematical Society, Providence, RI, 2009.

Emil Artin. Zur Theorie der hyperkomplexen Zahlen. *Abh. Math. Sem. Univ. Hamburg*, 5(1):251–260, 1927.

Ibrahim Assem, Daniel Simson, and Andrzej Skowroński. *Elements of the representation theory of associative algebras. Vol. 1: Techniques of representation theory*, volume 65 of *London Mathematical Society Student Texts*. Cambridge University Press, Cambridge, 2006.

Jean Bénabou. Catégories avec multiplication. *C. R. Acad. Sci. Paris*, 256:1887–1890, 1963.

Jean Bénabou. Algèbre élémentaire dans les catégories avec multiplication. *C. R. Acad. Sci. Paris*, 258: 771–774, 1964.

George M. Bergman. The diamond lemma for ring theory. *Adv. in Math.*, 29(2):178–218, 1978.

Gabriella Böhm. *Hopf algebras and their generalizations from a category theoretical point of view*, volume 2226 of *Lecture Notes in Mathematics*. Springer, Cham, 2018.

Paul Boisen. Graded Morita theory. *J. Algebra*, 164(1):1–25, 1994.

Mitya Boyarchenko and Vladimir Drinfeld. A duality formalism in the spirit of Grothendieck and Verdier. *Quantum Topol.*, 4(4):447–489, 2013.

Paul Bruillard, Siu-Hung Ng, Eric C. Rowell, and Zhenghan Wang. Rank-finiteness for modular categories. *J. Amer. Math. Soc.*, 29(3):857–881, 2016.

David A. Buchsbaum. Exact categories and duality. *Trans. Amer. Math. Soc.*, 80:1–34, 1955.

Sabine Burgdorf, Kristijan Cafuta, Igor Klep, and Janez Povh. The tracial moment problem and trace-optimization of polynomials. *Math. Program.*, 137(1-2, Ser. A):557–578, 2013.

Elie Cartan. Les groupes bilinéaires et les systèmes de nombres complexes. In *Annales de la Faculté des sciences de Toulouse: Mathématiques*, volume 12, pages B1–B64, 1898.

Henri Cartan and Samuel Eilenberg. *Homological algebra*. Princeton University Press, Princeton, N. J., 1956.

Arthur Cayley. II. A memoir on the theory of matrices. *Philosophical transactions of the Royal society of London*, (148):17–37, 1858.

Eugenia Cheng. *The joy of abstraction—an exploration of math, category theory, and life*. Cambridge University Press, Cambridge, 2022.

Calin Chindris, Ryan Kinser, and Jerzy Weyman. Module varieties and representation type of finite-dimensional algebras. *Int. Math. Res. Not. IMRN*, (3):631–650, 2015.

Miriam Cohen and Susan Montgomery. Group-graded rings, smash products, and group actions. *Trans. Amer. Math. Soc.*, 282(1):237–258, 1984.

Paul M. Cohn. *Further algebra and applications*. Springer-Verlag London, Ltd., London, 2003.

Severino C. Coutinho. *A primer of algebraic D-modules*, volume 33 of *London Mathematical Society Student Texts*. Cambridge University Press, Cambridge, 1995.

Alexei Davydov, Michael Müger, Dmitri Nikshych, and Victor Ostrik. The Witt group of non-degenerate braided fusion categories. *J. Reine Angew. Math.*, 677:135–177, 2013.

Leonard Eugene Dickson. Definitions of a linear associative algebra by independent postulates. *Trans. Amer. Math. Soc.*, 4(1):21–26, 1903.

Leonard Eugene Dickson. *Algebras and their arithmetics*, volume 15. University of Chicago Press, 1923.

Christopher L. Douglas, Christopher Schommer-Pries, and Noah Snyder. The balanced tensor product of module categories (extended preprint). *ArXiv preprint: math.QA/1406.4204v3. Retrieved October 6, 2023*, 2018.

Christopher L. Douglas, Christopher Schommer-Pries, and Noah Snyder. The balanced tensor product of module categories. *Kyoto J. Math.*, 59(1):167–179, 2019.

Christopher L. Douglas, Christopher Schommer-Pries, and Noah Snyder. Dualizable tensor categories. *Mem. Amer. Math. Soc.*, 268(1308):vii+88, 2020.

Vladimir Drinfeld, Shlomo Gelaki, Dmitri Nikshych, and Victor Ostrik. On braided fusion categories. I. *Selecta Math. (N.S.)*, 16(1):1–119, 2010.

Beno Eckmann and Peter J. Hilton. Group-like structures in general categories. I. Multiplications and comultiplications. *Math. Ann.*, 145:227–255, 1962.

Samuel Eilenberg. Abstract description of some basic functors. *J. Indian Math. Soc. (N.S.)*, 24:231–234 (1961), 1960.

Samuel Eilenberg and Saunders MacLane. General theory of natural equivalences. *Trans. Amer. Math. Soc.*, 58:231–294, 1945.

Pavel Etingof and Viktor Ostrik. Finite tensor categories. *Mosc. Math. J.*, 4(3):627–654, 782–783, 2004.

Pavel Etingof, Dmitri Nikshych, and Viktor Ostrik. On fusion categories. *Ann. of Math. (2)*, 162(2):581–642, 2005.

Pavel Etingof, Dmitri Nikshych, and Victor Ostrik. With an appendix by Ehud Meir. Fusion categories and homotopy theory. *Quantum Topol.*, 1(3):209–273, 2010.

Pavel Etingof, Shlomo Gelaki, Dmitri Nikshych, and Victor Ostrik. *Tensor categories*, volume 205 of *Mathematical Surveys and Monographs*. American Mathematical Society, Providence, RI, 2015.

Brendan Fong and David I. Spivak. *An invitation to applied category theory: Seven sketches in compositionality*. Cambridge University Press, Cambridge, 2019.

Adolf Fraenkel. Über die Teiler der Null und die Zerlegung von Ringen. *J. Reine Angew. Math.*, 145: 139–176, 1915.

Georg Frobenius. Über Matrizen aus nicht negativen Elementen. 1912.

Jürgen Fuchs, Ingo Runkel, and Christoph Schweigert. TFT construction of RCFT correlators. I. Partition functions. *Nuclear Phys. B*, 646(3):353–497, 2002.

Jürgen Fuchs, Gregor Schaumann, and Christoph Schweigert. A trace for bimodule categories. *Appl. Categ. Structures*, 25(2):227–268, 2017.

Jürgen Fuchs, Gregor Schaumann, and Christoph Schweigert. Eilenberg-Watts calculus for finite categories and a bimodule Radford S^4 theorem. *Trans. Amer. Math. Soc.*, 373(1):1–40, 2020.

Jürgen Fuchs, César Galindo, David Jaklitsch, and Christoph Schweigert. Spherical Morita contexts and relative Serre functors. *ArXiv preprint math.QA/2207.07031v2. Retrieved November 7, 2022*, 2022.

Jürgen Fuchs, Gregor Schaumann, Christoph Schweigert, and Simon Wood. Grothendieck-Verdier duality in categories of bimodules and weak module functors. *ArXiv preprint math.QA/2306.17668v1. Retrieved October 6, 2023*, 2023.

Roger Godement. *Topologie algébrique et théorie des faisceaux*. Publications de l'Institut de Mathématiques de l'Université de Strasbourg, No. 13. Hermann, Paris, 1958.

Kenneth R. Goodearl and Robert B. Warfield, Jr. *An introduction to noncommutative Noetherian rings*, volume 61 of *London Mathematical Society Student Texts*. Cambridge University Press, Cambridge, second edition, 2004.

Hermann Grassmann. *Die Wissenschaft der extensiven Größe oder die Ausdehnungslehre, eine neue mathematische Disciplin: Die lineale Ausdehnungslehre, eine neuer Zweig der Mathematik*, volume 1. Wigand, 1844.

Justin Greenough. *Bimodule categories and monoidal 2-structure*. Thesis (Ph.D.)–University of New Hampshire. ProQuest LLC, Ann Arbor, MI, 2010.

Justin Greenough. Relative centers and tensor products of tensor and braided fusion categories. *J. Algebra*, 388:374–396, 2013.

Alexander Grothendieck. Sur quelques points d'algèbre homologique. *Tohoku Math. J. (2)*, 9:119–221, 1957.

William Rowan Hamilton. *Elements of quaternions*. London: Longmans, Green, & Company, 1866.

Chris Heunen and Jamie Vicary. *Categories for quantum theory: An introduction*, volume 28 of *Oxford Graduate Texts in Mathematics*. Oxford University Press, Oxford, 2019.

Kurt A. Hirsch. A note on non-commutative polynomials. *Journal of the London Mathematical Society*, s1-12(4):264–266, 1937.

Otto Hölder. Zurückführung einer beliebigen algebraischen Gleichung auf eine Kette von Gleichungen. *Math. Ann.*, 34(1):26–56, 1889.

Charles Hopkins. Rings with minimal condition for left ideals. *Ann. of Math. (2)*, 40:712–730, 1939.

Peter J. Huber. Homotopy theory in general categories. *Math. Ann.*, 144:361–385, 1961.

Gordon James and Martin Liebeck. *Representations and characters of groups*. Cambridge University Press, New York, second edition, 2001.

Niles Johnson and Donald Yau. *2-dimensional categories*. Oxford University Press, Oxford, 2021.

Camille Jordan. *Traité des substitutions et des équations algébriques*. Les Grands Classiques Gauthier-Villars. Éditions Jacques Gabay, Sceaux, 1989. Reprint of the 1870 original.

André Joyal and Ross Street. Braided tensor categories. *Advances in Mathematics*, 102(1):20–78, 1993.

Daniel M. Kan. Adjoint functors. *Trans. Amer. Math. Soc.*, 87:294–329, 1958.

Gregory M. Kelly. On MacLane's conditions for coherence of natural associativities, commutativities, etc. *J. Algebra*, 1:397–402, 1964.

Gregory M. Kelly. Doctrinal adjunction. In *Category Seminar (Proc. Sem., Sydney, 1972/1973)*, Lecture Notes in Math., Vol. 420, pages 257–280. Springer, Berlin, 1974.

Gregory M. Kelly. Basic concepts of enriched category theory. *Repr. Theory Appl. Categ.*, (10):vi+137, 2005. Reprint of the 1982 original [Cambridge Univ. Press, Cambridge].

Masoud Khalkhali. *Basic noncommutative geometry*. EMS Series of Lectures in Mathematics. European Mathematical Society (EMS), Zürich, second edition, 2013.

Joachim Kock. *Frobenius algebras and 2D topological quantum field theories*, volume 59 of *London Mathematical Society Student Texts*. Cambridge University Press, Cambridge, 2004.

Wolfgang Krull. Über verallgemeinerte endliche Abelsche Gruppen. *Math. Z.*, 23(1):161–196, 1925.

Alexander S. Kuz'min, Viktor T. Markov, Alexander A. Mikhalëv, Alexandr V. Mikhalëv, and Alexandr A. Nechaev. Cryptographic algorithms on groups and algebras. *Fundam. Prikl. Mat.*, 20(1):205–222, 2015.

Jaroslaw Kwapisz. Cocyclic subshifts. *Math. Z.*, 234(2):255–290, 2000.

Stephen Lack and Ross Street. The formal theory of monads. II. volume 175, pages 243–265. 2002.

Jeanne LaDuke. The study of linear associative algebras in the United States, 1870–1927. In *Emmy Noether in Bryn Mawr (Bryn Mawr, Pa., 1982)*, pages 147–159. Springer, New York, 1983.

Tsit Yuen Lam. *Lectures on modules and rings*, volume 189 of *Graduate Texts in Mathematics*. Springer-Verlag, New York, 1999.

Robert Laugwitz and Chelsea Walton. Constructing non-semisimple modular categories with local modules. *Comm. Math. Phys.*, 403(3):1363–1409, 2023.

F. William Lawvere. Functorial semantics of algebraic theories. *Proc. Nat. Acad. Sci. U.S.A.*, 50:869–872, 1963.

Fernando Liu Lopez and Chelsea Walton. Twists of graded algebras in monoidal categories. *ArXiv preprint math.QA/2311.18105v1. Retrieved December 6, 2023*, 2023.

Martin Lorenz. *A tour of representation theory*, volume 193 of *Graduate Studies in Mathematics*. American Mathematical Society, Providence, RI, 2018.

Saunders Mac Lane. Natural associativity and commutativity. *Rice Univ. Stud.*, 49(4):28–46, 1963.

Saunders Mac Lane. *Categories for the working mathematician*, volume 5 of *Graduate Texts in Mathematics*. Springer-Verlag, New York, second edition, 1998.

Saunders MacLane. *Categories for the working mathematician*. Graduate Texts in Mathematics, Vol. 5. Springer-Verlag, New York-Berlin, 1971.

Matilde Marcolli and Walter D. van Suijlekom. Gauge networks in noncommutative geometry. *J. Geom. Phys.*, 75:71–91, 2014.

Heinrich Maschke. Beweis des Satzes, dass diejenigen endlichen linearen Substitutionsgruppen, in welchen einige durchgehends verschwindende Coefficienten auftreten, intransitiv sind. *Math. Ann.*, 52(2-3):363–368, 1899.

Volodymyr Mazorchuk and Vanessa Miemietz. Cell 2-representations of finitary 2-categories. *Compos. Math.*, 147(5):1519–1545, 2011.

Catherine Meusburger. State sum models with defects based on spherical fusion categories. *Adv. Math.*, 429:Paper No. 109177, 97, 2023.

Barry Mitchell. The full imbedding theorem. *Amer. J. Math.*, 86:619–637, 1964.

Barry Mitchell. *Theory of categories*. Pure and Applied Mathematics, Vol. XVII. Academic Press, New York-London, 1965.

Theodor Molien. Ueber systeme höherer complexer zahlen. *Mathematische Annalen*, 41(1):83–156, 1892.

Yiby Morales, Monique Müller, Julia Plavnik, Ana Ros Camacho, Angela Tabiri, and Chelsea Walton. Algebraic structures in group-theoretical fusion categories. *Algebras Represent. Theory*, pages 1–33, 2022.

Kiiti Morita. Duality for modules and its applications to the theory of rings with minimum condition. *Sci. Rep. Tokyo Kyoiku Daigaku Sect. A*, 6:83–142, 1958.

Sonia Natale. On the equivalence of module categories over a group-theoretical fusion category. *SIGMA Symmetry Integrability Geom. Methods Appl.*, 13:Paper No. 042, 9, 2017.

Emmy Noether. Idealtheorie in Ringbereichen. *Math. Ann.*, 83(1-2):24–66, 1921.

Emmy Noether. Hyperkomplexe Größen und Darstellungstheorie. *Math. Z.*, 30(1):641–692, 1929.

Gabriela Olteanu and Inneke Van Gelder. Construction of minimal non-abelian left group codes. *Des. Codes Cryptogr.*, 75(3):359–373, 2015.

Victor Ostrik. Fusion categories of rank 2. *Math. Res. Lett.*, 10(2-3):177–183, 2003a.

Victor Ostrik. Module categories over the Drinfeld double of a finite group. *Int. Math. Res. Not.*, (27):1507–1520, 2003b.

Victor Ostrik. Module categories, weak Hopf algebras and modular invariants. *Transform. Groups*, 8(2):177–206, 2003c.

Bodo Pareigis. Non-additive ring and module theory. I. General theory of monoids. *Publ. Math. Debrecen*, 24(1-2):189–204, 1977a.

Bodo Pareigis. Non-additive ring and module theory. II. $\mathcal{C}$-categories, $\mathcal{C}$-functors and $\mathcal{C}$-morphisms. *Publ. Math. Debrecen*, 24(3-4):351–361, 1977b.

Bodo Pareigis. Non-additive ring and module theory. III. Morita equivalences. *Publ. Math. Debrecen*, 25(1-2):177–186, 1978.

Giuseppe Peano. *Calcolo geometrico secondo l'Ausdehnungslehre di H. Grassmann: preceduto dalla operazioni della logica deduttiva*, volume 3. Fratelli Bocca Editori, Turin, 1888.

Benjamin Peirce. Linear Associative Algebra. *Amer. J. Math.*, 4(1-4):97–229, 1881.

Oskar Perron. Zur Theorie der Matrices. *Math. Ann.*, 64(2):248–263, 1907.

Richard S. Pierce. *Associative algebras*, volume 9 of *Studies in the History of Modern Science*. Springer-Verlag, New York-Berlin, 1982.

Nicolae Popescu. *Abelian categories with applications to rings and modules*. London Mathematical Society Monographs, No. 3. Academic Press, London-New York, 1973.

Leonid Positselski and Jan Šťovíček. Topologically semisimple and topologically perfect topological rings. *Publ. Mat.*, 66(2):457–540, 2022.

Birgit Richter. *From categories to homotopy theory*, volume 188 of *Cambridge Studies in Advanced Mathematics*. Cambridge University Press, Cambridge, 2020.

Emily Riehl. *Category theory in context*. Aurora Dover Modern Math Originals. Dover Publications, Inc., Mineola, NY, 2016.

Joseph J. Rotman. *An introduction to homological algebra*. Universitext. Springer, New York, second edition, 2009.

Neantro Saavedra Rivano. *Catégories Tannakiennes*. Lecture Notes in Mathematics, Vol. 265. Springer-Verlag, Berlin-New York, 1972.

Gregor Schaumann. Pivotal tricategories and a categorification of inner-product modules. *Algebr. Represent. Theory*, 18(6):1407–1479, 2015.

Otto Schmidt. Sur les produits directs. *Bull. Soc. Math. France*, 41:161–164, 1913.

Issai Schur. Neue begründung der theorie der gruppencharaktere. *Sitzungsberichte der Akademie der Wiss. zu Berlin, Physikalisch-Math. Kl*, pages 406–432, 1905.

Kenichi Shimizu. Relative Serre functor for comodule algebras. *ArXiv preprint math.CT/1904.00376v2. Retrieved November 7, 2022*, 2019.

Kenichi Shimizu. Further results on the structure of (co)ends in finite tensor categories. *Appl. Categ. Structures*, 28(2):237–286, 2020.

Karen E. Smith, Lauri Kahanpää, Pekka Kekäläinen, and William Traves. *An invitation to algebraic geometry*. Universitext. Springer-Verlag, New York, 2000.

Bo Stenström. *Rings of quotients: An introduction to methods of ring theory*. Die Grundlehren der mathematischen Wissenschaften, Band 217. Springer-Verlag, New York-Heidelberg, 1975.

Ross Street. The formal theory of monads. *J. Pure Appl. Algebra*, 2(2):149–168, 1972.

Daisuke Tambara and Shigeru Yamagami. Tensor categories with fusion rules of self-duality for finite abelian groups. *J. Algebra*, 209(2):692–707, 1998.

Vladimir Turaev and Alexis Virelizier. *Monoidal categories and topological field theory*, volume 322 of *Progress in Mathematics*. Birkhäuser/Springer, Cham, 2017.

Frank Vallentin. Symmetry in semidefinite programs. *Linear Algebra Appl.*, 430(1):360–369, 2009.

Chelsea Walton. An invitation to noncommutative algebra. In *A celebration of the EDGE program's impact on the mathematics community and beyond*, volume 18 of *Assoc. Women Math. Ser.*, pages 339–366. Springer, Cham, 2019.

Chelsea Walton and Harshit Yadav. Filtered Frobenius algebras in monoidal categories. *Int. Math. Res. Not. IMRN*, (24):21494–21535, 2023.

Charles E. Watts. Intrinsic characterizations of some additive functors. *Proc. Amer. Math. Soc.*, 11:5–8, 1960.

Joseph H. MacLagan Wedderburn. On Hypercomplex Numbers. *Proc. London Math. Soc. (2)*, 6:77–118, 1908.

Charles A. Weibel. *An introduction to homological algebra*, volume 38 of *Cambridge Studies in Advanced Mathematics*. Cambridge University Press, Cambridge, 1994.

Nobuo Yoneda. On the homology theory of modules. *J. Fac. Sci. Univ. Tokyo Sect. I*, 7:193–227, 1954.

James J. Zhang. Twisted graded algebras and equivalences of graded categories. *Proc. London Math. Soc. (3)*, 72(2):281–311, 1996.

Yuanzhao Zhang and Adilson E. Motter. Symmetry-independent stability analysis of synchronization patterns. *SIAM Rev.*, 62(4):817–836, 2020.

$\mathcal{E}_2(f, f')$, 268
$\mathcal{E}_2^{X,Y}(f, f')$, 268
$\mathcal{G}$, 72
$\mathcal{L}$, 88
$\mathcal{L}(X)$, 88
$\mathcal{M}$, 145
$\mathcal{M}^*$, 167
$^*\mathcal{M}$, 167
$\mathcal{N}^*$, 167
$^*\mathcal{N}$, 167
$\mathcal{O}$, 88, 97
$\mathcal{O}(G)$, 8
$\mathcal{O}_q(GL_2(\mathbb{C}))$, 7
$\mathcal{O}_q(X)$, 6
$\mathcal{O}(GL_2(\mathbb{C}))$, 7
$\mathcal{O}(X_q)$, 6
$\mathcal{O}(\mathbb{C}^2)$, 5
$\mathcal{V}$, 186
$\underline{\mathcal{Z}}(\mathcal{C})$, 129

$\mathfrak{BimCat}$, 271
$\mathfrak{Bim}(\mathcal{C})$, 270
$\mathfrak{Cat}$, 270

$\vec{0}_X$, 76
0, 76
$_X\vec{0}$, 76
1, 137
80sMusic, 74
(A, B)-Bimod, 73
(A, B)-FdBimod, 73
(B_1, B_2)-Bimod$(\mathcal{C})$, 227
(B_1, B_2)-Bimod$(\mathcal{C})_\square$, 236
$(\mathcal{C}, \mathcal{D})$-Bimod, 151
A-Bimod, 73, 136
A-Bimod$(\mathcal{C})$, 238
A-FdBimod, 73
A-FdMod, 73
A-Mod, 73
A-Mod$(\mathcal{C})$, 224
A-Mod$(\mathcal{C})_\square$, 236
G-Mod, 8, 73, 136
N-GrAlg, 72
R-Mod$_{ab}$, 97

V-Quiv, 138
$\mathbb{N}$-GrAlg, 72
$\mathbb{N}$-GrAlg$(\mathcal{C})$, 240
$\mathbb{N}$-GrMod$_A(\mathcal{C})$, 242
$\mathbb{N}$-GrAlg$(\mathcal{C})_A$, 242
$\mathbb{N}$-GrAlg$_{B_1}(\mathcal{C})_{B_2}$, 242
$\mathcal{C}$-Bimod, 151
$\mathcal{C}$-Mod, 148
nCob, 74, 138
Ab, 72, 136
Aff, 73, 97
Alg, 72
Alg$(\mathcal{C})$, 206
Alg$(\mathcal{C})_\square$, 235
Alg$_\circledast$, 136
Alg$_{\otimes_\Bbbk}$, 136
Aut$(\mathcal{A})$, 138
Aut$(\mathcal{C})$, 96
Ban, 74
Bim, 73, 99
Braid, 138
Cat, 73, 137
ComAlg, 72
ComRing, 72
Comod-$\mathcal{O}(G)$, 8
DirGraph, 74
End$(\mathcal{C})$, 90
End$(\mathcal{A})$, 138
End$_{\mathcal{C}\text{-Mod}}(\mathcal{M})$, 148
End$_{\text{Mod-}\mathcal{C}}(\mathcal{M})$, 148
FdAlg, 72
FdHilb, 74, 138
FdMod-A, 73
FdRep(A), 73
FdVec, 72, 136
FdVec$_\oplus$, 136
FgAlg, 72
FgRedComAlg, 97
Fib, 180
FinSet, 73
Fun$(\mathcal{C}, \mathcal{D})$, 90
Fun$_{\text{Mod-}\mathcal{C}}(\mathcal{M}, \mathcal{N})$, 150
Graph, 74
Group, 72

Group$(\mathcal{A}_\sqcap)$, 278
G$_{\text{cat}}$, 96
Hilb, 74, 138
Ising, 181
I, 76
Mat, 129
Meas, 74
Mfld, 73
Mod-A, 73
Mod-$A(\mathcal{C})$, 224
Mod-$A(\mathcal{C})_\square$, 236
Mod-G, 73
Mod-$\mathcal{C}$, 148
Monad$(\mathcal{A})$, 221
Monoid, 72
Perm, 150
Poset, 74
Quiv, 74
Rel, 73
Rep(A), 72
Rep(G), 73
Rex$_{\mathcal{C}\text{-Mod}}(\mathcal{M}, \mathcal{N})$, 167
Rex$_{\text{Mod-}\mathcal{C}}(\mathcal{M}, \mathcal{N})$, 167, 244
Ring, 72, 136
Rng, 72
SameTaste, 74
Scheme, 97
Set, 73, 137
Set$_*$, 73
SharePw, 74
Skel$(\mathcal{C})$, 94
TY(G, χ, τ), 181
Top, 74, 138
Top$_*$, 74
T, 76
Vec, 72, 136
Vec$'_G$, 137
Vec$_G$, 136
Vec$_G^\omega$, 137
Vec$_N$, 72, 136
Vec$_\mathbb{C}$, 8
Vec$_\oplus$, 136

INDEX OF TERMINOLOGY

(That's all, folks. See you later in Volumes 2 and 3.)

www.ingramcontent.com/pod-product-compliance
Lightning Source LLC
Chambersburg PA
CBHW080251030726
47593CB00009B/2439